Statistical Methods and Applications in Systems Assurance and Quality

In the era of big data and real-time monitoring, vast amounts of information are generated every second through sensors and advanced tracking systems. The information contained in the data needs to be carefully analyzed and accurate results must be delivered to the users as soon as possible. This book explores the critical role of statistics in modern technologies and applied sciences such as health, informatics, and production systems.

Statistical *Methods and Applications* in Systems Assurance and Quality provides a comprehensive overview of statistical methods for ensuring system quality and reliability. It discusses best practices for improving the quality of manufacturing products by using proper statistical methods. The book covers the current state of statistical methods used for system quality and conveys the importance of healthcare systems and employing statistics in the detection of epidemic outbreaks.

Designed for researchers, industry professionals, and academics, this reference is ideal for those working in manufacturing, healthcare, computer engineering, statistical process control, and machine learning applications. It serves as a valuable resource for anyone seeking to enhance system performance through data-driven insights.

Advanced Research in Reliability and System Assurance Engineering

Series Editor: Mangey Ram, Professor, Graphic Era University, Uttarakhand, India

Reliability Management and Engineering
Challenges and Future Trends
Edited by Harish Garg and Mangey Ram

Applied Systems Analysis
Science and Art of Solving Real-Life Problems
F. P. Tarasenko

Stochastic Models in Reliability Engineering
Lirong Cui, Ilia Frenkel, and Anatoly Lisnianski

Predictive Analytics
Modeling and Optimization
Vijay Kumar and Mangey Ram

Design of Mechanical Systems Based on Statistics
A Guide to Improving Product Reliability
Seong-woo Woo

Social Networks
Modeling and Analysis
Niyati Aggrawal and Adarsh Anand

Operations Research
Methods, Techniques, and Advancements
Edited by Amit Kumar and Mangey Ram

Statistical Modeling of Reliability Structures and Industrial Processes
Edited by Ioannis S. Triantafyllou and Mangey Ram

Industrial Reliability and Safety Engineering
Applications and Practices
Edited by Dilbagh Panchal, Mangey Ram, Prasenjit Chatterjee, and Anish Kumar Sachdeva

Reliability and Maintenance Modeling with Optimization
Advances and Applications
Edited by Mitsutaka Kimura, Satoshi Mizutani, Mitsuhiro Imaizumi, and Kodo Ito

Intelligent Prognostics for Engineering Systems with Machine Learning Techniques
Edited by Gunjan Soni, Om Prakash Yadav, Gaurav Kumar Badhotiya and Mangey Ram

Statistical Methods and Applications in Systems Assurance and Quality
Edited by Sotiris Bersimis, Polychronis Economou, and Athanasios Rakitzis

For more information about this series, please visit: https://www.routledge.com/Advanced-Research-in-Reliability-and-System-Assurance-Engineering/book-series/CRCARRSAE

Statistical Methods and Applications in Systems Assurance and Quality

Edited by

Sotiris Bersimis, Polychronis Economou, and Athanasios Rakitzis

CRC Press
Taylor & Francis Group
Boca Raton London New York

CRC Press is an imprint of the
Taylor & Francis Group, an **informa** business

Designed cover image: Shutterstock

First edition published 2026
by CRC Press
2385 NW Executive Center Drive, Suite 320, Boca Raton, FL 33431

and by CRC Press
4 Park Square, Milton Park, Abingdon, Oxon, OX14 4RN

CRC Press is an imprint of Taylor & Francis Group, LLC

ISBN: 978-1-032-66404-0 (hbk)
ISBN: 978-1-032-66432-3 (pbk)
ISBN: 978-1-032-66433-0 (ebk)

DOI: 10.1201/9781032664330

Typeset in Nimbus Roman
by KnowledgeWorks Global Ltd.

Access the Support Material: routledge.com/9781032664040

Contents

SECTION III Statistical Methods in Sports and Industrial Systems Assurance

SECTION IV Miscellaneous Applications of Statistical Methods in Systems Assurance

Preface

The Panhellenic Statistics Conference is the main scientific event in Statistics and Probability, in Greece. It is co-organized annually by the Greek Statistical Institute (GSI), a member of the Federation of European National Statistical Societies, and an Academic Institution in Greece or Cyprus, since 1985.

The 35th Panhellenic and 1st International Statistics Conference was organized jointly by the Greek Statistical Institute and the University of West Attica, in the city of Athens, in person and virtually, from May 25 to May 28, 2023. The conference brought together Statisticians, Data Analysts, Mathematicians, Computer Scientists, Econometricians as well as doctors and scientists from the wide area of health Sciences, who exchanged ideas and discussed the recent advances in Statistical Science, with a special focus of their applications in health Sciences.

In the big data era and the frequent use of sensors and detailed monitoring systems thousands of observations become available every second. The information contained in the data needs to be carefully analyzed and accurate results must be delivered to the users as soon as possible. Examples of systems that produce crucial data during their operation time include financial transactions tracking systems, traffic sensors, packaging processes in a manufacturing company, and healthcare monitoring systems of medical examinations. Statistical methods have nowadays a wide range of applications in several areas of applied science. Specifically, statistical methods for systems assurance have, among others, a direct application in network monitoring and the early detection of computer intrusion, financial transactions tracking systems, traffic sensors, packaging processes, healthcare monitoring systems, and sports analytics. In addition, statistical machine learning methods are necessary to handle complex systems and produce accurate and on-time crucial results that assure that the system works safely, and the produced products (or the delivered services) are of the highest quality.

To be more specific, data collected by the aforementioned systems are usually multivariate in nature, whereas very often are count data or even mixed data (multivariate data from continuous and discrete distributions). Thus, there are obvious violations of common statistical assumptions (e.g., normality, independence). Consequently, analytical results that allow the direct calculation of specific performance indices are not always possible, and numerical methods, improved algorithms, and other tailor-made solutions are necessary when the purpose is to assure the system's safety and/or its stability during operation time.

These improved and/or tailor-made algorithms and methods are highlighted by this book which brings into the reader's attention some of the most interesting topics related to the application of statistics in modern technologies and applied sciences such as health, informatics, production systems, and sports analytics.

The book consists of 13 papers that were carefully selected and reviewed by the scientific program committee. They are divided into four parts (sections), all related to the application of statistical methods in systems assurance.

SECTION I: STATISTICAL QUALITY CONTROL METHODS IN SYSTEMS ASSURANCE

Section I consists of the first four chapters, all devoted to particular statistical process control/monitoring (SPC/SPM) methods for systems assurance. Specifically:

In Chapter 1, new EWMA-type control charts for monitoring serially independent or autocorrelated count processes are introduced and studied. The proposed charts are based on the respective Stein identities for Poisson, binomial and negative binomial distributions. Their performance is investigated by simulations while their usefulness is highlighted via an application in health surveillance.

In Chapter 2, control charts for monitoring time-varying ARMA processes are proposed and studied. The focus is on processes that can be modelled as a time series model with a time-varying AR component and a static MA part. First, a model is fitted to the original process to take into account the dependence structure, and then the process is controlled by monitoring its residuals. The performance of the proposed control charts is investigated by simulations while special attention is paid on the estimation and identification of these time-varying ARMA models. A real-data example highlights their usefulness in practice.

In Chapter 3, the performance of one-sided Shewhart control charts for zero-inflated processes is evaluated when the process parameters are estimated. The study focuses on the two most common zero-inflated probability models (the zero-inflated Poisson and the zero-inflated binomial). Using simulation, the in-control and the out-of-control performance of the charts is investigated, under an unconditional perspective, and guidelines are provided regarding their statistical design.

In Chapter 4, the performance of two self-starting CUSUM charts is compared numerically. Specifically, the performance of a Bayesian self-starting CUSUM chart and its frequentist alternative is investigated for a wide variety of scenarios, with special attention on the detection of changes in the mean of normal data. The proposed charts can be used for the online monitoring of a process without the presence of a preliminary Phase I analysis.

SECTION II: STATISTICAL METHODS IN HEALTHCARE SYSTEMS ASSURANCE

Section II consists of the next four chapters, which are all devoted to applications of statistical methods in healthcare. Specifically:

In Chapter 5, a semi-parametric mixture model is used to capture the responder/non-responder effects in a treatment group during in a two-stage randomized clinical trial (RCT). The treatment effect is described by the proportion of treated patients that respond to the treatment and by the difference in the mean responses between responders and non-responders. Using the method of moment estimators the treatment effect as well as its size is estimated. Asymptotic confidence intervals are also given while a real-life application illustrates the usefulness of the theoretical results in practice.

In Chapter 6, an extensive overview on the use of polygenic risk scores (PRS) is provided and it is investigated if they can be used in clinical practice as informative predictive tools. PRS could be considered as a numerical expression of a person's genetic predisposition for developing specific phenotypic traits or diseases aiming to classify this person to a low, medium or high-risk category (Cardiovascular diseases, Cancer Types, etc.). Various types of PRS are presented and their pros and cons in applying them in clinical practice are discussed.

In Chapter 7, a literature review of predictive models based on machine learning techniques in the healthcare industry is provided. More precisely, the overview summarizes predictive models for no-shows phenomenon in the health sector and for hospital re-admissions between the years 2019–2023. The analysis of the existing articles shows that the use of machine learning algorithms improves significantly the quality of services provided in clinics and hospitals, reducing at the same time the enormous healthcare services expenditures. In Chapter 8, a comprehensive and adaptable approach for estimating healthcare cost associated with a specific disease is introduced. By integrating various aspects of disease management and leveraging diverse data sources (from defining treatment protocols to assessing expenses for medications, procedures, diagnostic tests, and consultations), the proposed framework enhances the accuracy and reliability of cost estimates. In addition, it acknowledges the importance of adjusting costs based on the target population size, and thus, by linking the estimated prevalence of a specific disease to the overall population, the methodology enables scalable cost projections, facilitating informed decision-making in resource allocation.

SECTION III: STATISTICAL METHODS IN SPORTS AND INDUSTRIAL SYSTEMS ASSURANCE

Section III consists of Chapters 9, 10, and 11 which contain applications of modern statistical methods in sports and industry. Specifically:

In Chapter 9, applied considerations about the prominent role of data and the need to focus on the data produced by machine learning algorithms are provided. With the aid of sequential learning, it is possible to optimally combine heterogeneous or multi-scale model outputs (also known as "second-order" data). The proposed methodology is illustrated with two examples from industry where machine learning algorithms are used to combine these "second-order" data and generate predictions, where the focus is not on selecting the best model but on how to the data can be useful.

In Chapter 10, a comprehensive overview of the research in the field of sports analytics is provided. Since this area received increased attention in recent years, this study identifies the emerging fields in sports industry, providing valuable insights regarding athletes' performance, training strategies, and game tactics across a wide range of sports. By applying analytical methods, numerous publications are analyzed and classified, thus providing a comprehensive systematic review.

In Chapter 11, a case study is presented, using data from NBA and Euroleague games from seasons 2000–2020, and the aim is to predict score outcomes. Using penalized Poisson and penalized negative binomial regressions, predictions are made regarding the different types of plays for each opposing team. Then, these predictions are used to simulate each game, using a compound Poisson process to predict the game outcome. The proposed methodology allows for an improved visualization and interpretation of the predicted outcomes, compared to other competitive methods.

SECTION IV: MISCELLANEOUS APPLICATIONS OF STATISTICAL METHODS IN SYSTEMS ASSURANCE

Section IV consists of Chapters 12 and 13 and is devoted to the application of statistical methods in systems assurance in cases that have not been covered by the previous chapters. Specifically:

In Chapter 12, a new methodology for constructing probability distributions for counts (counting distributions) is introduced. Specifically, the methodology is based on considering natural exponential families of probability distributions. Since these families are uniquely determined by their variance functions, the counting distributions can be constructed from variance functions that show some regularity conditions. The usefulness of these new counting distributions in practice is supported by real-world applications such as in mortality projection, and in insurance risk modelling.

In Chapter 13, a case study on analyzing and modelling data on contamination of river water in Uttarakhand, India, is presented. Several analytical measures of water quality are analyzed and practical problems such as the heavy tailed character of the distribution of the chemical contamination or the case of missing data, are discussed. In addition, the study outlines the necessity of appropriate data transformations for river water contamination models.

The high scientific level of the 35th Panhellenic and 1st International Statistics Conference is determined by the quality of its papers. We believe that this book reflects the high research quality of the participants in the conference as well as the significance of their research contributions in modern statistical methods for systems assurance and quality.

Before closing, we would like to express our deep gratitude to the members of the scientific program committee of the 35th Panhellenic and 1st International Statistics Conference, who carefully invited researchers from around the world as well as to all reviewers for their diligent work and commitment. It was a very pleasant experience corresponding with all the authors involved. Also, we would like to cordially thank all the authors who have contributed to this book and provided great support and cooperation during all the stages of this project. Finally, we would like to express our sincere thanks to the CRC Press production team for their valuable assistance, patience, and understanding during the preparation of this book.

January 2025,

Sotiris Bersimis
Department of Business Administration,
University of Piraeus, Greece

Polychronis Economou
Department of Civil Engineering,
University of Patras, Greece

Athanasios C. Rakitzis
Department of Statistics and Insurance Science,
University of Piraeus, Greece

Editors

Dr. Sotiris Bersimis is a Professor at the University of Piraeus, Department of Business Administration and an elected member of International Statistical Institute. He is a PhD holder in Statistics, from the Statistics and Insurance Science Department of the University of Piraeus. He also holds an MSc in Statistics from Athens University of Economics and Business and a BSc in Statistics and Insurance Science from the University of Piraeus. His research interests are mostly related to the development and implementation of stochastic models for processes monitoring (processes' quality control, patients' surveillance, quality of services, etc.), as well as with the theory of runs and patterns. He also works in the fields of data analytics and statistical machine learning. He has published more than 70 papers in international scientific while more than 120 of his papers have been presented in National (Greek) and International Conferences. In the past, during the period 9/2020–4/2022 he served as the Chair of the Department of Statistics and Insurance Science and as a member of the Senate of the University of Piraeus. In addition, during the period 12/2015–1/2019 he served as the President of the Hellenic Organization for Health Care Services Provision while during the period 4/2018–1/2019 he has been elected and served as the president of European Healthcare Fraud & Corruption Network (EHFCN).

Dr. Polychronis Economou is an Associate Professor in the Department of Civil Engineering at the University of Patras, Greece. He earned a BSc in Mathematics from the University of Ioannina in 2001 and an MSc in Applied Mathematics from the National Technical University of Athens (NTUA) in 2003. In 2007, he completed his doctoral thesis, Heterogeneity Models in Lifetime Data Analysis, also at NTUA. Dr. Economou's research spans several areas of applied statistics, including Survival and Reliability Models, Diagnostic Tests, Statistical Process Control and Monitoring, and Biased Sampling and Weighted Distributions. His work is characterized by methodological rigor and practical applications, often addressing complex real-world problems. He has contributed significantly to the field, with over 75 articles published in international peer-reviewed journals and conference proceedings, earning widespread recognition for their impact. Additionally, he has co-authored four books, providing valuable resources for students and researchers in applied statistics. Dr. Economou actively collaborates with academic and industry partners, reflecting his commitment to advancing both theoretical and applied aspects of his research areas. He serves as an associate editor for three prestigious scientific journals specializing in statistics, further highlighting his standing in the academic community. Beyond his research and editorial responsibilities, Dr. Economou is an experienced educator, dedicated to mentoring students and guiding them in the development of statistical expertise to address contemporary challenges.

Dr. Athanasios Rakitzis is an Assistant Professor in the Department of Statistics and Insurance Science at the University of Piraeus, in Greece. He holds a BSc in Mathematics from the National and Kapodestrian University of Athens, Greece and a MSc in Applied Statistics from the University of Piraeus, Greece. In 2008 he obtained his PhD in Statistics from the same department. He has served as a visiting professor (Lecturer) in the Department of Mathematics and Statistics, University of Cyprus, Cyprus, as an Assistant Professor at the Department of Statistics and Actuarial Mathematics, University of Aegean, Greece while he has also worked as a Post-doctoral researcher (Marie Curie Fellow) at the Université de Nantes, Institute Universitaire de Technologie de Nantes, France. He has published more than 70 research articles in international peer-review journals and conference proceedings and is the co-author of 2 books. He has participated, with announcement, in more than 100 National (Greek) and International Conferences. Also, he has supervised 2 PhD students in pursuing their PhD degree and 40 MSc theses. His research interests include statistical process control/monitoring and applied probability. Finally, he is a member of the Greek Statistical Institute and the European Network of Business and Industrial Statistics (ENBIS).

Contributors

Bagos, Pantelis
Department of Computer Science and Biomedical Informatics, University of Thessaly
E-Mail: pbagos@compgen.org
ORCID: 0000-0003-4935-2325

Bal, Kiran J.
FMA Finanzmarktaufsicht Österreich, Austria

Bar-Lev, Shaul K.
Holon Institute of Technology, Holon, Israel
E-Mail: barlev@stat.haifa.ac.il
ORCID: 0000-0002-1770-7091

Bersimis, Fragiskos
Department of Statistics and Insurance Science, University of Piraeus
E-Mail: fbersimis@unipi.gr
ORCID: 0000-0002-8450-0453

Bersimis, Sotiris
Department of Business Administration, University of Piraeus, Greece
E-Mail: sbersim@unipi.gr
ORCID: 0000-0002-7448-1172

Bourazas, Konstantinos
Department of Mathematics and Statistics & KIOS Research and Innovation Center of Excellence, University of Cyprus, Nicosia, Cyprus
E-Mail: kbourazas@aueb.gr
ORCID: 0000-0002-2691-4869

Economou, Polychronis
Department of Civil Engineering, University of Patras, Greece
E-Mail: peconom@upatras.gr
ORCID: 0000-0001-6452-5920

Georgakellos, Dimitrios
University of Piraeus, Department of Business Administration, Piraeus, Greece
E-Mail: dgeorg@unipi.gr
ORCID: 0000-0002-7956-4727

Jeske, Daniel R.
University of California, Riverside, Department of Statistics
E-Mail: daniel.jeske@ucr.edu

Karaminas, Antonios
Department of Business Administration, University of Piraeus, Greece

Ley, Christophe
University of Luxembourg
E-Mail: christophe.ley@uni.lu
ORCID: 0000-0002-2290-8437

Mamzeridou, Eftychia
University of the Aegean, Department of Statistics and Actuarial-Financial Mathematics
E-Mail: emam@aegean.gr

Manisera, Marica
University of Brescia
E-Mail: marica.manisera@unibs.it
ORCID: 0000-0002-2982-0243

Maravelakis, Petros E.
University of Piraeus, Department of Business Administration
E-Mail: maravel@unipi.gr
ORCID: 0000-0002-7477-5947

Nika, Elisavet
University of Piraeus, Department of Business Administration, Piraeus, Greece
E-Mail: elisavet.nika@yahoo.gr

Pant, Kumud
Department of Biotechnology, Graphic Era University, Dehradun, India

Platjouw, Emiel
Signpost Belgium

Poggi, Jean-Michel
LMO, University Paris-Saclay, Orsay, France; University Paris Cité, Dept SD, IUT Paris-Rives de Seine, Paris, France
E-Mail: jean-michel.poggi@universite-paris-saclay.fr
ORCID: 0000-0002-8222-1653

Rakitzis, Athanasios C.
University of Piraeus, Department of Statistics and Insurance Science
E-Mail: arakitz@unipi.gr
ORCID: 0000-0002-3054-4004

Ridder, Ad
Vrije Universiteit Amsterdam, Netherlands
E-Mail: ad.ridder@vu.nl
ORCID: 0000-0002-2414-0548

Sadeghkhani, Abdolnasser
North Carolina Agricultural and Technical State University
E-Mail: asadeghkhani@ncat.edu
ORCID: 0000-0002-3074-0443

Soza, Ludy Núñez
Departamento de Matemática, Facultad de Ciencias, Universidad de Tarapacá, Arica, Chile
ORCID: 0000-0003-4858-1900

Spyropoulos, Christos
Department of Business Administration, University of Piraeus, Greece

Stehlík, Milan
Institute of Statistics, University of Valparaíso, Valparaíso, Chile; Facultad de Ingeniería, Universidad Andrés Bello, Valparaíso, Chile; University of Applied Sciences Upper Austria, Campus Wels, Austria
E-Mail: mlnstehlik@gmail.com
ORCID: 0000-0003-2749-5990

Stehlíková, Silvia
Independent Researcher, Valparaíso, Chile

Stillman, Eleanor
School of Mathematical and Physical Sciences
E-Mail: e.c.stillman@sheffield.ac.uk
ORCID: 0000-0002-5971-0985

Triantafyllopoulos, Kostas
School of Mathematical and Physical Sciences
E-Mail: k.triantafyllopoulos@sheffield.ac.uk
ORCID: 0000-0002-4144-4092

Tsiampalis, Thomas
Harokopio University of Athens, School of Health Sciences & Education, Department of Nutrition and Dietetics, Athens, Greece; University of Thessaly, School of Physical Education, Sports and Dietetics, Department of Nutrition and Dietetics, Trikala, Greece
E-Mail: ttsiampalis@gmail.com
ORCID: 0000-0002-4625-460X

Weiss, Christian H.
Helmut Schmidt University, Department of Mathematics and Statistics, Hamburg, Germany
E-Mail: weissc@hsu-hh.de
ORCID: 0000-0001-8739-6631

Zuccolotto, Paola
University of Brescia
E-Mail: paola.zuccolotto@unibs.it
ORCID: 0000-0003-4399-7018

Section I

Statistical Quality Control Methods in Systems Assurance

1 Stein EWMA Control Charts for Count Processes

Christian H. Weiss
Helmut Schmidt University, Department of Mathematics and Statistics, Hamburg, Germany.
E-Mail: weissc@hsu-hh.de. ORCID: 0000-0001-8739-6631.

1.1 INTRODUCTION

The sequential monitoring of count processes $(X_t)_{t\in\mathbb{N}=\{1,2,...\}}$ (i.e., where the X_t have a quantitative range contained in $\mathbb{N}_0 = \{0, 1, ...\}$) is of utmost importance in many application areas, such as the quality control of manufactured items (counts of defects or non-conformities) or health surveillance (counts of infections or hospital admissions); see Montgomery, 2009; Weiss, 2015, 2018 for examples and references. The primary tool for managing statistical process control (SPC) for (X_t) is the use of attributes control charts. These charts involve plotting specific statistics sequentially until signals indicate the necessity for corrective action. These statistics are computed in an online manner from the sequence of incoming counts (X_t). The control chart triggers a signal ("alarm") if the plotted statistic violates the specified control limits (CLs). The CLs are derived from the so-called in-control model of (X_t), i.e., a stochastic model that assumes (X_t) to operate under stable conditions. Hence, an alarm (a violation of the CLs) is interpreted as an indication of a possible process deterioration (out-of-control situation). Therefore, if the alarm occurs when the process (X_t) is genuinely out of control, it is considered a true alarm. On the other hand, an alarm under in-control conditions is regarded as a false alarm. It stands to reason that CLs should be chosen to avoid false alarms for as long as possible, and to receive a true alarm as soon as possible. The default metric for evaluating these durations until alarm is the average run length (ARL), i.e., the expected number of plotted statistics until the first alarm. For a more detailed description of the aforementioned terms and concepts as well as for further references, the reader may consult SPC textbooks such as Montgomery, 2009; Qiu, 2014.

Various control charts for diverse count processes have been developed during the last decades, covering serially independent or autocorrelated counts, and counts having full $\mathbb{N}_0$ as their range (unbounded counts) or just the finite subset $\{0, ..., n\}$ with specified $n \in \mathbb{N}$ (bounded counts). The simplest type of control chart is obtained by plotting the counts X_t themselves against appropriately chosen CLs, which is called a *c*-chart if monitoring unbounded counts and an *np*-chart for bounded counts. Such types of Shewhart control chart can be classified as being memory-less as the tth plotted statistic does not comprise information about earlier counts (at least not beyond the mere effect of autocorrelation). Consequently, while Shewhart charts might be quick in detecting a sudden strong process shift, they are slow in detecting small process changes. Therefore, also several memory-type control charts for counts have been proposed, whereas in the present research, our focus is on exponentially weighted moving average (EWMA) charts. Some recent references on EWMA-type control charts for counts are Anastasopoulou & Rakitzis, 2022a, 2022b; Borror *et al.*, 1998; Gan, 1990; Morais & Knoth, 2020; Morais *et al.*, 2018; Rakitzis *et al.*, 2015; Weiss,

DOI: 10.1201/9781032664330-1

2011. Further references (also on other types of count control charts) can be found in Alevizakos & Koukouvinos, 2020; Weiss, 2015, 2018. The default version of the EWMA chart plots the statistics

$$Z_0 = \mu_0, \quad Z_t = \lambda \cdot X_t + (1-\lambda) \cdot Z_{t-1} \quad \text{for } t = 1, 2, \ldots \tag{1.1}$$

against a specified lower and upper CL (LCL and UCL, respectively), where $\mu_0 > 0$ denotes the in-control mean of (X_t). The smoothing parameter $\lambda \in (0,1]$ in (1.1) controls the strength of the memory (the smaller λ, the stronger the memory). If $\lambda = 1$, then (1.1) reduces to the c- or np-chart, respectively.

All the aforementioned count EWMA charts are mainly designed to detect shifts in the process mean, although they may sometimes ("accidentally") also react to increases in variance or changes in the autocorrelation structure. In the present research, however, our focus is on "more sophisticated" out-of-control scenarios, namely where the mean changes together with further distributional properties (i.e., where the process change cannot be traced back to a sole change in the mean parameter), or where "purely distributional changes" (not affecting the mean) happen. Such "distributional changes" might be increases or decreases in dispersion compared to the in-control model (overdispersion or underdispersion, respectively), or an excessive number of zero counts (zero inflation), to mention those being most relevant in practice. The basic idea of our approach is as follows. Many common count distributions can be characterized by a type of moment identity, referred to as *Stein identity*, which has to hold for a large class of functions (e.g., all bounded functions on $\mathbb{N}_0$). The idea to develop such identities dates back to Stein, 1972, 1986, and further contributions and references can be found in Landsman & Valdez, 2016; Sudheesh, 2009; Sudheesh & Tibiletti, 2009. Recently, such Stein identities were successfully used to develop powerful goodness-of-fit (GoF) tests for counts, see Betsch *et al.*, 2022; Weiss *et al.*, 2023 for Stein-type GoF-tests for independent and identically distributed (i. i. d.) counts, and Aleksandrov, Weiss & Jentsch, 2022; Aleksandrov, Weiss, Jentsch & Faymonville, 2022 for tests for count time series. Thus, it suggests itself to utilize these Stein identities also for developing sequential test procedures, namely count control charts for relevant types of in-control models. In a first research Weiss, 2024, this idea was tried for the special case of i. i. d. Poisson counts, and the achieved ARL performance was quite appealing. This motivates to develop and investigate Stein-based control charts for count data on a much broader scale, namely for various different count distributions and not only for i. i. d. but also for time series data. More precisely, we focus on the three most common count distributions in practice, namely Poisson (Poi), negative binomial (NB), and binomial (Bin), and we include first-order autoregressive (AR(1)) models having either Poi-, NB-, or Bin-distributed marginal distributions in our research.

The outline of this chapter is as follows. In Section 1.2, we briefly present the count models used for this research, and we provide the Stein identities for the Poi-, NB-, and Bin-distribution. These are used in Section 1.3 to construct novel EWMA-type control charts for counts, where the chart design with respect to diverse out-of-control scenarios is discussed in detail. In Section 1.4, results from a simulation study are presented, which allow to analyze the Stein EWMA charts' ARL performance. Section 1.5 then investigates a real-world data example on registrations in the emergency department of a children's hospital, which illustrates the application and interpretation of the novel Stein EWMA charts in practice. Finally, Section 1.6 concludes the article and discusses possible directions for future research.

1.2 COUNT MODELS AND STEIN IDENTITIES

The two most common distributions for unbounded counts are the Poi- and NB-distribution, while the Bin-distribution is the default choice for bounded counts; see Johnson *et al.*, 2005 for details and properties. The Poi(μ)-distribution with mean $\mu > 0$ is known to be equidispersed; i.e., its variance σ^2 satisfies $\sigma^2 = \mu$. By contrast, NB($\nu, \frac{\nu}{\nu+\mu}$) with $\nu, \mu > 0$ exhibits overdispersion relative to the Poi-distribution because its variance $\sigma^2 = (1+\frac{\mu}{\nu})\mu$ always exceeds the mean. The Bin$(n,\mu/n)$-distribution with $n \in \mathbb{N}$ and $\mu \in (0,n)$, in turn, refers to counts having the bounded range $\{0,\ldots,n\}$.

There is a huge variety of time series models related to the Poi-, NB-, or Bin-distribution; see Weiss, 2018 for a comprehensive overview. Here, many models use a *conditional* Poi-, NB-, or Bin-distribution (regression models), while the marginal distribution does not belong to any parametric model family. However, to be able to apply a Stein identity to given data, we need to specify the corresponding *marginal* distribution. Regarding the count time series models proposed so far Weiss, 2018, only a considerably smaller number of models have a Poi-, NB-, or Bin-*marginal* distribution. For the sake of studying the possible effects of serial dependence on the charts' performance, we shall consider three of the AR(1)-type processes surveyed by Weiss, 2008, whose model definitions use so-called thinning operators as integer substitutes of the multiplication:

- the Poi-INAR(1) process (integer AR) defined by

$$X_t = \rho \circ X_{t-1} + \varepsilon_t \quad \text{with i. i. d. } \varepsilon_t \sim \text{Poi}\big(\mu(1-\rho)\big), \tag{1.2}$$

- the NB-IINAR(1) process (iterated-thinning INAR) defined by

$$X_t = \rho \circledast_\pi X_{t-1} + \varepsilon_t \quad \text{with i. i. d. } \varepsilon_t \sim \text{NB}(\nu,\pi) \text{ and } \pi = \tfrac{\nu}{\mu(1-\rho)+\nu}, \tag{1.3}$$

- the BinAR(1) process defined by

$$X_t = \alpha \circ X_{t-1} + \beta \circ (n - X_{t-1}) \quad \text{with } \beta = (1-\rho)\tfrac{\mu}{n},\ \alpha = \beta + \rho. \tag{1.4}$$

Definitions (1.2) and (1.4) use the binomial thinning operator, which is defined by requiring a conditional Bin-distribution, namely $\theta \circ X | X \sim \text{Bin}(X,\theta)$ for $\theta \in (0,1)$. Definition (1.3), in turn, uses iterated thinning defined as $\rho \circledast_\pi X = \sum_{i=1}^{(\pi\rho)\circ X} Y_i$, where the counting series (Y_i) is i. i. d. according to $Y_i \sim 1 + \text{NB}(1,\pi)$. The crucial point for the subsequent research is the following.

Models (1.2)–(1.4) lead to stationary Markov chains with AR(1)-like autocorrelation function (ACF) $\rho(h) = \rho^h$ for time lags $h \in \mathbb{N}$, where $\rho \in (0,1)$ for all three models. The stationary marginal distributions are Poi(μ), NB$(\nu, \frac{\nu}{\nu+\mu})$, and Bin$(n,\mu/n)$, respectively; see Weiss, 2008. The parameter ρ allows to control the extent of serial dependence, where the boundary case $\rho \to 0$ leads to i. i. d. counts.

Independent of the value of ρ, we are always concerned with a Poi-, NB-, or Bin-marginal distribution. Any of these three distributions is uniquely characterized by a corresponding Stein identity; see Sudheesh & Tibiletti, 2009:

- $X \sim \text{Poi}(\mu)$ if and only if

$$E\big[X\,f(X)\big] = \mu\,E\big[f(X+1)\big] \tag{1.5}$$

holds for all bounded functions $f : \mathbb{N}_0 \to \mathbb{R}$;

- $X \sim \text{NB}(\nu, \frac{\nu}{\nu+\mu})$ if and only if

$$(\nu+\mu)\,E\big[X\,f(X)\big] = \mu\,E\big[(\nu+X)\,f(X+1)\big] \tag{1.6}$$

holds for all bounded functions $f : \mathbb{N}_0 \to \mathbb{R}$;

- $X \sim \text{Bin}(n,\mu/n)$ if and only if

$$(n-\mu)\,E\big[X\,f(X)\big] = \mu\,E\big[(n-X)\,f(X+1)\big] \tag{1.7}$$

holds for all bounded functions $f : \mathbb{N}_0 \to \mathbb{R}$.

Identity (1.5) is commonly referred to as the Stein–Chen identity (Chen, 1975). Note that the identities (1.5)–(1.7) constitute non-trivial statements only if f is not constant on $\mathbb{N}_0$ and if f is not identical to zero on $\mathbb{N}$.

In the subsequent Section 1.3, we shall derive control charts from the identities (1.5)–(1.7). As these identities have to hold for all bounded functions f under in-control conditions, we can select any choice of f for defining the control charts. This degree of freedom can be used to achieve particular sensitivity regarding a specified type of out-of-control scenario.

1.3 STEIN EWMA CHARTS FOR COUNTS

The three identities (1.5)–(1.7) always depend on three types of moment: the mean μ, the moment $E\big[X\,f(X)\big]$, and a moment involving $f(X+1)$. The idea of the GoF-tests in Aleksandrov, Weiss & Jentsch, 2022; Aleksandrov, Weiss, Jentsch & Faymonville, 2022; Weiss *et al.*, 2023 as well as of the Poisson EWMA charts in Weiss, 2024 was to derive a statistic by solving the identities in a certain way and to substitute the involved population moments by appropriate types of sample moments. In Weiss, 2024, two types of constructing an EWMA control chart were analyzed, and it turned out that one of these types is clearly superior, namely the one called "ABC-EWMA chart". Building on this experience, we propose the following Stein EWMA charts for sequentially monitoring Poi-, NB-, or Bin-counts, respectively, where $E_0[\cdot]$ expresses that the expectation is computed with respect to the in-control model. For all three Stein EWMA charts, we compute

$$\begin{aligned} A_0 &= E_0\big[X\,f(X)\big], \qquad A_t = \lambda\cdot X_t\,f(X_t) + (1-\lambda)\cdot A_{t-1},\\ C_0 &= \mu_0, \qquad C_t = \lambda\cdot X_t + (1-\lambda)\cdot C_{t-1}, \quad \text{for } t=1,2,\ldots \end{aligned} \tag{1.8}$$

Furthermore, we compute for the

- $\text{Poi}(\mu_0)$ in-control model:

$$\begin{aligned} B_0 &= E_0\big[f(X+1)\big], \qquad B_t = \lambda\cdot f(X_t+1) + (1-\lambda)\cdot B_{t-1},\\ Z_0^{\text{S}} &= 1, \qquad Z_t^{\text{S}} = \frac{A_t}{B_t\,C_t}, \quad \text{for } t=1,2,\ldots; \end{aligned} \tag{1.9}$$

- $\text{NB}(\nu, \frac{\nu}{\nu+\mu_0})$ in-control model:

$$\begin{aligned} B_0 &= E_0\big[(\nu+X)\,f(X+1)\big],\\ B_t &= \lambda\cdot(\nu+X_t)\,f(X_t+1) + (1-\lambda)\cdot B_{t-1},\\ Z_0^{\text{S}} &= 1, \quad Z_t^{\text{S}} = \frac{(\nu+C_t)\,A_t}{B_t\,C_t}, \quad \text{for } t=1,2,\ldots; \end{aligned} \tag{1.10}$$

- $\text{Bin}(n, \mu_0/n)$ in-control model:

$$\begin{aligned} B_0 &= E_0\big[(n-X)\,f(X+1)\big],\\ B_t &= \lambda\cdot(n-X_t)\,f(X_t+1) + (1-\lambda)\cdot B_{t-1},\\ Z_0^{\text{S}} &= 1, \quad Z_t^{\text{S}} = \frac{(n-C_t)\,A_t}{B_t\,C_t}, \quad \text{for } t=1,2,\ldots \end{aligned} \tag{1.11}$$

We define the respective *Stein EWMA chart* by plotting the statistics Z_t^{S} against appropriately chosen $\text{LCL} < 1 < \text{UCL}$, where we expect Z_t^{S} to vary closely around 1 under in-control assumptions.

The actual chart design now comprises two steps. First, we have to select the function f involved in (1.8)–(1.11). Here, one could generally choose any bounded function on $\mathbb{N}_0$, recall (1.5)–(1.7), except trivial choices like f being constant on $\mathbb{N}_0$, or $f \equiv 0$ on $\mathbb{N}$. But not any choice of f will lead to an appealing chart performance. Instead, f has to be chosen with respect to the anticipated out-of-control scenario, in an analogous way as proposed by Aleksandrov, Weiss & Jentsch, 2022; Aleksandrov, Weiss, Jentsch & Faymonville, 2022; Weiss *et al.*, 2023; Weiss, 2024. The idea is to interpret f as a weight function within the "A- and B-moments", which puts unequal weight on the integers in $\mathbb{N}_0$. For a given anticipated out-of-control scenario, one should put most weight on those regions of $\mathbb{N}_0$ where one gets the strongest departures from the in-control model. If we want to uncover zero inflation, for example, we should choose an f with a relatively large weight close to zero counts, whereas for "general" overdispersion (i.e., if the probability mass function (PMF) is flattened compared to the in-control model), it is advantageous to put more weight on large counts. So in accordance to the recommendations in Weiss, 2024, the following choices of f are considered in these two cases:

- for uncovering overdispersion relative to the in-control model, we use the linear weights $f(x) = |x-1|$;
- for uncovering zero inflation relative to the in-control model, we use the root weights $f(x) = |x-1|^{1/4}$.

The case of underdispersion has not been investigated by Weiss, 2024. Thus, as a starting point, we follow the findings of Weiss *et al.*, 2023 on GoF-tests and

- for uncovering underdispersion relative to the in-control model, we use the inverse weights $f(x) = 1/(x+1)$.

The described approach for choosing the weight function is later illustrated in some more detail when discussing Figure 1.1. The ARL performance of the aforementioned choices of f is analyzed in Section 1.4. As we shall see, the underdispersion scenario is much more demanding than overdispersion or zero inflation. Therefore, further choices for f shall be considered later in Section 1.4.2.

After having specified f, the second step of chart design is the choice of the triple $(\lambda, \mathrm{LCL}, \mathrm{UCL})$. Assume for the moment that the smoothing parameter λ has already been fixed. Then, LCL and UCL are determined based on ARL considerations. While different ARL concepts exist in the literature Knoth, 2006, it is common to use the zero-state ARL for evaluating the in-control performance. The simplest solution (the one used here) is symmetric CLs of the form $\mathrm{LCL} = 1-L$ and $\mathrm{UCL} = 1+L$, where L is chosen such that the desired in-control ARL (ARL_0) is met in close approximation (the textbook choice is the target value 370). Alternatively, one could define asymmetric CLs by also considering the out-of-control ARL performance, but this is only reasonable if a particular out-of-control scenario has been fixed. For example, if one assumes that the out-of-control models differ from the in-control one solely in terms of the mean μ, then asymmetric CLs may allow to obtain an unbiased ARL performance with respect to μ, i.e., the ARL as a function of μ is maximal in the in-control mean μ_0 and decreases symmetrically around μ_0; see Section 4 in Morais & Knoth, 2020. But as we shall consider a broad variety of out-of-control scenarios, we restrict to symmetric CLs here, i.e., $\mu_0 \mp L$ for ordinary EWMA and $1 \mp L$ for Stein EWMA charts.

At this point, let us recall that the above approach for choosing f assumes that we have specified a *single* relevant out-of-control scenario, such as "zero inflation". If the application context allows for *various* out-of-control scenarios (e.g., if underdispersion could also be possible), one can run multiple Stein EWMA charts in parallel, each designed for a different type of deterioration. This is later done in Section 1.5 when monitoring the emergency counts. Therefore, one Stein EWMA chart is designed to detect overdispersion, another one for zero inflation, and a third one for underdispersion. Then, the observed pattern of alarms enables a kind of targeted diagnosis. For example, if the "zero inflation"-chart is the first to trigger an alarm, we conclude that we might be confronted with zero inflation (rather than with underdispersion, etc.). Certainly, if running many charts in parallel, the risk increases that one of these charts gives a false alarm ("multiple testing").

Generally, the risk of false alarms is controlled by setting a target value ARL_0 for the in-control ARL. As described before, in our simulation study, we set $\text{ARL}_0 \approx 370$ for each single chart for comparability. But if multiple control charts are applied simultaneously, it would also be possible to determine their CLs such that the *joint* in-control ARL meets a target value, i.e., the ARL would be defined as the expected time until one of the charts triggers the first alarm.

Another remark refers to the choice of the smoothing parameter λ. To ensure sufficient memory, one often chooses small values for λ, such as $\lambda \in \{0.25, 0.10, 0.05\}$. The actual choice of λ might be done based on out-of-control ARLs. For example, we could first compute several candidate triples $(\lambda_i, \text{LCL}_i, \text{UCL}_i)$ having the same ARL_0. Then, we fix a specified out-of-control scenario and select that triple where the corresponding out-of-control ARL is closest to a given target value. In this way, one might end up with different λ for different Stein EWMA charts. But as we choose $f(x)$ with respect to different out-of-control scenarios, comparable chart designs are most easily obtained by a unique λ. Furthermore, all Stein EWMA charts follow the same type of recursive scheme, which makes it plausible to use a unique value for λ. Thus, in the sequel, we follow the practice in Weiss, 2024 and set $\lambda = 0.10$ throughout our analyses.

Let us conclude this section with a note on out-of-control ARLs. As mentioned before, there are a couple of competing ARL concepts for evaluating a control chart's out-of-control performance; see Knoth, 2006, which are typically defined with respect to different positions of the change point. Here, the change point τ expresses the time when the process turns out of control, i.e., the process is in control (out of control) for $t < \tau$ ($t \geq \tau$). The out-of-control zero-state ARL assumes that the process change happens right at the beginning of process monitoring, i.e., the change point equals $\tau = 1$. The conditional expected delay $\text{CED}(\tau)$, by contrast, assumes a later change point $\tau > 1$, where the limit $\tau \to \infty$ leads to the steady-state ARL. While $\tau = 1$ would happen for a mis-specified in-control model, late change points are often more realistic in practice (but one does not know the true value of τ in advance). Fortunately, it turned out in the initial analyses of Weiss, 2024 that the exact position of τ does not have a substantial effect on the computed out-of-control ARL values, i.e., the Stein EWMA charts showed roughly the same performance for early and late process changes. Since the charts of the present study are defined in complete analogy to those of Weiss, 2024, we expect the value of the zero-state ARL to be representative also for $\tau > 1$ (for "problematic" types of control chart, see the discussion in Knoth *et al.*, 2023). For this reason, we restrict our subsequent analyses to zero-state ARLs, and these are computed based on simulations. So one simulates the considered process for R times (we use $R = 10^4$), applies the specified control chart to it, and determines the time until the first alarm. In this way, one gets R run lengths $l_1, \ldots, l_R$, and the sample mean thereof provides an approximate value of the actual ARL.

1.4 SIMULATION-BASED PERFORMANCE ANALYSES

1.4.1 OVERDISPERSION OR ZERO INFLATION

Let us start our performance analyses with the case of overdispersion or zero inflation (compared to the in-control model). Strictly speaking, a zero-inflated distribution also exhibits increased dispersion, but this increase is caused by a single-point mass in zero. With overdispersion, by contrast, we refer to a flattened PMF compared to the in-control model's PMF. For unbounded counts, we use the NB-distribution for generating overdispersion, and the zero-inflated Poisson (ZIP) distribution for zero inflation. In the bounded-counts case, we use the beta-binomial (BB) distribution for overdispersion and the zero-inflated binomial (ZIB) distribution for zero inflation; see Appendix A in Weiss, 2018 for details on these distributions. For the sake of a unique representation, we specify any of the aforementioned models by the mean μ and by an appropriate dispersion index, namely by $I_P = \sigma^2/\mu$ for unbounded counts and $I_B = n\sigma^2/(\mu(n-\mu))$ for bounded counts. Note that $I_P = 1$ for the Poi-distribution while $I_P > 1$ for NB and ZIP. Analogously, $I_B = 1$ for the Bin-distribution and $I_B > 1$ for BB and ZIB. The in-control means are set at either

Table 1.1

Simulated ARLs of ordinary EWMA and Stein EWMA chart for Bin-counts with $n = 10$. BB and ZIB with $I_B = 5/3$.

μ_0	$\mu =$	$\mu_0 - 0.25$	μ_0	$\mu_0 + 0.25$	$\mu_0 - 0.25$	μ_0	$\mu_0 + 0.25$	$\mu_0 - 0.25$	μ_0	$\mu_0 + 0.25$
(a)	i. i. d. counts									
		EWMA		*0.7805*	EWMA^S	$\lvert x-1\rvert$	*0.534*	EWMA^S	$\lvert x-1\rvert^{1/4}$	*0.4235*
2	ZIB	69.2	87.9	51.6	22.7	26.1	29.2	16.6	19.1	21.9
	Bin	171.5	**370.2**	99.3	240.2	**369.5**	550.6	191.5	**370.6**	671.0
	BB	71.5	90.0	51.8	25.9	29.2	33.3	30.8	40.5	55.6
		EWMA		*0.974*	EWMA^S	$\lvert x-1\rvert$	*0.2115*	EWMA^S	$\lvert x-1\rvert^{1/4}$	*0.0511*
5	ZIB	69.2	87.9	51.6	18.9	19.9	20.6	12.9	14.0	15.5
	Bin	162.5	**369.5**	164.1	307.2	**370.1**	443.4	311.5	**369.5**	417.2
	BB	66.7	88.2	66.3	25.1	26.9	29.0	27.4	28.9	30.3
(b)	AR(1) counts with $\rho = 0.5$									
		EWMA		*1.191*	EWMA^S	$\lvert x-1\rvert$	*0.639*	EWMA^S	$\lvert x-1\rvert^{1/4}$	*0.568*
2	ZIB	77.9	73.8	55.8	22.9	24.3	25.3	23.8	26.9	30.0
	Bin	384.8	**370.1**	158.0	247.1	**369.7**	554.4	211.9	**371.2**	634.3
	BB	105.0	105.7	77.2	33.0	39.5	47.0	44.7	58.9	77.1
		EWMA		*1.493*	EWMA^S	$\lvert x-1\rvert$	*0.225*	EWMA^S	$\lvert x-1\rvert^{1/4}$	*0.0528*
5	ZIB	30.8	29.9	28.0	7.9	7.7	7.3	7.1	6.9	6.5
	Bin	258.0	**369.1**	257.3	316.7	**370.9**	424.1	293.9	**370.9**	421.8
	BB	86.3	96.1	85.9	35.3	37.6	39.5	37.9	39.4	40.7

Notes: In-control ARL printed in bold font, CL L shown in italic font. "EWMA" = ordinary EWMA; "EWMA^S" = Stein EWMA, where weight functions $f(x) = |x-1|$ and $|x-1|^{1/4}$.

Interpretation:			
	sole mean change	pure distrib. change	zero inflation / overdispersion

$\mu_0 = 2$ (low counts) or $\mu_0 = 5$ (medium counts). The data-generating process (DGP) produces either i. i. d. counts ($\rho = 0$) or AR(1)-like counts with $\rho = 0.5$, recall (1.2)–(1.4). The simulated ARLs rely on 10^4 replications, the target value for ARL_0 is 370.

As the case of overdispersion or zero inflation relative to an in-control Poi-distribution was already analyzed by Weiss, 2024, we restrict our subsequent analyses to in-control NB- and Bin-models; see Tables 1.1–1.2 for the obtained results. The Bin-results are quite close to the Poi-results anyway, which is not surprising as these distributions are related to each other by the Poisson limit theorem. Let us start our discussion with the Bin-case in Table 1.1. The design of this table (and of any further table) is as follows. It consists of twelve 3×3-blocks with a negative (positive) mean shift on the left (right) and no mean shift in the centre column. In the Bin-rows, the binomial distribution is preserved, whereas we have distributional changes in the ZIB- and BB-rows. Note that at the bottom of each table, there is a scheme for interpreting the ARLs within each block. Part (a) of each table refers to i. i. d. counts, part (b) to AR(1)-like counts. Note that the BB- and ZIB-AR(1) models are constructed as in (3.27) of Weiss, 2018 by replacing the Bin-thinnings in (1.4) by BB- or ZIB-thinnings, respectively. Finally, the first column of blocks always refers to the ordinary EWMA chart (1.1), whereas the remaining two columns refer to Stein EWMA charts, i.e., to (1.11) in case of Table 1.1.

Part (a) of Table 1.1 just confirms the findings of Weiss, 2024 for the Poi-case. If we are concerned with the textbook situation, i.e., if there is a change solely in the mean parameter (the distribution is preserved otherwise), then the ordinary EWMA chart is clearly the best choice. But if there is an additional distributional change, or even only a distributional change, while

Table 1.2
Simulated ARLs of ordinary EWMA and Stein EWMA chart for NB-counts with $I_P = 5/3$. ZIP with $I_P = 5/3$, and overdispersed NB ("oNB") with $I_P = 5/2$.

μ_0	$\mu =$	$\mu_0 - 0.25$	μ_0	$\mu_0 + 0.25$	$\mu_0 - 0.25$	μ_0	$\mu_0 + 0.25$	$\mu_0 - 0.25$	μ_0	$\mu_0 + 0.25$
(a)	i. i. d. counts									
		EWMA		*1.156*	EWMA^S	$\lvert x-1\rvert$	*0.349*	EWMA^S	$\lvert x-1\rvert^{1/4}$	*0.3146*
2	ZIP	506.6	462.0	149.7	139.4	257.2	481.5	54.2	81.0	124.9
	NB	605.8	**370.7**	133.1	172.1	**370.9**	892.2	154.0	**369.9**	1001.2
	oNB	171.9	135.1	75.6	45.2	67.2	103.8	52.2	86.9	163.2
		EWMA		*1.805*	EWMA^S	$\lvert x-1\rvert$	*0.1554*	EWMA^S	$\lvert x-1\rvert^{1/4}$	*0.0883*
5	ZIP	342.6	407.5	261.9	116.2	143.9	170.2	22.7	24.7	26.8
	NB	444.7	**370.8**	205.3	267.3	**369.8**	522.7	260.0	**370.1**	537.2
	oNB	130.2	124.7	94.8	42.9	51.3	59.1	55.5	68.7	87.3
(b)	AR(1) counts with $\rho = 0.5$									
		EWMA		*1.855*	EWMA^S	$\lvert x-1\rvert$	*0.45*	EWMA^S	$\lvert x-1\rvert^{1/4}$	*0.4415*
2	ZIP	1287.7	505.7	238.0	108.3	167.0	266.3	90.6	139.7	219.4
	NB	770.9	**369.7**	200.4	187.4	**370.7**	710.6	182.7	**370.7**	768.8
	oNB	288.9	178.8	115.2	61.1	93.5	141.8	75.4	123.1	210.1
		EWMA		*2.78*	EWMA^S	$\lvert x-1\rvert$	*0.177*	EWMA^S	$\lvert x-1\rvert^{1/4}$	*0.1105*
5	ZIP	502.7	408.0	284.2	147.8	178.4	214.4	93.6	110.2	131.1
	NB	510.8	**369.6**	242.5	278.0	**370.8**	494.2	276.7	**370.2**	492.8
	oNB	183.8	156.4	124.4	60.1	71.6	83.0	78.5	97.1	122.0

Notes: In-control ARL printed in bold font, CL L shown in italic font. "EWMA" = ordinary EWMA; "EWMA^S" = Stein EWMA, where weight functions $f(x) = |x-1|$ and $|x-1|^{1/4}$.

Interpretation:			
	sole mean change	pure distrib. change	zero inflation / overdispersion

the mean is kept fixed, then the Stein EWMA charts become superior. More precisely, if we are concerned with overdispersion (BB-model), then the weight function $f(x) = |x-1|$ leads to the lowest ARLs, whereas zero inflation is best detected using $f(x) = |x-1|^{1/4}$. It should be noted that we draw the same conclusions from part (a) of Table 1.2, which refers to the NB-case (in-control level of dispersion is $I_P = 5/3$) with Stein EWMA chart (1.10). There, overdispersion (relative to the in-control model) was generated by an NB-distribution with higher dispersion, namely $I_P = 5/2$, whereas we used the ZIP-distribution with $I_P = 5/3$ for causing zero inflation. Thus, we generally conclude that if monitoring i. i. d. counts and if being concerned with overdispersion under out-of-control conditions, the Stein EWMA chart with $f(x) = |x-1|$ is the best choice, whereas $f(x) = |x-1|^{1/4}$ performs best under zero inflation. If the type of out-of-control situation is not clear in advance, it is recommended to run all three charts of Tables 1.1–1.2 simultaneously. As these charts show a rather different ARL performance, they can be used for a kind of targeted diagnosis, i.e., one can conclude from the observed pattern of alarms on the type of out-of-control situation. For example, if the chart with $f(x) = |x-1|^{1/4}$ signals first, we conclude on zero inflation.

Next, let us look at parts (b) of Tables 1.1–1.2, i.e., on the case of positively correlated counts. Note that in Table 1.2, the overdispersed NB-counts are still generated by the NB-IINAR(1) model (1.3), whereas the ZIP-INAR(1) model was used for zero inflation, i.e., model (1.2) with ZIP-innovations (ε_t). In most cases in parts (b), CLs have to be increased and the charts' ARL performances get worse. But it still holds that the Stein EWMA charts are superior under overdispersion or zero inflation. As the only difference, the ARL performances of both Stein EWMA charts are now quite similar under zero inflation in the Bin-case, but not for the NB-case.

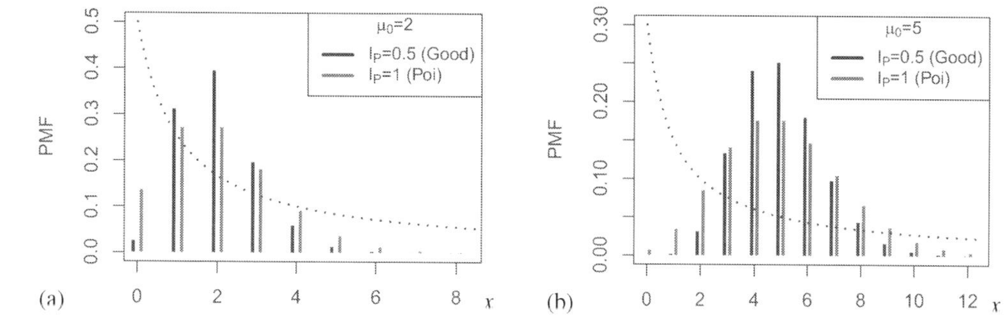

Figure 1.1 PMF plots for Good with $l_{\mathrm{P}} = 0.5$ vs. Poi, for (a) $\mu_0 = 2$ and (b) $\mu_0 = 5$. Dotted line proportional to $1/(x+1)$.

So altogether, the conclusions done for i. i. d. counts still apply, but with generally increased out-of-control ARLs due to the serial dependence.

1.4.2 UNDERDISPERSION

Underdispersion is the opposite phenomenon to overdispersion, i.e., the PMF is concentrated more closely around the mean. While there are hardly any models for underdispersion of bounded counts, the underdispersion of unbounded counts received more interest in the literature. For the subsequent simulations, we use the Good distribution Weiss, 2018, Appendix A for generating underdispersion relative to the Poi-distribution (see Table 1.3), whereas for underdispersion relative to an NB-distribution, we use either an NB-distribution with less dispersion or the Poi-distribution (see Table 1.4). It quickly gets clear from Tables 1.3–1.4 that underdispersion is quite demanding for process monitoring. If looking first at the ordinary EWMA chart (first columns of blocks), we recognize severely increasing ARLs in the presence of underdispersion. This is reasonable as fluctuations of the DGP are reduced such that small mean shifts (we have ± 0.25) hardly lead to a violation of the CLs. The second columns in Tables 1.3–1.4 refer to the Stein EWMA charts (1.9) and (1.10), where we used $f(x) = 1/(x+1)$ as recommended by Weiss *et al.*, 2023 in the context of GoF-tests. While this chart indeed performs well for the low mean $\mu = 2$, the ARL performance is very bad for $\mu = 5$, again demonstrating that process monitoring in the presence of underdispersion is quite demanding. To better understand these difficulties and to find a solution, let us look at Figure 1.1, where we visualize the effect of underdispersion (black) compared to the in-control model (grey) by PMF plots. It can be seen that the strongest deviations are in a small area below μ_0. For $\mu_0 = 2$ in (a), the main deviations happen for $x \in \{0, 1, 2\}$. In fact, $f(x) = 1/(x+1)$ (symbolized by the dotted graphs in Figure 1.1) puts most weight to this region, which explains the good ARL performance in this case. For $\mu_0 = 5$ in (b), however, the main deviations are for $x \in \{2, \ldots, 5\}$. As $f(x) = 1/(x+1)$ puts relatively low weight into this region, we now get a poor ARL performance.

To get a good performance regarding underdispersion, Figure 1.1 suggests to put most weight somewhat below the in-control mean, which leads to the following idea. The in-control PMF itself (plotted in grey in Figure 1.1), interpreted as a weight function, puts most weight around the mean of the in-control distribution. Thus, we may just shift this PMF downwards to move most weight below the mean. Inspired by Figure 1.1 and some simulation experiments, the idea was developed to define $f(x)$ by simply shifting the in-control PMF by 2. More precisely, for the in-control Poi-model of Table 1.3, we use $p_{\mathrm{P}}(x+2)$ as the weight function and $p_{\mathrm{N}}(x+2)$ for the NB-model of Table 1.4. Here, $p_{\mathrm{P}}(\cdot)$ and $p_{\mathrm{N}}(\cdot)$ abbreviate the PMFs of the in-control Poi- and NB-distribution, respectively. The computed chart designs and ARLs are summarized in the respective third column of Tables 1.3–1.4. We now get notably decreasing ARLs for increasing underdispersion in any scenario, both under i. i. d. and AR(1)-like counts.

Table 1.3
Simulated ARLs of ordinary EWMA and Stein EWMA chart for Poi-counts. Good with $I_P = 3/4$ (first row) and $I_P = 1/2$ (second row).

μ_0	$\mu =$	$\mu_0 - 0.25$	μ_0	$\mu_0 + 0.25$	$\mu_0 - 0.25$	μ_0	$\mu_0 + 0.25$	$\mu_0 - 0.25$	μ_0	$\mu_0 + 0.25$
(a)	i. i. d. counts									
		EWMA		*0.877*	EWMA^S	$1/(x+1)$	*0.223*	EWMA^S	$p_P(x+2)$	*0.608*
2	Poi	252.6	**369.1**	106.1	274.6	**368.9**	470.8	538.9	**370.3**	271.7
	Good	622.4	948.9	158.5	96.9	142.3	249.5	90.7	71.4	60.5
	Good	3611.8	6346.9	380.6	32.3	42.2	63.0	29.1	26.2	24.3
		EWMA		*1.388*	EWMA^S	$1/(x+1)$	*0.1775*	EWMA^S	$p_P(x+2)$	*0.293*
5	Poi	309.9	**371.4**	185.1	352.9	**370.5**	398.1	526.1	**368.7**	268.9
	Good	1006.0	1015.9	327.2	$> 10^4$	$> 10^4$	$> 10^4$	228.4	149.3	106.4
	Good	8154.5	7882.4	1168.9	$> 10^4$	$> 10^4$	$> 10^4$	52.5	40.1	33.2
(b)	AR(1) counts with $\rho = 0.5$									
		EWMA		*1.351*	EWMA^S	$1/(x+1)$	*0.2467*	EWMA^S	$p_P(x+2)$	*0.7235*
2	Poi	627.2	**371.0**	162.0	274.8	**370.0**	478.4	530.0	**370.5**	273.2
	Good	2177.5	814.6	261.3	153.7	242.0	439.2	143.5	108.9	89.6
	Good	$> 10^4$	2866.0	581.6	47.6	62.4	92.3	43.7	36.7	33.2
		EWMA		*2.123*	EWMA^S	$1/(x+1)$	*0.1707*	EWMA^S	$p_P(x+2)$	*0.345*
5	Poi	432.2	**369.8**	236.3	332.3	**370.1**	395.3	514.5	**370.1**	280.7
	Good	1369.3	909.2	446.1	2652.2	4029.5	5159.4	226.6	161.8	122.0
	Good	$> 10^4$	4264.3	1402.0	1856.0	6087.1	$> 10^4$	69.8	53.9	44.3
Notes:	In-control ARL printed in bold font, CL L shown in italic font. "EWMA" = ordinary EWMA; "EWMA^S" = Stein EWMA, where weight functions $f(x) = 1/(x+1)$ and $p_P(x+2)$.									
Interpre-tation:		sole mean change				pure distrib. change		increasing underdispersion		

1.5 AN ILLUSTRATIVE DATA APPLICATION

Weiss & Testik, 2015 analyzed a large set of daily count time series referring to registrations in the emergency department of a children's hospital. More precisely, these emergency counts were determined per 5-min interval from 08:00:00 to 23:59:59 on a day (so length $T = 192$), and the full set of time series covers the period from February 13 to August 13, 2009. Weiss & Testik, 2015 used the sixteen time series from February 13 to 28 to develop the in-control model (Phase-I analysis), namely a Poi-INAR(1) model with in-control mean $\mu_0 = 2.1$ and dependence parameter $\rho_0 = 0.78$. Control charts based on this model were then applied to prospective process monitoring (Phase-II application). While most Phase-II series did not contradict the in-control model, Weiss & Testik, 2015 recognized a few unusual days, two of which shall now serve as illustrative data examples.

Let us start our analyses by designing the control charts based on the in-control model. To make the results consistent with Section 1.4, the CLs of each chart are chosen such that the in-control ARL is close to 370. On the one hand, we consider the c-chart and the ordinary EWMA chart (1.1) as competitors. On the other hand, the new Stein EWMA chart (1.9) was used together with four weight function: $f(x) = |x-1|$ and $|x-1|^{1/4}$ for indicating possible overdispersion or zero inflation, and $f(x) = 1/(x+1)$ and $p_P(x+2)$ for underdispersion. While the ARLs of the c-chart are computed numerically exactly by using the Markov-chain approach Weiss, 2018, Section 8.2.2, we again used simulations with 10^4 replications for the remaining charts. Here, the c-chart is most difficult to design due to discreteness; the best design has LCL = 0 and UCL = 6 (so an only one-sided design, where counts $X_t > 6$ cause an alarm) and $\text{ARL}_0 \approx 326.2$. The EWMA chart designs (again with $\lambda = 0.1$) are truly two-sided, namely

Table 1.4

Simulated ARLs of ordinary EWMA and Stein EWMA chart for NB-counts with $I_P = 5/3$. Underdispersed NB ("uNB") with $I_P = 4/3$.

μ_0	$\mu =$	$\mu_0 - 0.25$	μ_0	$\mu_0 + 0.25$	$\mu_0 - 0.25$	μ_0	$\mu_0 + 0.25$	$\mu_0 - 0.25$	μ_0	$\mu_0 + 0.25$
(a)	i. i. d. counts									
		EWMA		*1.156*	EWMA^S	$1/(x+1)$	*0.2215*	EWMA^S	$p_N(x+2)$	*0.4163*
2	NB	605.8	**370.7**	133.1	240.3	**371.5**	590.6	367.7	**370.3**	325.2
	uNB	1531.2	799.1	202.7	280.0	380.5	568.9	315.1	213.5	152.7
	Poi	5998.4	3237.7	429.8	106.1	128.8	179.9	93.7	70.8	57.3
		EWMA		*1.805*	EWMA^S	$1/(x+1)$	*0.165*	EWMA^S	$p_N(x+2)$	*0.22*
5	NB	444.7	**370.8**	205.3	301.9	**369.1**	452.9	399.3	**369.1**	328.4
	uNB	1072.6	840.9	372.4	1185.9	1531.8	1921.6	455.4	313.1	226.8
	Poi	4236.6	3623.7	986.3	7006.1	$> 10^4$	$> 10^4$	132.6	96.9	76.5
(b)	AR(1) counts with $\rho = 0.5$									
		EWMA		*1.855*	EWMA^S	$1/(x+1)$	*0.2412*	EWMA^S	$p_N(x+2)$	*0.4626*
2	NB	770.9	**369.7**	200.4	266.3	**369.8**	514.0	440.5	**369.8**	293.9
	uNB	1797.1	685.4	315.4	233.3	319.1	441.7	248.6	177.5	137.2
	Poi	$> 10^4$	2329.1	713.9	103.7	136.9	190.5	87.6	68.6	58.2
		EWMA		*2.78*	EWMA^S	$1/(x+1)$	*0.1727*	EWMA^S	$p_N(x+2)$	*0.247*
5	NB	510.8	**369.6**	242.5	308.7	**370.7**	442.5	414.8	**370.0**	306.6
	uNB	1148.0	700.0	420.9	787.5	973.5	1222.5	329.4	241.2	191.5
	Poi	4918.5	2317.8	1082.8	1689.4	2741.1	4019.6	124.0	96.6	78.9
Notes:		In-control ARL printed in bold font, CL L shown in italic font. "EWMA" = ordinary EWMA; "EWMA^S" = Stein EWMA, where weight functions $f(x) = 1/(x+1)$ and $p_N(x+2)$.								
Interpretation:		sole mean change				pure distrib. change		increasing underdispersion		

- EWMA (1.1) with $L = 1.851$ and $\text{ARL}_0 \approx 370.3$;
- Stein EWMA (1.9) with $f(x) = |x-1|$, $L = 0.848$, $\text{ARL}_0 \approx 370.5$;
- Stein EWMA (1.9) with $f(x) = |x-1|^{1/4}$, $L = 0.829$, $\text{ARL}_0 \approx 370.5$;
- Stein EWMA (1.9) with $f(x) = 1/(x+1)$, $L = 0.2994$, $\text{ARL}_0 \approx 370.5$;
- Stein EWMA (1.9) with $f(x) = p_P(x+2)$, $L = 0.9594$, $\text{ARL}_0 \approx 370.2$.

As a first illustrative example, we apply these control charts to the emergency counts collected on March 28, 2009, see Figure 1.2. The c-chart triggers an alarm at $t = 23$, the ordinary EWMA only rather late at $t = 171$, both indicating that the in-control model is violated. The situation gets more clear if looking at the four Stein EWMA charts in Figure 1.2. The charts in (c) and (d) both trigger a very early alarm at $t = 6$, whereas those of (e) and (f) do not trigger an alarm at all. This indicates that we are confronted with overdispersion and zero inflation. In fact, comparing the sample properties of the emergency series from March 28, 2009, to the in-control model, we note a slight increase in the mean μ (from 2.1 to ≈ 2.323) and a substantial increase in dispersion I_P (from 1 to ≈ 1.312). Furthermore, the number of zeros equals 27, being larger than 23.5 as expected under the in-control model. So the Stein EWMA charts gave a clear diagnosis of the type of out-of-control situation. In addition, their alarm at $t = 6$ was not only much faster than those of c-chart and ordinary EWMA chart, but also than those of the control charts in Weiss & Testik, 2015 (these trigger at $t = 48$ and $t = 50$, respectively).

As the second illustrative example, see Figure 1.3, we consider the emergency counts from July 16, 2009, where no results are reported by Weiss & Testik, 2015. This time, neither c-chart nor ordinary EWMA chart trigger an alarm, so the user would conclude that the emergency series is in control. Looking at the Stein EWMA charts, however, the ones in (e) and (f) signal at times $t = 35$

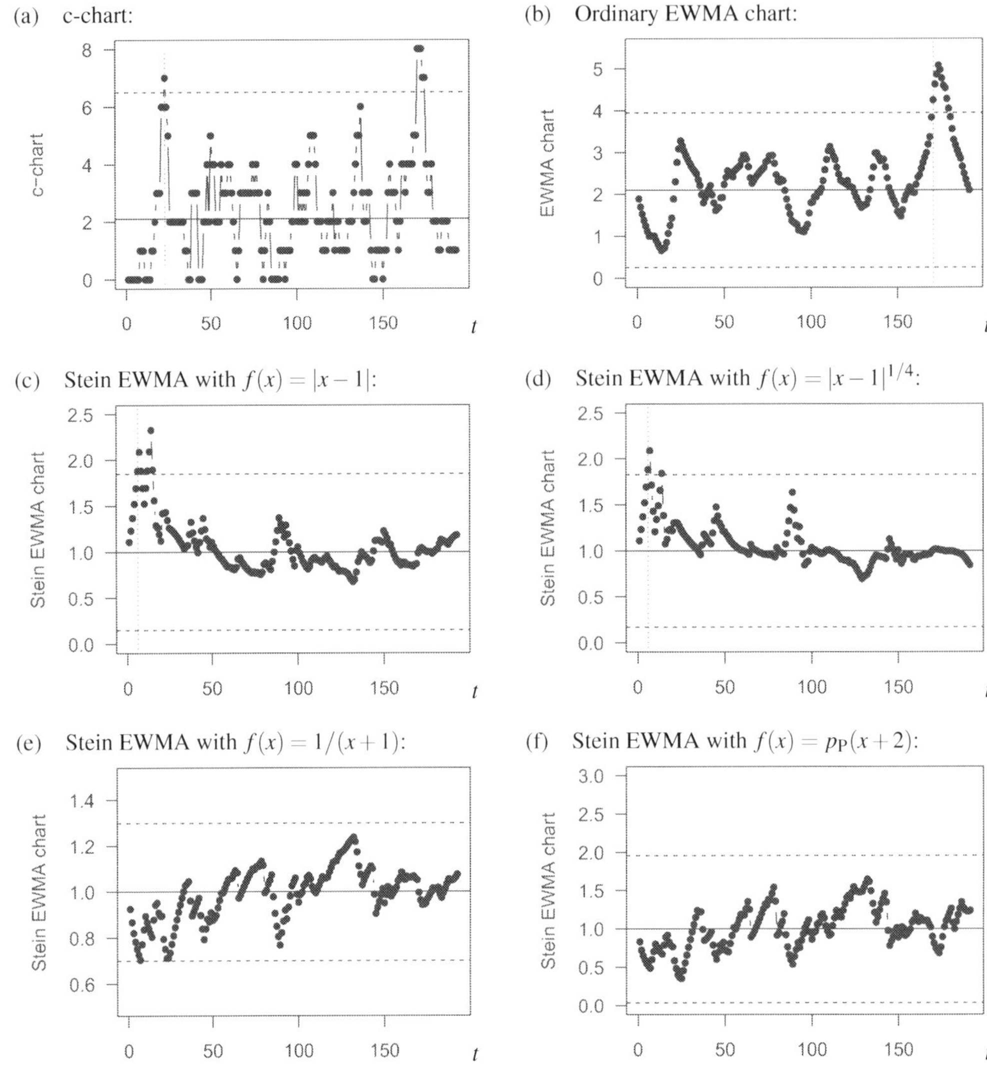

Figure 1.2 Emergency counts from March 28, 2009, see Section 1.5: c-chart in (a), ordinary EWMA chart in (b), and different Stein EWMA charts in (c)–(f). CLs as dashed lines, solid centre line, and first alarm at dotted line.

and $t = 74$, respectively, so we seem to be confronted with underdispersion. In fact, the sample value of I_P is ≈ 0.710 being notably smaller than 1, but also the mean has decreased from 2.1 to ≈ 1.911. As we know from Section 1.4.2, the ordinary EWMA chart is hardly able to detect mean shifts in the presence of underdispersion, so the novel Stein EWMA charts constitute a welcome complement for this scenario.

1.6 CONCLUSIONS AND FUTURE RESEARCH

For the monitoring of either Poi-, NB-, or Bin-counts, corresponding Stein EWMA charts were proposed, which are constructed by utilizing the respective Stein identity. Their ARL performance was investigated by simulations, both for i. i. d. and AR(1)-like counts. It turned out that

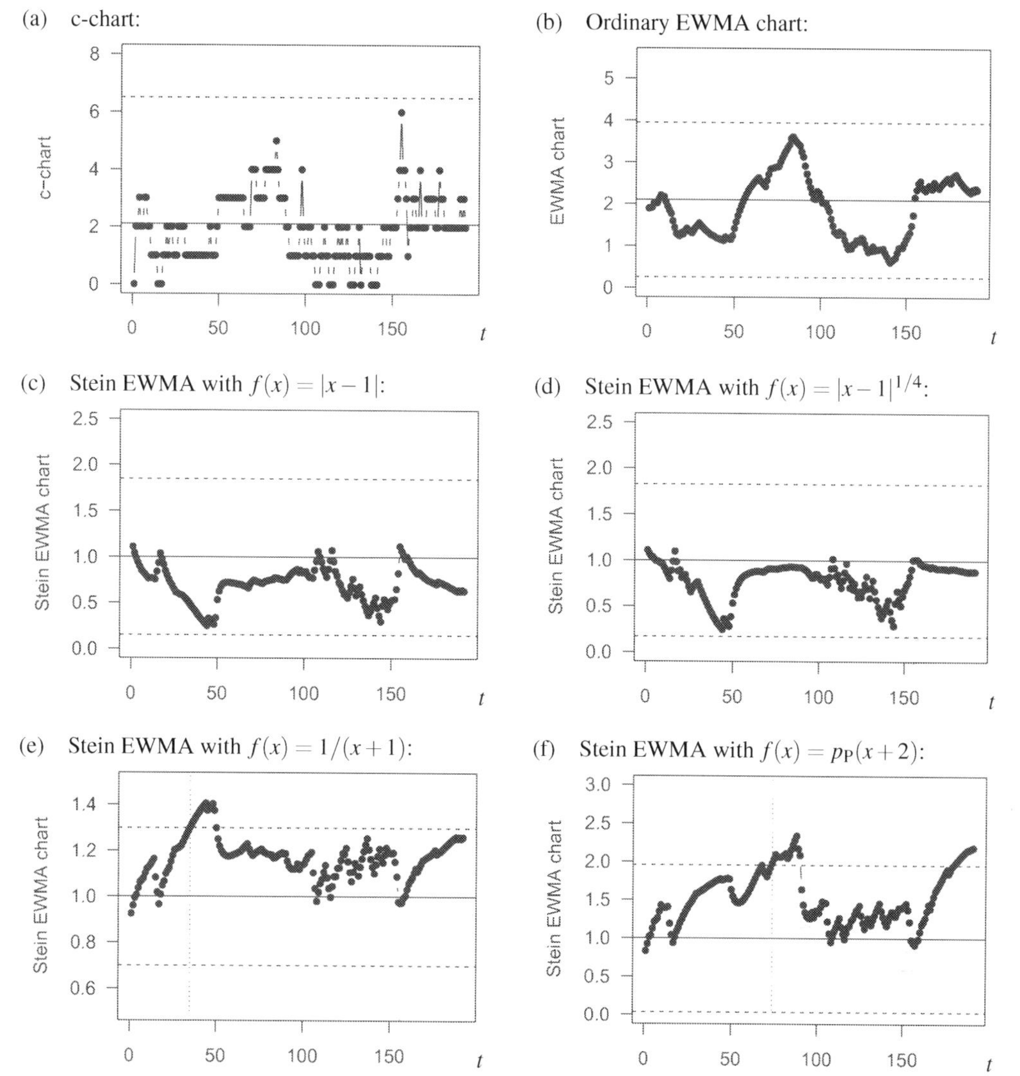

Figure 1.3 Emergency counts from July 16, 2009, see Section 1.5: c-chart in (a), ordinary EWMA chart in (b), and different Stein EWMA charts in (c)–(f). CLs as dashed lines, solid centre line, and first alarm at dotted line.

overdispersion is best detected by using the linear weights $f(x) = |x-1|$, and zero inflation by the root weights $f(x) = |x-1|^{1/4}$. Count monitoring in the presence of underdispersion, however, turned out to be quite demanding. While the ordinary EWMA charts performs very poorly in this case, the Stein EWMA chart with inverse weights $f(x) = 1/(x+1)$ has appealing out-of-control ARLs for low counts. Even more promising is the weight function obtained by a downward shift of the in-control PMF. These findings were also confirmed by the data example on emergency counts. Nevertheless, it appears that the monitoring of underdispersion requires additional research activity.

There are further directions for future research. As cumulative sum (CUSUM) charts often show a better out-of-control performance than EWMA charts, it would be relevant to develop and investigate Stein CUSUM charts for count processes. Here, a residuals-based approach in analogy to Weiss & Testik, 2015 would be attractive, as this would not only be applicable to Poi-, NB-, or Bin-marginals, but also to in-control models having a Poi-, NB-, or Bin-conditional distribution. Furthermore, Stein EWMA or CUSUM charts would also be interesting for continuously distributed variables data; for this case, Stein identities can be found in Landsman & Valdez, 2016; Sudheesh, 2009, among others.

Acknowledgements

The author thanks the two referees for their useful comments on an earlier draft of this article. This research was funded by the Deutsche Forschungsgemeinschaft (DFG, German Research Foundation) – Projektnummer 437270842.

REFERENCES

Aleksandrov, B., Weiss, C. H., & Jentsch, C. (2022). Goodness-of-fit tests for Poisson count time series based on the Stein–Chen identity. *Statistica Neerlandica, 76*(1), 35–64.

Aleksandrov, B., Weiss, C. H., Jentsch, C., & Faymonville, M. (2022). Novel goodness-of-fit tests for binomial count time series. *Statistics, 56*(5), 957–990.

Alevizakos, V., & Koukouvinos, C. (2020). A comparative study on Poisson control charts. *Quality Technology & Quantitative Management, 17*(3), 354–382.

Anastasopoulou, M., & Rakitzis, A. C. (2022a). EWMA control charts for monitoring correlated counts with finite range. *Journal of Applied Statistics, 49*(3), 553–573.

Anastasopoulou, M., & Rakitzis, A. C. (2022b). *Monitoring a BAR(1) process with EWMA and DEWMA control charts.* In Tran (Ed.), *Control charts and machine learning for anomaly detection in manufacturing* (77–103). Springer.

Betsch, S., Ebner, B., & Nestmann, F. (2022). Characterizations of non-normalized discrete probability distributions and their application in statistics. *Electronic Journal of Statistics, 16*(1), 1303–1329.

Borror, C. M., Champ, C. W., & Ridgon, S. E. (1998). Poisson EWMA control charts. *Journal of Quality Technology, 30*(4), 352–361.

Chen, L. H. Y. (1975). Poisson approximation for dependent trials. *Annals of Probability, 3*(3), 534–545.

Gan, F. F. (1990). Monitoring Poisson observations using modified exponentially weighted moving average control charts. *Communications in Statistics—Simulation and Computation, 19*(1), 103–124.

Johnson, N. L., Kemp, A. W., & Kotz, S. (2005). *Univariate discrete distributions* (*3rd*). John Wiley & Sons, Inc.

Knoth, S. (2006). *The art of evaluating monitoring schemes—how to measure the performance of control charts?* In Lenz & Wilrich (Eds.), *Frontiers in statistical quality control 8* (74–99). Physica-Verlag.

Knoth, S., Saleh, N. A., Mahmoud, M. A., Woodall, W. H., & Tercero-Gómez, V. G. (2023). A critique of a variety of "memory-based" process monitoring methods. *Journal of Quality Technology, 55*(1), 18–42.

Landsman, Z., & Valdez, E. A. (2016). The tail Stein's identity with applications to risk measures. *North American Actuarial Journal, 20*(4), 313–326.

Montgomery, D. C. (2009). *Introduction to statistical quality control* (*6th*). John Wiley & Sons, Inc.

Morais, M. C., & Knoth, S. (2020). Improving the ARL profile and the accuracy of its calculation for Poisson EWMA charts. *Quality and Reliability Engineering International, 36*(3), 876–889.

Morais, M. C., Knoth, S., & Weiss, C. H. (2018). An ARL-unbiased thinning-based EWMA chart to monitor counts. *Sequential Analysis, 37*(4), 487–510.

Qiu, P. (2014). *Introduction to statistical process control.* CRC Press, Taylor & Francis Group.

Rakitzis, A. C., Castagliola, P., & Maravelakis, P. E. (2015). A new memory-type monitoring technique for count data. *Computers & Industrial Engineering*, *85*, 235–247.

Stein, C. (1972). A bound for the error in the normal approximation to the distribution of a sum of dependent random variables. *Proceedings of the Sixth Berkeley Symposium on Mathematical Statistics and Probability*, *2*, 583–602.

Stein, C. (1986). *Approximate computation of expectations* (7). Institute of Mathematical Statistics.

Sudheesh, K. K. (2009). On Stein's identity and its applications. *Statistics & Probability Letters*, *79*(12), 1444–1449.

Sudheesh, K. K., & Tibiletti, L. (2009). Moment identity for discrete random variable and its applications. *Statistics*, *46*(6), 767–775.

Weiss, C. H. (2024). Control charts for Poisson counts based on the Stein—Chen identity. In *Advanced statistical methods in process monitoring, finance, and environmental science: Essays in honour of wolfgang schmid*, 195–209.

Weiss, C. H. (2008). Thinning operations for modelling time series of counts—a survey. *AStA Advances in Statistical Analysis*, *92*(3), 319–341.

Weiss, C. H. (2011). Detecting mean increases in Poisson INAR(1) processes with EWMA control charts. *Journal of Applied Statistics*, *38*(2), 383–398.

Weiss, C. H. (2015). SPC methods for time-dependent processes of counts—a literature review. *Cogent Mathematics*, *2*(1), 1111116.

Weiss, C. H. (2018). *An introduction to discrete-valued time series*. John Wiley & Sons, Inc.

Weiss, C. H., & Testik, M. C. (2015). Residuals-based CUSUM charts for Poisson INAR(1) processes. *Journal of Quality Technology*, *47*(1), 30–42.

Weiss, C., Puig, P., & Aleksandrov, B. (2023). Optimal Stein-type goodness-of-fit tests for count data. *Biometrical Journal*, *65*(2), 2200073.

2 Time-varying ARMA Models for Residual Control Charts

Kostas Triantafyllopoulos
School of Mathematical and Physical Sciences, University of Sheffield
E-Mail: k.triantafyllopoulos@sheffield.ac.uk.
ORCID: 0000-0002-4144-4092.

Eleanor Stillman
School of Mathematical and Physical Sciences, University of Sheffield
E-Mail: e.c.stillman@sheffield.ac.uk.
ORCID: 0000-0002-5971-0985.

2.1 INTRODUCTION

Statistical process control (SPC) and its numerous advances and extensions are popular tools for process monitoring and control. Following the ground-breaking work of Walter A. Shewhart (see e.g., Montgomery, 2009), the control chart plays an important role in the detection of process anomalies in both natural and industrial systems. In recent years, advances in high-frequency sampling and data extraction methods, have meant that control charts for individual observations have been developed. Such charts include the exponentially weighted moving average (EWMA) and cumulative sum (CUSUM) (see e.g., Montgomery, 2009 and references therein). However, these charts are designed assuming the observations are independent (often with additional distributional assumptions, typically Normality).

In many situations, the assumption of independent observations is invalid. This is particularly true in more recent non-industrial, observational settings, such as health surveillance or environmental monitoring, examples of which are studied in detail in Frisén *et al.*, 2011 and in Bersimis & Sachlas, 2022. In contrast to large-scale manufacturing processes where one can assess compliance of batches of items well-spaced in time, in observational or high-frequency applications there is little opportunity to design a sampling scheme to limit the effect of sequential correlation in the underlying process. One standard option in this case is to use a modified chart, in which the control limits have been adjusted to take account of the autocorrelation, see, for example, Osei-Aning *et al.*, 2017, Yao *et al.*, 2023 and references therein. An alternative is to adopt a model-based approach. Here one begins by fitting a model which is able to accommodate the identified structural complexities of the data and proceeds to base the control charting procedure on its residuals, or standardized residuals, which are almost invariably assumed to follow an (independent) Gaussian white noise process.

We discuss the concept of residual charting further in Section 2.2, but our conclusion is that very often the models proposed fail to take account of the full structure of the data and the residuals retain a degree of autocorrelation sufficient to undermine the monitoring and control function (see examples in Section 2.3). This is particularly true when the original process exhibits only local

DOI: 10.1201/9781032664330-2

rather than global (weak or second order) stationarity. For the remainder of the chapter, we turn our attention to seeking more sophisticated models which can make suitable allowance for such behaviour. An obvious class is the time-varying Auto Regressive Moving Average (tv-ARMA) models (e.g., Dahlhaus, 2012; Prado *et al.*, 2021; West & Harrison, 1997) in which one can allow for the dependence of successive terms in the main series and in the innovation process and permit the extent of that dependence to vary slowly over time.

It is suggested that tv-ARMA models might offer a parsimonious solution to the problem if efficient means can be found to estimate their parameters. In Section 2.4 this possibility is explored. We decide, in fact, to restrict attention to the tv-ARMA with only AR elements time-varying (and MA coefficients fixed or static), which appears to offer sufficient flexibility while limiting the number of parameters to be estimated and the complexity of that estimation process. The Expectation-Maximization (EM) algorithm is shown to provide a practical estimation option for such models.

Section 2.5 illustrates the use of tv-ARMA models with an analysis of a real-time series (Pan & Jarrett, 2004) from a production process, which has been studied by several authors previously, see e.g., Pan & Jarrett, 2004 and Triantafyllopoulos & Bersimis, 2016. The chapter closes with some concluding remarks in Section 2.6.

2.2 RESIDUAL CHARTS

Residual charting, introduced in Alwan & Roberts, 1988, is widely employed, as reported in the review Psarakis & Papaleonida, 2007. Under this approach, the residuals from a model fitted to so-called Phase I data are used to establish a control charting procedure for ongoing monitoring in Phase II.

In principle, any standard chart can be used for residual charting. There may, however, need to be further modifications to the charts to allow for the extra uncertainty introduced by the need for parameter estimation during the model-fitting procedure (Testik, 2005). In fact, new forms of charts have been introduced to take advantage of the model-based approach. In this setting, one can view the detection of process aberrations as a model comparison procedure, which attempts to detect whether the null model of Phase I has been superseded by an alternative at some point in Phase II. This admits the opportunity of using methods based on the likelihood ratio. This approach is adopted via the concept of "sufficient reduction" in a frequentist setting by Frisen and co-workers (see e.g., Frisén *et al.*, 2011), or via Bayes factors (Triantafyllopoulos & Bersimis, 2016) under the Bayesian paradigm.

Frisen *et al.* appeal to the sufficiency principle to derive their test procedures (Frisén *et al.*, 2011). This approach has the advantage that asymptotic results for the distribution of the likelihood ratio allow the provision of control limits (approximately) guaranteeing specified levels of performance. It also extends naturally to multivariate settings and to modelling discrete distributions for counting non-conforming items or defectives, etc. (Frisén *et al.*, 2011, Siripanthana & Stillman, 2016).

In contrast, Triantafyllopoulos and Bersimis Triantafyllopoulos & Bersimis, 2016 adopt a Bayesian viewpoint. An appeal to core Bayesian philosophy leads to consideration of the Bayes factor. Its definition as the ratio of average likelihoods under the competing models has a similar intuitive interpretation and possibilities for multivariate extension. An advantage over the likelihood ratio method is that it can be applied to any two models, not necessarily nested, but, in contrast, distributional results are not available and one is forced to reference the informal "grades" suggested by Jeffreys Jeffreys, 1961; see also the review article on Bayes factors Kass & Raftery, 1995.

The residual charting approach is probably more commonly utilized than modified control charting since there is clear value in modelling the process itself, but it does not offer a complete solution. We address some of its deficiencies below.

Firstly, there are practical problems in applying model-based solutions which have attracted surprisingly little attention. One key issue is the definition of the residuals themselves. To be specific, having selected a suitable model and estimated its parameters based on the Phase I data, the model is assumed to continue to Phase II, and the residuals are regarded as the deviations of the observations from its predicted values, but the precise mechanism by which the Phase II predictions are made is rarely precisely specified. Natural options are either to make all predictions from the Phase I data only (as 1,2,... step ahead forecasts), or to use sequential updating as each confirmed Phase II point is encountered and predict only a single point in advance (e.g., Box *et al.*, 2009). It is well-known that k-step ahead forecasts rapidly revert to the sample mean, so this version seems untenable, but the computational burden of repeated model refitting, or search for efficient updating schemes does not seem to be widely explored. In addition, the distinction between Phases I and II becomes blurred once Phase II data is utilized in the model-fitting. Further complications ensue if one wishes to provide a scheme which is robust, or easily computationally adaptable, to interventions made once an out-of-control alarm has been signalled; such interventions would obviously occur in practice.

Leaving aside these practical problems of the charts themselves, we turn now to the issue of establishing independent residuals. This central requirement is demanded by all standard charts, but it is easy to envisage situations where the assumption of residual independence is invalid. This can occur for two key reasons; the residuals might be autocorrelated if:

1. the model fit is poor, even if the "correct" model has been used;
2. the structural complexity of the data has been inadequately accounted for by the chosen model (model misspecification).

Problem (1) is usually limited by looking at measures of goodness of fit and revising the model choice until acceptable diagnostics are found. However, these diagnostics themselves are based on a null assumption of independence, and their behaviour when the null fails by dependence (as opposed to some other route) is not widely discussed. Given we are working with a single realization of a potentially complicated process and that, with real data under consideration, we always fit the "wrong model", autocorrelated residuals are commonly seen. Even when an apparently suitable model has been selected and the fit appears reasonable, autocorrelated residuals can still occur if, in fact, the true model is not well-represented by the very limited Phase I data. Therefore, it is always possible that the wrong model might be adopted from Phase I and that might not represent well the underlying process.

Model misspecification (2) is increasingly likely to occur as the true data structure becomes increasingly complex. In modern, closely-sampled, processes, it is almost inevitable that adjacent observations are dependent and so time series are frequently used as candidate models. However, even this rich class can frequently fail to capture the full complexity of the process. A particular example is a non-stationary, or only locally stationary, process; for a review of locally stationary processes, the reader is referred to Dahlhaus, 2012. We illustrate such problems with examples in Section 2.3.

We note that with non-stationary processes, standard control charts (for the mean) are doomed to failure since, by definition, the mean is a not constant baseline against which to check for change. Nonetheless, in many practical situations, changes in mean level can easily be accommodated by adapting processes, as long as the change is not too rapid. In these cases, the real necessity is to discern points which indicate "a more marked change than expected under these manageable slow drifts". It is suggested that one should routinely consider models which assume only local stationarity, since these a) represent genuinely manageable conditions for process operation in many cases and b) decrease the chance of autocorrelation in the residuals since time-dependent changes can be accommodated. A natural class of models to consider is the time-varying Auto Regressive Moving Average (tv-ARMA) family (e.g., Prado *et al.*, 2021; Triantafyllopoulos, 2021; West & Harrison, 1997) and this is considered further in Section 2.4.

2.3 AUTOCORRELATED RESIDUALS

In this section, we demonstrate the points made above using two simple experiments. More specifically: we simulate data, we fit a time series model and we discuss whether the residuals of each model are independent or not.

In the first example, we simulate 50 observations from a stationary autoregressive moving average model of orders 1 and 2, abbreviated as ARMA(1,2), so that observations y_t are generated by

$$y_t = 0.8y_{t-1} + \varepsilon_t + 0.5\varepsilon_{t-1} - 0.7\varepsilon_{t-2},$$

where ε_t is white noise with variance 2. This model is weakly stationary and causal, since $|0.8| < 1$ and it is also invertible since the MA coefficients lie inside the invertibility region; the invertibility region of this model is given by $\rho_1 + \rho_2 < 1$, $-\rho_1 + \rho_2 < 1$ and $|\rho_2| < 1$, where ρ_1, ρ_2 are the MA coefficients and here $\rho_1 = 0.5$ and $\rho_2 = -0.7$ satisfy these conditions; for more discussion on the invertibility region of this model the reader is referred to the classic textbook of Box *et al.* Box *et al.*, 2008, Section 3.3.4. The invertibility of the model guarantees that the autocorrelation function is unique. Fitting to this data an ARMA(1,2), whose parameters are estimated by the data, results in residuals that are uncorrelated and hence support the assertion of independence; the left panel of Figure 2.1 shows the residuals. Suppose now that the wrong model is fitted to this data; this is known as model misspecification. In this case we fit a ARMA(1,1) model, or

$$y_t = \phi y_{t-1} + \varepsilon_t + \rho\varepsilon_{t-1},$$

where ϕ is the AR coefficient and ρ is the MA coefficient. Hence, here we have omitted the second MA coefficient. Fitting this model results in residuals that are autocorrelated; this is depicted in the right panel of Figure 2.1. Similar results apply if we fit models such as ARMA(2,1), MA(2), AR(2), and so forth.

The example above gives some evidence of autocorrelated residuals being the result of fitting the "wrong" model. It may be argued that such a scenario may not be met in practice, the standard Box-Jenkins procedure Box *et al.*, 2008 can identify an appropriate model. In practice, considering fitting models to real data, there is not a "correct" model, and a number of so called "appropriate" models may result in autocorrelated residuals. This may be due to the use of a penalized likelihood model selection procedures, such as the Akaike or Bayesian information criteria, which can build models with fewer parameters than those needed to give uncorrelated residuals. For an example of this see Section 2.5 and in particular Table 2.1.

In our second example, we illustrate how autocorrelated residuals can arise in the face of non-stationarity, even when the "correct" model is fitted. We also take this opportunity to introduce methods for handling time-varying time series which we will extend later.

We begin by simulating 50 values from the time-varying AR(2) model

$$y_t = \phi_{1.t}y_{t-1} + \phi_{2.t}y_{t-2} + \varepsilon_t, \tag{2.1}$$

where $\phi_{i.t}$ are random walk processes, i.e., $\phi_{i.t} = \phi_{i.t-1} + w_{i.t}$, for $i = 1,2$ and $w_{i.t}$ and ε_t are independent white noises with variances 0.05 and 1, respectively. It is also assumed that $\phi_{1.t}$ and $\phi_{2.t}$ are independent for any t. Figure 2.2 shows the simulated data (left panel). We have fitted model 2.1 to the simulated data using the Kalman filter, as described in Triantafyllopoulos & Bersimis, 2016. Indeed, we cast 2.1 in state space form as

$$y_t = [y_{t-1}, y_{t-2}] \begin{bmatrix} \phi_{1.t} \\ \phi_{2.t} \end{bmatrix} + \varepsilon_t = x_t^\top \phi_t + \varepsilon_t,$$

$$\phi_t = \phi_{t-1} + w_t,$$

where $x_t = [y_{t-1}, y_{t-2}]^\top$ is the design vector, $\phi_t = [\phi_{1.t}, \phi_{2.t}]^\top$ is the state vector consisting of the time-varying AR coefficients and the innovation vector $w_t = [w_{1.t}, w_{2.t}]^\top$ is Gaussian white noise

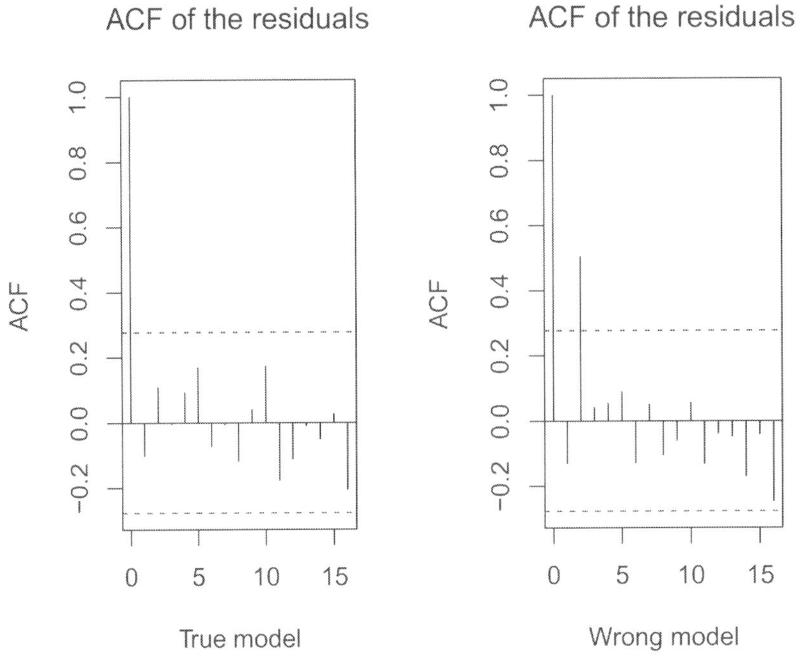

Figure 2.1 Correlograms of the residuals of simulated data from an ARMA(1,2) model: left panel shows values of the ACF of the residuals after fitting the true model and right panel after fitting the wrong model.

with covariance matrix $0.05\mathbf{I}_2$, where $\mathbf{I}_2$ denotes the 2×2 identity matrix. Here $\phi_{1,t}$ and $\phi_{2,t}$ are independent and are also independent of the assumed prior $\phi_2 \sim N(m_2, \mathbf{P}_2)$, for some vector mean m_2 and covariance matrix $\mathbf{P}_2$ (note that because y_{t-2} appears in x_t, the system is initialized at $t = 2$). Here we have chosen a weakly informative prior

$$m_2 = \begin{bmatrix} 1 \\ 1 \end{bmatrix} \quad \text{and} \quad \mathbf{P}_2 = 1000\mathbf{I}_2,$$

so that the precision $\mathbf{P}_2^{-1}$ is close to the zero matrix, which reflects our ignorance regarding the prior specification. Using the Kalman filter as discussed in Triantafyllopoulos & Bersimis, 2016 we can obtain the posterior distribution of ϕ_t, given $y_1, \ldots, y_t$, the one-step ahead forecast means $\hat{y}_{t+1} = E(y_{t+1} \mid y_1, \ldots, y_t)$ and the residuals or one-step ahead forecast errors as $e_{t+1} = y_{t+1} - \hat{y}_{t+1}$, for $t = 2, \ldots, n-1$ and some fixed value of $n > 2$. Kalman filters and Bayesian inference for state space models are discussed in detail by West & Harrison, 1997 and Triantafyllopoulos, 2021 among other authors.

We now consider whether residuals $e_3, e_4, \ldots, e_n$, are independent or not. In the right panel of Figure 2.2 the correlogram of the residuals is plotted together with confidence limits. It is clear from the plot that the sample autocorrelation function value at lag 1 is significant, which suggests the residuals are not uncorrelated and hence not independent. In particular, this plot suggests that an MA(1) evolution for the residuals might be appropriate.

Although this phenomenon (of presence of autocorrelation in the residuals of a correctly specified model) might not always occur, in our experience it occurs quite frequently. Sometimes residuals might indicate a particular clear time-dependence structure (as in this example), while at other times we may have a number of marginally significant residuals, which hints at a more generalized structural deficiency, but again casts doubt on the assumption of independence of the residuals. When we deal with real data, for which the time series model is not the "true model", the occurrence of autocorrelated residuals seems even more likely. In both examples discussed in this

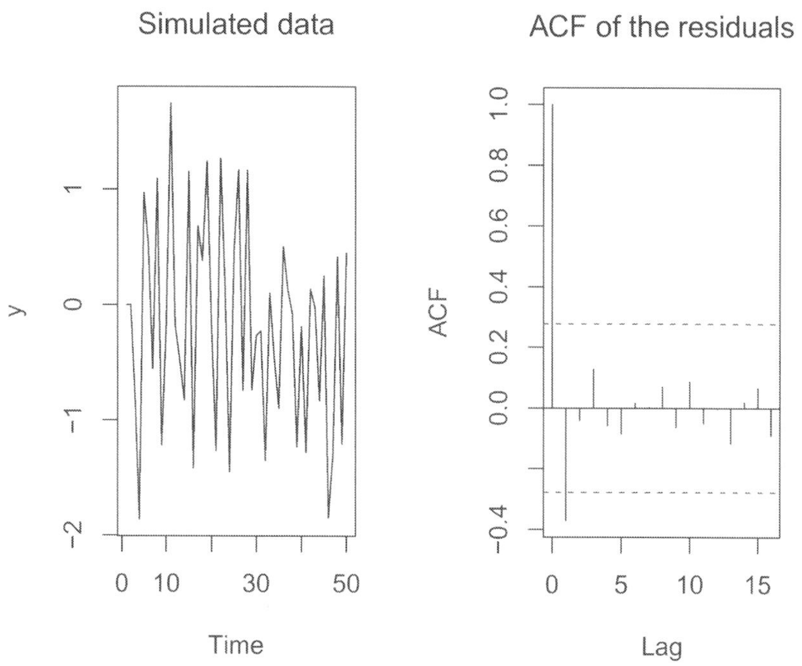

Figure 2.2 Simulated data from a time-varying AR(2) model (left panel) and correlogram of the residuals (right panel).

section we have suggested that autocorrelated residuals may be resulting either by fitting the"wrong" model or by the presence of non-stationarity, which sometimes is not easy to detect. Our experiments are based on single simulations of length 50 and are more suggestive than conclusive. For definitive results, a more thorough simulation study including Monte Carlo experiments would be conducted.

The conclusion of this short experiment is that it would be wrong to use standard residual control charts and thresholds. One other possible way forward might be to adopt a recursive approach and model the residuals of the original process with another time series model, but this seems inefficient and undesirable. Instead we seek a richer class of models (for the original series) to improve our chance of capturing the full model structure, leaving us with uncorrelated residuals which can be monitored using standard charts.

2.4 TIME-VARYING AUTO REGRESSIVE MOVING AVERAGE MODELS

2.4.1 MOTIVATION

In order to address the need for models of processes which are only locally stationary, Triantafyllopoulos & Bersimis, 2016 considered a time-varying autoregressive model. However, this might not be an appropriate model in many situations. For example, suppose that a process $\{y_t\}$ is generated by the time-varying moving average process of Triantafyllopoulos & Nason, 2007, described as

$$y_t = \xi_t + \gamma_t \xi_{t-1}, \tag{2.2}$$

where the time-varying parameter γ_t follows a suitable stochastic process and ξ_t is an i.i.d Gaussian innovation with a time-varying variance V_t. For more details of the definition and Bayesian inference of this model the reader is referred to Triantafyllopoulos & Nason, 2007. The point we make in this chapter is that if y_t is generated by 2.2, then it is unlikely that y_t would be modelled well with the time-varying autoregressive model of Triantafyllopoulos & Bersimis, 2016. Noting the correspondence between autoregressive and moving average processes (see e.g., Brockwell

& Davis, 1991) one might wonder whether a higher-order autoregressive model might suffice. However, a very high-order AR model would be needed to capture the dynamics of the first-order MA model 2.2, and this large number of AR time-varying coefficients might introduce excessive parameter uncertainty and result in a poor fit. As a result, we expect that residuals are likely to be autocorrelated.

This observation motivates a more general procedure than that proposed in Triantafyllopoulos & Bersimis, 2016 in order to consider the possibility that the process is driven by the time-varying autoregressive structure of Triantafyllopoulos & Bersimis, 2016.

2.4.2 MODEL DEFINITION

Suppose that for time t, observations y_t are generated by the time series model

$$y_t = \phi_{1.t} y_{t-1} + \cdots + \phi_{p.t} y_{t-p} + v_t, \quad t \geq \max(p,q)+1, \tag{2.3}$$

$$v_t = \varepsilon_t + \rho_1 \varepsilon_{t-1} + \cdots + \rho_q \varepsilon_{t-q}, \tag{2.4}$$

$$\phi_{i.t} = \alpha_i \phi_{i.t-1} + w_{i.t}, \quad i = 1, \ldots, p, \tag{2.5}$$

where $\phi_{i.t}$ are the time-varying coefficients, ε_t are the autocorrelated innovations with corresponding coefficients ρ_i, ε_t is Gaussian white noise with variance σ_ε^2 and $w_{i.t}$ is Gaussian white noise with variance σ_w^2; it is also assumed that $w_{i.t}$ is independent of $w_{j.t}$, for any $i \neq j$ and that $w_{i.t}$ is independent of ε_t, for any $i = 1, \ldots, p$. The time-varying coefficients $\phi_{i.t}$ are allowed to vary according to an autoregressive-type evolution with AR coefficients α_i, for $i = 1, \ldots, p$. Here we may assume $|\alpha_i| < 1$, if we wish $\phi_{i.t}$ to follow an AR(1) process, but we may allow $\alpha_i = 1$ to cater for the random walk evolution (as in section 2.3).

Our model formulation relates to the time-varying autoregressive state-space models discussed in Prado *et al.*, 2021, Chapter 5 and in references therein; see also West *et al.*, 1999 and Grenier, 1983 among others. The above model formulation also extends the standard ARMA model to incorporate time-varying AR coefficients. The autoregressive order is $p \geq 1$ and there are q lags of correlation in the innovations. We note that if the innovations are uncorrelated, then $\rho_1 = \cdots = \rho_q = 0$ and $v_t = \varepsilon_t$. Hence, by estimating ρ_j one can determine the presence of autocorrelation of the innovations. We can see that model 2.3–2.5 is a time-varying ARMA(p,q) model, more specifically an ARMA model with time-varying AR coefficients and time-invariant MA coefficients. Model 2.3–2.5 extends the time-varying model of Triantafyllopoulos & Bersimis, 2016. Clearly, the model of Triantafyllopoulos & Bersimis, 2016 is obtained by 2.3–2.5 if we set $\alpha_1 = \cdots = \alpha_p = 1$ and $\rho_1 = \cdots = \rho_q = 0$. Hence, model 2.3–2.5 allows for wider application and is capable of dealing with the issue of autocorrelated innovations. The possibility of extending the model further to permit time-varying MA coefficients also is discussed in Section 2.6.

Model 2.3–2.5 can be cast in state space form by defining the $(p+q) \times 1$ (column) state vector $\beta_t = [\phi_{1.t}, \phi_{2.t}, \ldots, \phi_{p.t}, \varepsilon_{t-1}, \varepsilon_{t-2} \ldots, \varepsilon_{t-q}]^\top$. The state or evolution equation is given by

$$\begin{bmatrix} \phi_{1.t} \\ \phi_{2.t} \\ \vdots \\ \phi_{p.t} \\ \varepsilon_{t-1} \\ \varepsilon_{t-2} \\ \vdots \\ \varepsilon_{t-q} \end{bmatrix} = \begin{bmatrix} \alpha_1 & 0 & \cdots & 0 & 0 & \cdots & 0 \\ 0 & \alpha_2 & \cdots & 0 & 0 & \cdots & 0 \\ \vdots & \vdots & \ddots & \vdots & \vdots & \ddots & \vdots \\ 0 & 0 & \cdots & \alpha_p & 0 & \cdots & 0 \\ 0 & 0 & \cdots & 0 & 0 & \cdots & 0 \\ 0 & 0 & \cdots & 0 & 0 & \cdots & 0 \\ \vdots & \vdots & \ddots & \vdots & \vdots & \ddots & \vdots \\ 0 & 0 & \cdots & 0 & 0 & \cdots & 0 \end{bmatrix} \begin{bmatrix} \phi_{1.t-1} \\ \phi_{2.t-1} \\ \vdots \\ \phi_{p.t-1} \\ \varepsilon_{t-2} \\ \varepsilon_{t-3} \\ \vdots \\ \varepsilon_{t-q-1} \end{bmatrix} + \begin{bmatrix} w_{1.t} \\ w_{2.t} \\ \vdots \\ w_{p.t} \\ \varepsilon_{t-1} \\ \varepsilon_{t-2} \\ \vdots \\ \varepsilon_{t-q} \end{bmatrix}$$

$$\beta_t = \mathbf{F} \beta_{t-1} + \zeta_t, \tag{2.6}$$

where ζ_t follows a $(p+q)$-multivariate Gaussian distribution with zero mean vector and covariance matrix

$$\mathbf{Z} = \begin{bmatrix} \sigma_w^2 & 0 & \cdots & 0 & 0 & 0 & \cdots & 0 \\ 0 & \sigma_w^2 & \cdots & 0 & 0 & 0 & \cdots & 0 \\ \vdots & \vdots & \ddots & \vdots & \vdots & \vdots & \ddots & \vdots \\ 0 & 0 & \cdots & \sigma_w^2 & 0 & 0 & \cdots & 0 \\ 0 & 0 & \cdots & 0 & \sigma_\varepsilon^2 & 0 & \cdots & 0 \\ 0 & 0 & \cdots & 0 & 0 & \sigma_\varepsilon^2 & \cdots & 0 \\ \vdots & \vdots & \ddots & \vdots & \vdots & \vdots & \ddots & \vdots \\ 0 & 0 & \cdots & 0 & 0 & 0 & \cdots & \sigma_\varepsilon^2 \end{bmatrix}.$$

We note that $\mathbf{Z}$ is a $(p+q)\times(p+q)$ diagonal and positive definite matrix.

The observation equation is

$$y_t = [y_{t-1}, y_{t-2}, \cdots, y_{t-p}, \rho_1, \ldots, \rho_q] \begin{bmatrix} \phi_{1,t} \\ \phi_{2,t} \\ \vdots \\ \phi_{p,t} \\ \varepsilon_{t-1} \\ \vdots \\ \varepsilon_{t-q} \end{bmatrix} + \varepsilon_t = x_t^\top \beta_t + \varepsilon_t. \tag{2.7}$$

The model is initiated at $t = t_0 = \max(p,q)$ and a multivariate Gaussian prior is set for β_{t_0}, or

$$\beta_{t_0} \sim N(m_{t_0}, \mathbf{P}_{t_0}), \tag{2.8}$$

for some vector mean m_{t_0} and some covariance matrix $\mathbf{P}_{t_0}$. To represent no prior knowledge, a weakly informative prior might take $m_{t_0} = [1/p, 1, \ldots, 1/p, 0, 0, \ldots, 0]^\top$ and $\mathbf{P}_{t_0} = 1000\mathbf{I}_{p+q}$, so that the precision $\mathbf{P}_{t_0}^{-1}$ is close to the $(p+q)\times(p+q)$ zero matrix. For the prior mean vector the setting suggests equating the innovations ε_j's equal to zero and ϕ_i's equal to $1/p$, so that a prior expectation of y_{t_0+1} is equal to the sample mean $\bar{y}(t_0) = p^{-1}(y_{t_0-p} + y_{t_0-p+1} + \ldots + y_{t_0})$. The prior distribution of β_{t_0} needs to be Gaussian, in order to allow for the derivation of the posterior distribution of β_t, using the Kalman filter, discussed briefly in Section 2.4.3. Likewise, the EM algorithm, which uses forward estimation (posterior distribution of β_t, given data up to time t) and smoothed estimation (the distribution of β_t, given all data), requires the Gaussian assumption on the prior of β_{t_0}. There is flexibility though over the prior mean vector m_{t_0} and the prior covariance matrix $\mathbf{P}_{t_0}$; above we have given a weakly information prior setup for these moments. For more discussion of prior specification for state space models the reader is referred to West & Harrison, 1997, Triantafyllopoulos, 2021 and in references therein.

The working model is the linear and Gaussian state space model 2.6–2.7, equipped with prior 2.8. In the next section, we discuss inference for this model.

2.4.3 MAXIMUM LIKELIHOOD ESTIMATION

Consider the linear and Gaussian state space model 2.6–2.7, with the prior 2.8. Suppose that observations $y_{t_0+1}, y_{t_0+2}, \ldots, y_n$ are collected in Phase I and write $y = \{y_{t_0+1}, y_{t_0+2}, \ldots, y_n\}$, for some $n > t_0$. Write $\theta = [\alpha_1, \ldots, \alpha_p, \rho_1, \ldots, \rho_q, \sigma_w^2, \sigma_\varepsilon^2]^\top$ the vector of static parameters subject to estimation. This section discusses maximum likelihood estimation of θ based on the collection of observations y. Once θ is estimated in Phase I, then any residual control chart can proceed in Phase II in a standard manner.

Conditionally on θ, model 2.6–2.7 is a standard state space model, with known design vector x_t, evolution matrix $\mathbf{F}$, and known innovation variance and covariance matrices σ_ε^2 and $\mathbf{Z}$ (for the definition of these components see the previous section). Hence, conditionally on θ we can obtain the smoothed distribution of β_t, given all observations, i.e., $\beta_t \mid y \sim N(\hat{\beta}_{t|n}, \mathbf{P}_{t|n})$, where the smoothed mean vector $\hat{\beta}_{t|n}$ and covariance matrix $\mathbf{P}_{t|n}$ are routinely obtained as an extension of the Kalman filter, as reported in many textbooks; see e.g., West & Harrison, 1997, Chapter 4 and Triantafyllopoulos, 2021, Section 3.3, Theorem 3.4. In order to estimate θ we shall employ maximum likelihood estimation, in particular, we employ the Expectation Maximization (EM) algorithm set-out in the seminal paper of Dempster *et al.*, 1977. The EM algorithm for state space models was originally proposed in Shumway & Stoffer, 1982 and is discussed in Shumway & Stoffer, 2017, Section 6.3, Fahrmeir & Tutz, 2001, Section 8.1.2 and in Triantafyllopoulos, 2021, Section 4.3 among others.

We shall assume that the prior of the state vector β_{t_0} is specified as in equation 2.8, hence the EM algorithm will not be used to estimate the prior mean and covariance matrix of β_{t_0}. That approach may be used, as reported in Durbin & Koopman, 2012, but in our experience estimation of the initial moments slows the convergence of the EM algorithm without necessarily offering an advantage in the accuracy in the estimation of θ.

The log-likelihood function of θ, based on observations $y = \{y_{t_0+1}, \dots, y_n\}$ and states $\beta = \{\beta_{t_0+1}, \dots, \beta_n\}$ is

$$\begin{aligned}
\ell(\theta; y, \beta) &= \log p(y_{t_0+1}, \dots, y_n, \beta_{t_0+1}, \dots, \beta_n \mid \theta) \\
&= \log \prod_{t=t_0+1}^{n} p(y_t \mid \beta_t, \theta) \prod_{t=t_0+1}^{n} p(\beta_t \mid \beta_{t-1}, \theta) \\
&= -\frac{n-t_0}{2} \log \sigma_\varepsilon^2 - \frac{1}{2\sigma_\varepsilon^2} \sum_{t=t_0+1}^{n} (y_t - x_t^\top \beta_t)^2 \\
&\quad -\frac{n-t_0}{2} |\mathbf{Z}| - \frac{1}{2} \sum_{t=t_0+1}^{n} (\beta_t - \mathbf{F}\beta_{t-1})^\top \mathbf{Z}^{-1} (\beta_t - \mathbf{F}\beta_{t-1}).
\end{aligned}$$

It is noted here that $\mathbf{F} = \mathbf{F}(\theta)$ is a function of $\alpha_1, \dots, \alpha_p$, $x_t = x_t(\theta)$ is a function of $\rho_1, \dots, \rho_q$, and $\mathbf{Z} = \mathbf{Z}(\theta)$ is a function of σ_w^2 and σ_ε^2. We assume that at iteration $j \geq 1$ of the algorithm $\theta^{(j)}$ denotes the estimated vector of θ. The algorithm proceeds by obtaining the estimated vector $\theta^{(j+1)}$ at iteration $j+1$ as follows.

In the **E-step** (Expectation) we need to calculate the conditional expectation

$$Q(\theta \mid \theta^{(j)}) = E[\ell(\theta; y, \beta) \mid y, \theta^{(j)}],$$

where $\theta^{(j)}$ is the current estimate of θ. Note that the expectation is taken over β, while θ is a deterministic vector.

Define the expectations

$$S_1 = E\left[-\frac{n-t_0}{2} \log \sigma_\varepsilon^2 - \frac{1}{2\sigma_\varepsilon^2} \sum_{t=t_0+1}^{n} (y_t - x_t^\top \beta_t)^2 \mid y, \theta^{(j)} \right],$$

$$S_2 = E\left[-\frac{n-t_0}{2} |\mathbf{Z}| - \frac{1}{2} \sum_{t=t_0+1}^{n} (\beta_t - \mathbf{F}\beta_{t-1})^\top \mathbf{Z}^{-1} (\beta_t - \mathbf{F}\beta_{t-1}) \mid y, \theta^{(j)} \right],$$

so that $Q(\theta \mid \theta^{(j)}) = S_1 + S_2$.

Following standard results, these expectations are calculated as follows (for complete derivations the reader is referred to Triantafyllopoulos, 2021, Section 4.3).

$$S_1 = -\frac{n-t_0}{2} \log \sigma_\varepsilon^2 - \frac{1}{2\sigma_\varepsilon^2} \sum_{t=t_0+1}^{n} \left[(y_t - x_t^\top \hat{\beta}_{t|n})^2 + x_t^\top \mathbf{P}_{t|n} x_t \right]$$

and

$$S_2 = -\frac{n-t_0}{2}|\mathbf{Z}| - \frac{1}{2}\sum_{t=t_0+1}^{n} \text{trace}\Big[\{(\hat{\beta}_{t|n} - \mathbf{F}\hat{\beta}_{t-1|n})(\hat{\beta}_{t|n} - \mathbf{F}\hat{\beta}_{t-1|n})^\top + \mathbf{P}_{t|n}$$
$$-\mathbf{P}_{t,t-1|n}\mathbf{F} - \mathbf{F}\mathbf{P}_{t,t-1|n}^\top + \mathbf{F}\mathbf{P}_{t-1|n}\mathbf{F}\}\mathbf{Z}^{-1}\Big],$$

where $\mathbf{P}_{t,t-1|n} = \text{Cov}(\beta_t, \beta_{t-1} \mid y)$ is known as the lag-one covariance smoother, computed by the recursions of $\mathbf{P}_{t|n}$, details of which can be found in Shumway & Stoffer, 2017, Section 6.2, Property 6.3 and Triantafyllopoulos, 2021, Section 3.3.2, Theorem 3.5. This completes the E-step.

Proceeding now to the **M-step** (Maximization), we need to find the maximizing vector θ of $Q(\theta \mid \theta^{(j)})$. The partial derivatives of S_1 and S_2, with respect to the components of $\theta = [\alpha_1, \ldots, \alpha_p, \rho_1, \ldots, \rho_q, \sigma_w^2, \sigma_\varepsilon^2]^\top$ are

$$\frac{\partial S_1}{\partial \sigma_\varepsilon^2} = -\frac{n-t_0}{2\sigma_\varepsilon^2} + \frac{1}{2\sigma_\varepsilon^4}\left[\sum_{t=t_0+1}^{n} (y_t - x_t^\top \hat{\beta}_{t|n})^2 + x_t^\top \mathbf{P}_{t|n} x_t\right]. \tag{2.9}$$

Define $e_j = [0, \ldots, 0, 1, 0, \ldots, 0]^\top$ to be the $(p+q) \times 1$ vector with a unit in the j-th place and 0 elsewhere, with $j = 1, \ldots, p+q$. For $i = 1, \ldots, q$ we have

$$\frac{\partial S_1}{\partial \rho_i} = \frac{1}{\sigma_\varepsilon^2}\sum_{t=t_0+1}^{n}\left[(y_t - x_t^\top \hat{\beta}_{t|n})\{\hat{\beta}_{t|n}^\top - x_t^\top \mathbf{P}_{t|n}\} e_{p+i}\right]. \tag{2.10}$$

Noting that $\mathbf{Z}$ is a diagonal matrix, the determinant of $\mathbf{Z}$ is $|\mathbf{Z}| = \sigma_w^{2p}\sigma_\varepsilon^{2q}$ and the inverse of $\mathbf{Z}$ is the diagonal matrix with elements $\sigma_w^{-2}, \ldots, \sigma_w^{-2}, \sigma_\varepsilon^{-2}, \ldots, \sigma_\varepsilon^{-2}$. Hence

$$\frac{\partial S_2}{\partial \sigma_w^2} = -\frac{(n-t_0)p}{2\sigma_w^2} + \frac{1}{2\sigma_w^4}\sum_{t=t_0+1}^{n} \text{trace}(\mathbf{A}_t\mathbf{J}), \tag{2.11}$$

where

$$\mathbf{A}_t = (\hat{\beta}_{t|n} - \mathbf{F}\hat{\beta}_{t-1|n})(\hat{\beta}_{t|n} - \mathbf{F}\hat{\beta}_{t-1|n})^\top + \mathbf{P}_{t|n} - \mathbf{P}_{t,t-1|n}\mathbf{F} - \mathbf{F}\mathbf{P}_{t,t-1|n}^\top + \mathbf{F}\mathbf{P}_{t-1|n}\mathbf{F}$$

and $\mathbf{J}$ is the diagonal matrix with units for the first p elements of its diagonal and zeroes elsewhere, or

$$\mathbf{J} = \begin{bmatrix} \mathbf{I}_p & \mathbf{0}_{p\times q} \\ \mathbf{0}_{q\times p} & \mathbf{0}_{q\times q} \end{bmatrix},$$

where $\mathbf{0}_{k\times m}$ denotes the $k \times m$ zero matrix and $\mathbf{I}_p$ is the $p \times p$ identity matrix.

Likewise, for σ_ε^2 we have

$$\frac{\partial S_2}{\partial \sigma_\varepsilon^2} = -\frac{(n-t_0)q}{2\sigma_\varepsilon^2} + \frac{1}{2\sigma_\varepsilon^4}\sum_{t=t_0+1}^{n} \text{trace}(\mathbf{A}_t\mathbf{J}'), \tag{2.12}$$

where

$$\mathbf{J}' = \begin{bmatrix} \mathbf{0}_{p\times p} & \mathbf{0}_{p\times q} \\ \mathbf{0}_{q\times p} & \mathbf{I}_q \end{bmatrix}.$$

For the partial derivative of S_2, with respect to α_i, first we note that $\frac{\partial \mathbf{F}}{\partial \alpha_i} = \mathbf{K}_i$, where $\mathbf{K}_i$ is a $(p+q) \times (p+q)$ matrix with a unit in its (i,i) element and zero otherwise, for $i = 1, \ldots, p$. Then from the definition of S_2 above we have

$$\frac{\partial S_2}{\partial \alpha_i} = -\frac{1}{2}\sum_{t=t_0+1}^{n} \text{trace}\Big[\{-\mathbf{K}_i\hat{\beta}_{t-1|n}(\hat{\beta}_{t|n} - \mathbf{F}\hat{\beta}_{t-1|n})^\top - (\hat{\beta}_{t|n} - \mathbf{F}\hat{\beta}_{t-1|n})\hat{\beta}_{t-1|n}\mathbf{K}_i$$
$$-\mathbf{P}_{t,t-1|n}\mathbf{K}_i - \mathbf{K}_i\mathbf{P}_{t,t-1|n}^\top + \mathbf{K}_i\mathbf{P}_{t-1|n}\mathbf{F} + \mathbf{F}\mathbf{P}_{t-1|n}\mathbf{K}_i\}\mathbf{Z}^{-1}\Big]. \tag{2.13}$$

From 2.9 and 2.12 we have that $\partial S_1/\partial\sigma_\varepsilon^2 + \partial S_2/\partial\sigma_\varepsilon^2 = 0$ implies the maximum value for σ_ε^2 at iteration $j+1$ as

$$\left(\sigma_\varepsilon^2\right)^{(j+1)} = \frac{1}{(n-t_0)(q+1)} \sum_{t=t_0+1}^{n} \left[(y_t - x_t^\top \hat{\beta}_{t|n})^2 + x_t^\top \mathbf{P}_{t|n} x_t + \text{trace}(\mathbf{A}_t \mathbf{J}')\right]. \tag{2.14}$$

where it is understood that the values of x_t, $\hat{\beta}_{t|n}$, $\mathbf{P}_{t|n}$ and $\mathbf{A}_t$ on the right hand side use $\theta = \theta^{(j)}$, being computed at iteration j.

Following a similar calculation, from 2.11, the maximum value of σ_w^2 is obtained by $\partial S_2/\partial\sigma_w^2 = 0$ as

$$\left(\sigma_w^2\right)^{(j+1)} = \frac{1}{p(n-t_0)} \sum_{t=t_0+1}^{n} \text{trace}(\mathbf{A}_t \mathbf{J}). \tag{2.15}$$

Now we turn our attention to the maximization of ρ_i, $i = 1,\ldots,q$. From equation 2.10 we need to solve $\partial S_1/\partial\rho_i = 0$. From 2.10 we obtain the normal equation

$$\sum_{t=t_0+1}^{n} x_t^\top (\hat{\beta}_{t|n}\hat{\beta}_{t|n}^\top + \mathbf{P}_{t|n}) e_{p+i} = \sum_{t=t_0+1}^{n} y_t. \tag{2.16}$$

Write $\mathbf{B}_t = \hat{\beta}_{t|n}\hat{\beta}_{t|n}^\top + \mathbf{P}_{t|n}$ and consider the following partition of x_t, $\mathbf{B}_t$ and e_{p+i}:

$$x_t = \begin{bmatrix} y_t^* \\ \rho \end{bmatrix}, \quad \mathbf{B}_t = \begin{bmatrix} \mathbf{B}_{11.t} & \mathbf{B}_{12.t} \\ \mathbf{B}_{12.t}^\top & \mathbf{B}_{22.t} \end{bmatrix}, \quad e_{p+i} = \begin{bmatrix} 0_p \\ e_i^* \end{bmatrix},$$

where $y_t^* = [y_{t-1},\ldots,y_{t-p}]^\top$, $\rho = [\rho_1,\ldots,\rho_q]^\top$, $0_p = [0,\ldots,0]^\top$, $e_i^* = [0,\ldots,0,1,0,\ldots,0]^\top$ ($q\times 1$ vector with a unit in its i-th position and zero elsewhere), $\mathbf{B}_{11.t}$ a $p\times p$ submatrix of $\mathbf{B}_t$, $\mathbf{B}_{22.t}$ a $q\times q$ submatrix of $\mathbf{B}_t$ and $\mathbf{B}_{12.t}$ a $p\times q$ submatrix of $\mathbf{B}_t$.

With this partition the normal equation 2.16 implies

$$\rho^\top \sum_{t=t_0+1}^{n} \mathbf{B}_{22.t} e_i^* = \sum_{t=t_0+1}^{n} y_t - \sum_{t=t_0+1}^{n} (y_t^*)^\top \mathbf{B}_{12.t} e_i^*.$$

Given that $\rho_k^{(j)}$ is available at iteration j for $k \neq i$, we can solve this equation for ρ_i to get

$$\rho_i^{(j+1)} = \frac{\sum_{t=t_0+1}^{n} y_t - \sum_{t=t_0+1}^{n} (y_t^*)^\top \mathbf{B}_{12.t} e_i^* - \sum_{k=1,k\neq i}^{q} \sum_{t=t_0+1}^{n} \rho_k^{(j)} b_{22.t}^{(k,i)}}{\sum_{t=t_0+1}^{n} b_{22.t}^{(i,i)}}, \tag{2.17}$$

where $b_{22.t}^{(k,i)}$ is the (k,i)-th element of the matrix $\mathbf{B}_{22.t}$, for $i = 1,\ldots,q$.

Finally, for α_i, we need to solve $\partial S_2/\partial\alpha_i$, for $i = 1,\ldots,p$. From equation 2.13 we have

$$\begin{aligned} &\sum_{t=t_0+1}^{n} \text{trace}\left[(\mathbf{K}_i \hat{\beta}_{t-1|n}\hat{\beta}_{t-1|n}^\top \mathbf{F} + \mathbf{F}\hat{\beta}_{t-1|n}\hat{\beta}_{t-1|n}^\top \mathbf{K}_i + \mathbf{K}_i \mathbf{P}_{t-1|n}\mathbf{F} + \mathbf{F}\mathbf{P}_{t-1|}\mathbf{K}_i)\mathbf{Z}^{-1}\right] \\ &= \sum_{t=t_0+1}^{n} \text{trace}\left[(\mathbf{K}_i \hat{\beta}_{t-1|n}\hat{\beta}_{t|n}^\top + \hat{\beta}_{t|n}\hat{\beta}_{t-1|n}^\top \mathbf{K}_i + \mathbf{P}_{t,t-1|n}\mathbf{K}_i + \mathbf{K}_i \mathbf{P}_{t,t-1|n}^\top)\mathbf{Z}^{-1}\right]. \end{aligned}$$

Because $\mathbf{F}$ and $\mathbf{K}_i$ are diagonal matrices, the above equation simplifies to

$$\begin{aligned} &\sum_{t=t_0+1}^{n} \text{trace}\left[(2\hat{\beta}_{t-1|n}\hat{\beta}_{t-1|n}^\top + 2\mathbf{P}_{t-1|n})\mathbf{F}\mathbf{K}_i\mathbf{Z}^{-1}\right] \\ &= \sum_{t=t_0+1}^{n} \text{trace}\left[(\hat{\beta}_{t-1|n}\hat{\beta}_{t|n}^\top + \hat{\beta}_{t|n}\hat{\beta}_{t-1|n}^\top + \mathbf{P}_{t,t-1|n} + \mathbf{P}_{t,t-1|n}^\top)\mathbf{K}_i\mathbf{Z}^{-1}\right] \end{aligned}$$

or

$$\sum_{t=t_0+1}^{n} \text{trace}(\mathbf{C}_t \mathbf{F} \mathbf{K}_i \mathbf{Z}^{-1}) = \sum_{t=t_0+1}^{n} \text{trace}(\mathbf{D}_t \mathbf{K}_i \mathbf{Z}^{-1}), \tag{2.18}$$

where $\mathbf{C}_t = 2\hat{\beta}_{t-1|n}\hat{\beta}_{t-1|n}^\top + 2\mathbf{P}_{t-1|n}$ and $\mathbf{D}_t = \hat{\beta}_{t-1|n}\hat{\beta}_{t|n}^\top + \hat{\beta}_{t|n}\hat{\beta}_{t-1|n}^\top + \mathbf{P}_{t,t-1|n} + \mathbf{P}_{t,t-1|n}^\top$. From the definitions of $\mathbf{K}_i$ and $\mathbf{Z}$ equation 2.18 we obtain the solution

$$\alpha_i^{(j+1)} = \frac{\sum_{t=t_0+1}^{n} d_{ii,t}}{\sum_{t=t_0+1}^{n} c_{ii,t}}, \quad i = 1, \dots, p, \tag{2.19}$$

where $c_{ii,t}$ is the (i,i) element of $\mathbf{C}_t$ and $d_{ii,t}$ is the (i,i) element of $\mathbf{D}_t$.

Putting together equations 2.14, 2.15, 2.17, 2.19, we obtain the estimate of the vector θ as

$$\theta^{(j+1)} = \left[\alpha_1^{(j+1)}, \dots, \alpha_q^{(j+1)}, \rho_1^{(j+1)}, \dots, \rho_q^{(j+1)}, (\sigma_w^2)^{(j+1)}, (\sigma_\varepsilon^2)^{(j+1)}\right]^\top$$

and this completes iteration $j+1$ of the EM algorithm. The algorithm continues to the next iteration until convergence, which typically requires the Euclidean distance $|\theta^{(m)} - \theta^{(m-1)}|$ to be smaller than some tolerance value, for some iteration m. Then the algorithm terminates and $\theta^{(m)}$ is the approximate MLE of θ. More details about stopping criteria for the EM algorithm can be found in Shumway & Stoffer, 2017, Section 6.3 and Triantafyllopoulos, 2021, Section 4.3. When the EM algorithm terminates, this completes Phase I. The residuals can be used to validate that the process is in control and establish a monitoring procedure for Phase II. Of course, while achieving uncorrelated residuals is vital for the application of control charting, it may also be needed in other applications of time series forecasting and modelling.

2.5 TEMPERATURE DATA

In this section, we illustrate the proposed methodology using part of the Production data set, reported in Pan & Jarrett, 2004. Temperatures at one location on a plastic mould are recorded. As reported in Pan & Jarrett, 2004, temperatures are not dynamically controlled over time and is not possible to intervene in the process, e.g., as in Box *et al.*, 2009. Interest is focused on the design of a control chart in order to detect any extreme values in the temperatures around the point; high temperatures can overheat the mould or low temperatures may increase heat diffusion around the point and result in poor production. The data set we consider is recorded over 276 equally spaced time points, the values of which are tabulated in Appendix C of Pan & Jarrett, 2004 (Variable 2).

The data, plotted in Figure 2.3, exhibit local stationarity. Indeed there seems initially to be some fluctuation with lower values being more persistent, followed by a period of stable fluctuations (being aligned more to stationarity) and finally after around time point 180 there seems to be an upward and downward local trend, reminiscent of a random walk.

In order to explore the structure of the data and to identify suitable time series models the plots of the autocorrelation function (ACF) and the partial autocorrelation function (PACF) are shown in Figure 2.4. Both the ACF and PACF indicate that the time series is stationary (lags are largely within the confidence intervals). However, there is some moderate evidence from the ACF that the time series might only be locally stationary (there are a few lags which are significant or marginally significant).

First, we shall consider fitting some (static) ARMA(p,q) models, for AR and MA orders $p = 0,1,2,3,4,5$ and $q = 0,1,2,3,4,5$; models with higher orders seem to be undesirable, as the number of parameters will increase uncertainty around the estimates. The ACF and PACF show evidence for a model with autoregressive terms of order 2 (since the ACF suggests a geometric decay and the PACF has a cut-off at lag 2). Brief results for a number of ARMA models are tabulated in Table 2.1. A few comments are in order. First of all, the model with the smallest AIC is ARMA(1,1) (AIC value

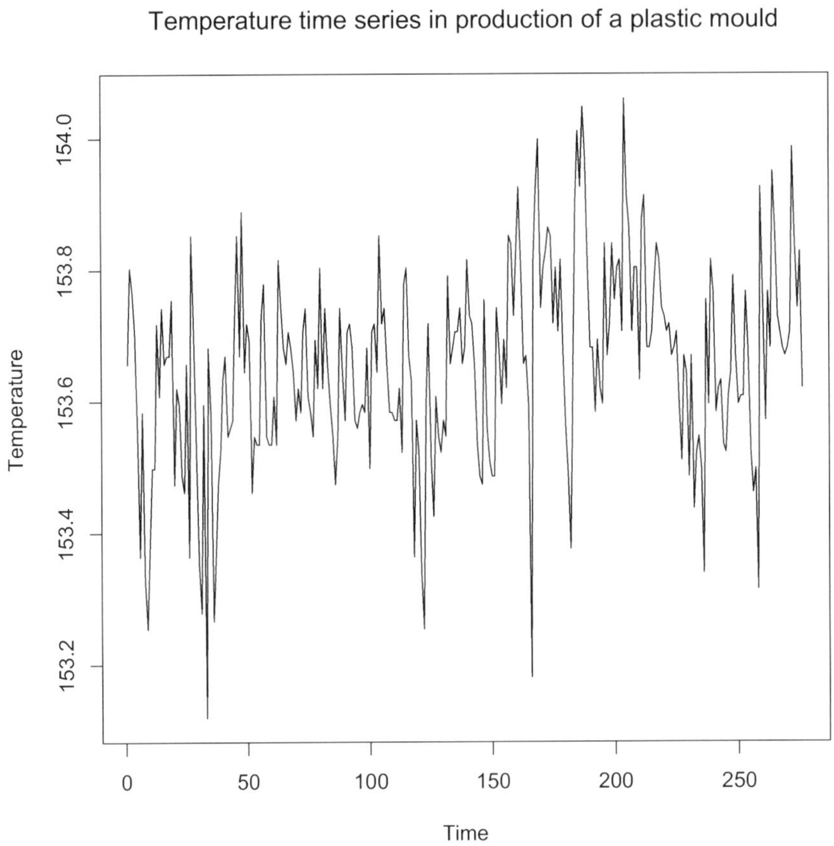

Figure 2.3 Temperature time series data

is −335.67), but this model still produces residuals which are autocorrelated (there is one significant lag in the residuals). The MA models have the worst performance, in particular MA(1) and MA(2); this is expected as the ACF and PACF in Figure 2.4 suggested an AR(2). Indeed, the AR(2) seems to do better, with a good value of AIC, but still results in correlated residuals. There are six models, with uncorrelated residuals; these models are: ARMA(1,4), ARMA(2,4), ARMA(5,1), ARMA(1,5), ARMA(5,2) and ARMA(5,5). Of these models, ARMA(1,4) has the smallest AIC (−334.18). The largest log-likelihood is obtained by the model with most parameters, i.e., ARMA(5,5), but the increase in the likelihood between ARMA(1,4) and ARMA(5,5) is small. Hence, we conclude here that considering the models which result in uncorrelated residuals ARMA(1,4) is the best. It may be argued that for this model the number of parameters to be estimated is quite large (1 AR and 4 MA terms) and that it would be desirable to consider a model with fewer parameters. In general the conclusion we reach is that standard model selection in time series may result in model fitting with autocorrelated residuals. It may be that the modeller will have to consider fitting higher order models in order to obtain uncorrelated residuals.

In order to circumvent the issues discussed above we consider the time-varying autoregressive models of Triantafyllopoulos & Bersimis, 2016. The AR structure is somewhat supported by the significant lags in the PACF of Figure 2.4 and the local stationarity of the time series, visually depicted in Figures 2.3 and 2.4. The time-varying AR models of Triantafyllopoulos & Bersimis, 2016 are essentially model 2.6–2.7, with $\alpha_1 = \cdots = \alpha_p = 1$ and $\rho_1 = \cdots = \rho_q = 0$, i.e., the AR coefficients $\phi_{i,t}$ are driven by random walks and there are no MA terms. Inference is based on the

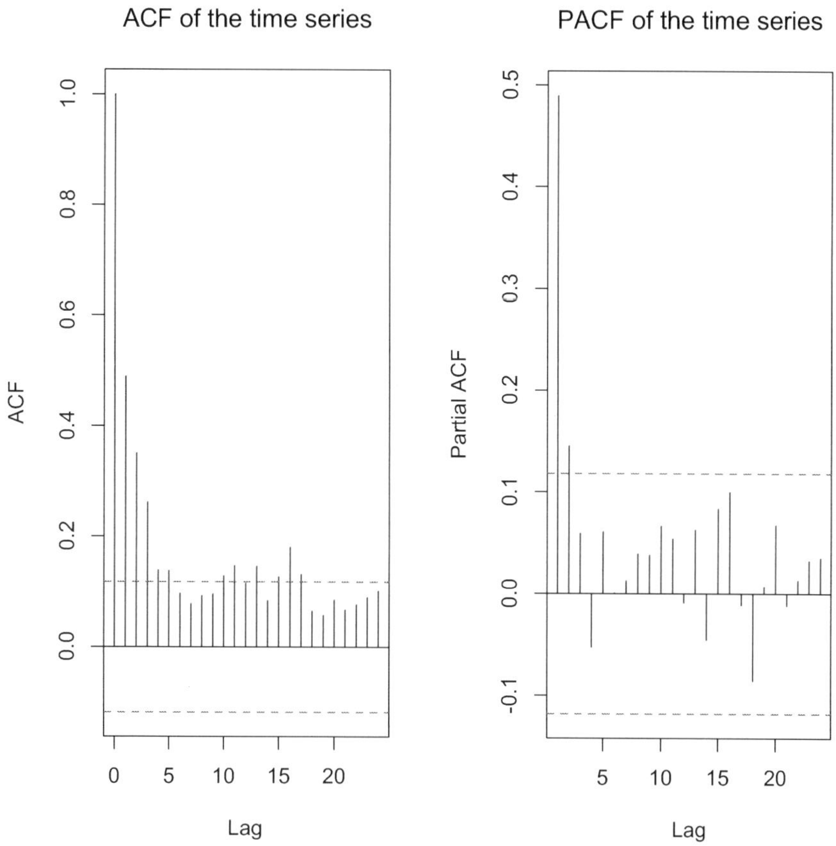

Figure 2.4 Temperature time series; left panel shows the ACF and right panel shows the PACF

Kalman filter together with a discount factor δ, which controls the variance of each random walk for $\phi_{i,t}$, for $i = 1,\ldots,p$. A lower discount factor results in erratic evolutions for $\phi_{i,t}$ and high discount factors close to 1 result in smooth evolutions; $\delta = 1$ implies $\phi_{i,t} = \phi_{i,t-1}$ for all t and hence the model reduces to a static AR model. Here, following suggestions of West & Harrison, 1997 we have used $\delta = 0.9$, which allows for $\phi_{i,t}$ to vary considerably being adaptive to new information, but also retain reasonable smoothness. In general, values close to, but less than, one are recommended, but the choice is problem specific. One possibility for choosing the discount factor is to maximize the likelihood function, but other approaches may include making the discount factor dynamic or placing a uniform or beta distribution on δ. For a detailed discussion the reader is referred to West & Harrison, 1997 and Triantafyllopoulos, 2021.

We have fitted nine successive time-varying AR models and we report on the maximized likelihood function, mean of squared standardized residuals (MSSR) and number of significant residuals together with the actual time point(s) on Table 2.2. For a perfect fit, the MMSR value should be equal to 1. Of these nine models, the tv-AR(3) has MSSR which is closest to 1 (1.162). The only model with uncorrelated residuals is the tv-AR(9). Despite its very large number of parameters, this model does not even have the best likelihood function of the models considered; the MSSR of this model is quite high (1.307) suggesting overall a poor performance. It would be desirable to accommodate time-varying AR behaviour without needing so many parameters to obtain uncorrelated residuals.

Table 2.1

Fitting of several AR, MA and ARMA (static) models for the temperature time series. Shown are the model, maximized log-likelihood, value of the AIC and the number of significant outliers in the ACF of the residuals; the number in parentheses indicates the time point(s) of the outlier(s).

Model	maximized log-likelihood	AIC	Significant residuals
AR(1)	168.59	−331.19	1 (16)
AR(2)	171.53	−335.07	1 (16)
AR(3)	172.02	−334.04	1 (16)
AR(4)	172.4	−332.8	1 (16)
AR(5)	172.91	−331.82	1 (16)
MA(1)	157.73	−309.45	4 (2,3,13,16)
MA(2)	164.79	−321.57	3 (3, 13, 16)
MA(3)	170.99	−331.97	1 (16)
MA(4)	171.23	−330.46	1 (16)
MA(5)	172.28	−330.57	1 (16)
ARMA(1,1)	171.84	−335.67	1 (16)
ARMA(2,1)	171.84	−333.69	1 (16)
ARMA(1,2)	171.85	−333.69	1 (16)
ARMA(2,2)	172.66	−333.31	1 (16)
ARMA(3,1)	172.74	−333.48	1 (16)
ARMA(1,3)	172.02	−332.03	1 (16)
ARMA(3,2)	172.83	−331.66	1 (16)
ARMA(3,3)	172.85	−329.69	1 (16)
ARMA(4,1)	172.73	−331.46	1 (16)
ARMA(1,4)	174.09	−334.18	0
ARMA(4,2)	174.00	−331.99	1 (16)
ARMA(2,4)	174.13	−332.25	0
ARMA(4,3)	174.00	−330.00	1 (16)
ARMA(4,4)	175.23	−330.46	1 (16)
ARMA(5,1)	174.28	−332.57	0
ARMA(1,5)	174.11	−332.23	0
ARMA(5,2)	174.35	−330.7	0
ARMA(2,5)	174.06	−330.11	1 (16)
ARMA(5,3)	176.86	−333.72	1 (16)
ARMA(3,5)	174.51	−329.02	1 (16)
ARMA(5,4)	174.61	−327.23	1 (16)
ARMA(5,5)	177.96	−331.93	0

The observations above lead us naturally to combine the time-varying AR models, which do not have MA terms, with the static ARMA models, which do not have time-varying AR structure into one model. We fit a number of tv-ARMA(p,q) models of the form of 2.6–2.7. For all models we use the prior 2.8 as described on page 25; in particular we use $\beta_{t_0} \sim N(0, 1000\mathbf{I}_{p+q})$. The zero mean vector reflects a standard belief assuming no previous data are available, whilst the low precision $1000^{-1}\mathbf{I}_{p+q}$ reflects our uncertainty around that belief. Note that this prior mean is reasonable, as

Table 2.2

Fitting of several time-varying AR models for the temperature time series. Shown are the model, log-likelihood, mean of squared standardized residuals (MSSR), and the number of significant outliers in the ACF of the residuals; the number in parentheses indicates the time point(s) of the outlier(s).

Model	log-likelihood	MSSR	Significant residuals
tv-AR(1)	−402762.1	2936.777	2 (1,16)
tv-AR(2)	744.3869	1.995	1 (1)
tv-AR(3)	754.5277	1.162	2 (1,2)
tv-AR(4)	851.3823	1.311	2 (1,2)
tv-AR(5)	889.2897	1.309	2 (2, 15)
tv-AR(6)	920.3376	1.623	2 (1, 18)
tv-AR(7)	955.4090	1.456	1 (13)
tv-AR(8)	965.9083	1.310	2 (2,8)
tv-AR(9)	949.9469	1.307	0

Table 2.3

Fitting of several time-varying ARMA models for the temperature time series. Shown are the model, maximized log-likelihood, mean of squared standardized residuals (MSSR), the number of significant outliers in the ACF of the residuals, and the number of iterations when the EM algorithm has converged; the number in parentheses indicates the time point(s) of the outlier(s).

Model	maximized log-likelihood	MSSR	significant residuals	Iterations
tv-ARMA(1,1)	−402624.3.1	1765.3	1 (16)	12
tv-ARMA(2,1)	−260.8692	1.023	0	15
tv-ARMA(1,2)	−402624.3	0.875	1 (16)	16
tv-ARMA(2,2)	−260.8682	1.235	0	17
tv-ARMA(3,1)	−442.1149	1.110	0	16
tv-ARMA(1,3)	−402624.3	1644.8	1 (16)	15
tv-ARMA(3,2)	−442.1133	1.025	0	17
tv-ARMA(2,3)	−260.8095	1.001	1 (3)	20
tv-ARMA(3,3)	−442.1101	0.998	0	19

the ε_j elements of β have zero mean and we essentially place a zero mean on the AR coefficients ϕ_j at time t_0.

Table 2.3 tabulates the models, the maximized log-likelihood, the mean of squared standardized residuals (MSSR), number of significant residuals and number of iterations required for the EM algorithm to converge. The stopping criterion is that the Euclidean distance $|\theta^{(m)} - \theta^{(m-1)}|$ to be smaller than some tolerance value $T = 1/1000$ so that the EM algorithm terminates at iteration m. After some experimentation we found that $T = 1/1000$ works well for all models, whilst lower values of T do not result in better convergence. The number of uncorrelated residuals do not lower, if the value of T is lowered, since tolerance is related to convergence of the parameters and does not

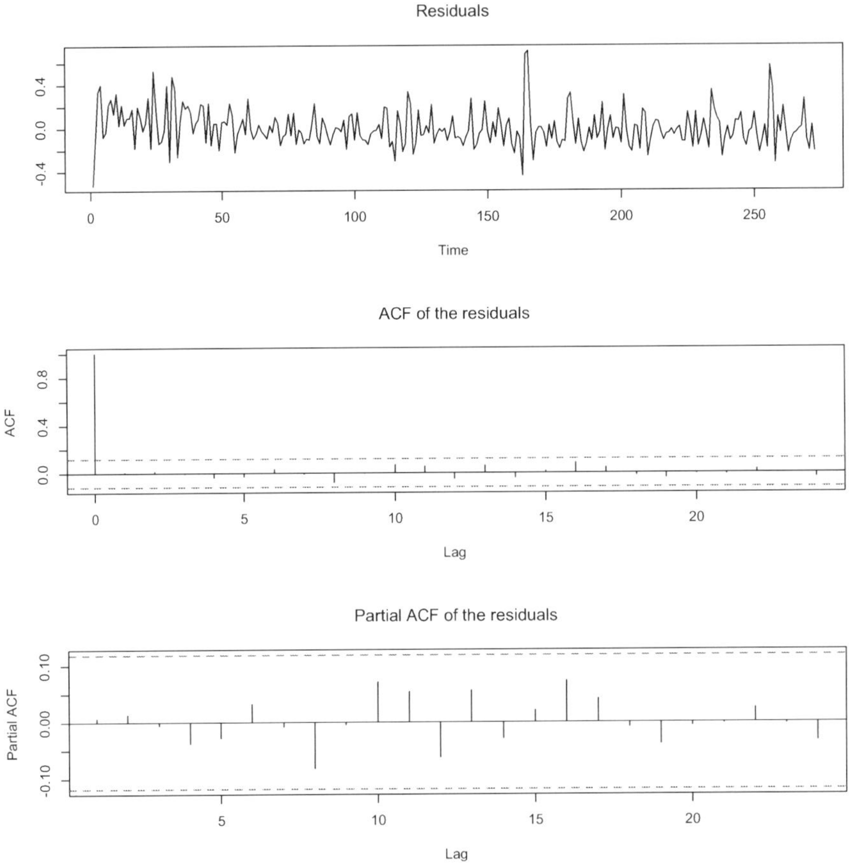

Figure 2.5 Residuals of the tv-ARMA(2,1) model; top panel shows the residuals, the middle panel shows the ACF of the residuals and the lower panel shows the partial ACF of the residuals.

improve the model fit. We observe that models tv-ARMA(1,1), tv-ARMA(1,2) and tv-ARMA(1,3) have too small a likelihood, indicating a poor fit; this agrees with the values of MSSR not being close to 1. Models tv-ARMA(2,1), tv-ARMA(2,2), tv-ARMA(3,1), tv-ARMA(3,2), tv-ARMA(3,3) give uncorrelated residuals. Of these five models tv-ARMA(2,2) has the largest log-likelihood (-260.8682), but the log-likelihood of tv-ARMA(2,1) is very close to it (-260.8692). For these models the value of MSSR is close to 1 indicating good overall performance. Overall, we favour the tv-ARMA(2,1) because it has fewer parameters yet produces almost the same log-likelihood as the tv-ARMA(2,2) and has MSSR value close to 1. In comparison to both the static ARMA and the time-varying AR models of Tables 2.1 and 2.2, the tv-ARMA(2,1) requires fewer parameters to produce uncorrelated residuals (note that we need a static ARMA(1,4) or a time-varying AR(9) to have no significant lags in the ACF of the residuals). The fitted model is

$$y_t = \phi_{1,t} y_{t-1} + \phi_{2,t-1} y_{t-1} + \varepsilon_t - 0.25\varepsilon_{t-1}, \varepsilon_t \sim N(0, 0.019),$$
$$\phi_{1,t} = 0.955\phi_{1,t-1} + w_{1,t}, \quad \phi_{2,t} = 0.823\phi_{2,t-1} + w_{2,t},$$
$$w_{1,t} \sim N(0, 1.453), \quad w_{2,t} \sim N(0, 1.453).$$

Figure 2.5 shows the residuals from the tv-ARMA(2,1) model, together with the ACF and partial ACF of the residuals. The residuals are uncorrelated as the ACF and PACF lie within the $\pm 1.96/\sqrt{276} = \pm 0.11798$ bounds for all lags.

2.6 DISCUSSION

In this section, we discuss some issues related to the design and application of time series-based residual control charts. Such charts have the advantage of utilizing the independence, or at least uncorrelatedness, of the residuals, thus avoiding the need to modify control limits to cater for autocorrelation of the original measurements. However, in applications, it is not uncommon to find that the residuals themselves are still autocorrelated, despite standard model selection procedures and checks being followed. A particular situation in which autocorrelated residuals can arise is when the process is only locally, not weakly, stationary.

We propose a flexible time-varying autoregressive moving average (tv-ARMA) model, which is capable of dealing with local stationarity as it is adaptable to new information and it includes autocorrelated innovations. The model offers a time-varying autoregressive structure, while the moving average terms are static. Casting the model in state space form permits use of well-established Bayesian inference and results in maximum likelihood estimation of hyperparameters using the EM algorithm; convergence of the algorithm is fast. We demonstrate that the class can give parsimonious, well-fitting models, resulting in uncorrelated residuals suitable for standard charting for monitoring purposes.

We have made a case for the adoption of tv-ARMA time series models for process modelling and residual production in Phase I. Time-varying ARMA models offer the flexibility of adaptive filtering of time-varying AR components while at the same time including the necessary MA static terms, which account for autocorrelated residuals. When the process subject to control is clearly weakly stationary, then a static ARMA model might be appropriate. However, in situations where the process might exhibit local stationarity, then this static model might not be suitable since high orders p and q might be needed to capture the dynamics of the data, or the residuals might be autocorrelated. Even if high orders are selected, local stationarity may result in more complex dynamics in Phase II, hence the static model might lead to poor fit, at least locally.

One may question the motivation behind the MA component being static. The MA coefficients of the lagged innovations are not likely to vary dynamically in practice, as the structure of autocorrelated errors (if no MA terms were included) is not likely to be dynamic. In addition to this, as pointed out in Triantafyllopoulos & Nason, 2007, time-varying moving average models introduce identifiability and estimation difficulties, which make these models impractical for routine application; indeed the time-varying models of West & Harrison, 1997 and Prado *et al.*, 2021 focus primarily on time-varying autoregressive structures.

In real series, it is quite hard to identify periods of static evolution of the process and periods of local stability, which indicate some small shifts from stationarity. Our proposed tv-ARMA model is able to deal both with stationary time series and with local stationarity. The estimation of either form is carried out automatically, without the need to make explicit assumptions or perform testing to select one form or the other, or to resort to complex simulation-based estimation methods. As a result, the proposed model 2.3–2.5 together with the estimation strategy of Section 2.4.3 offers a speedy and easy-to-design implementation, which is suitable to model the complexity of real time series and is accessible to practitioners.

REFERENCES

Alwan, L., & Roberts, H. V. (1988). Time-series modelling for statistical process control. *Journal of Business and Economic Statistics*, *6*, 87–95.

Bersimis, S., & Sachlas, A. (2022). Surveilling public health through statistical process monitoring: A literature review and a unified framework. *Communications in Statistics: Case Studies, Data Analysis and Applications*, *8*(3), 515–543.

Box, G. E. P., Jenkins, G. M., & Reinsel, G. C. (2008). *Time series analysis: Forecasting and control* (*4th*). Wiley.

Box, G. E. P., Luceño, A., & del C. Paniagua-Quinones, M. (2009). *Statistical control by monitoring and adjustment* (*2nd*). Wiley.

Brockwell, P. J., & Davis, R. A. (1991). *Time series: Theory and methods*. Springer.

Dahlhaus, R. (2012). *13 - locally stationary processes*. In T. Subba Rao, S. Subba Rao, & C. Rao (Eds.). Elsevier.

Dempster, P., Laird, N. M., & Rubin, D. B. (1977). Maximum likelihood from incomplete data via the EM algorithm. *Journal of the Royal Statistical Society Series B*, *39*(1), 1–38.

Durbin, J., & Koopman, S. J. (2012). *Time series analysis by state space methods* (*2nd*). Oxford University Press.

Fahrmeir, L., & Tutz, G. (2001). *Multivariate statistical modelling based on generalized linear models*. Springer.

Frisén, M., Anderson, E., & Schiöler, L. (2011). Sufficient reduction in multivariate surveillance. *Communications in Statistics: Theory and Methods*, *40*(10), 1821–1838.

Grenier, Y. (1983). Time-dependent ARMA modeling of nonstationary signals. *IEEE Transactions on Acoustics, Speech, and Signal Processing*, *31*(4), 899–911.

Jeffreys, H. (1961). *Theory of probability* (*3rd*). Oxford University Press.

Kass, R. E., & Raftery, A. E. (1995). Bayes factors. *Journal of the American Statistical Association*, *90*(430), 773–795.

Montgomery, D. C. (2009). *Introduction to Statistical Quality Control* (*6th*). John Wiley & Sons.

Osei-Aning, R., Abbasi, S. A., & Riaz, M. (2017). Optimization design of the CUSUM and EWMA charts for autocorrelated processes. *Quality and Reliability Engineering International*, *33*(8), 1827–1841.

Pan, X., & Jarrett, J. (2004). Applying state space to spc: Monitoring multivariate time series. *Journal of Applied Statistics*, *31*(4), 397–418.

Prado, R., Ferreira, M. A. R., & West, M. (2021). *Time series: Modelling, computation and inference* (*2nd*). Chapman; Hall/CRC.

Psarakis, S., & Papaleonida, G. E. A. (2007). Spc procedures for monitoring autocorrelated processes. *Quality Technology and Quantitative Management*, *4*, 501–540.

Shumway, R. H., & Stoffer, D. S. (1982). An approach to time series smoothing and forecasting using the EM algorithm. *Journal of Time Series Analysis*, *3*, 253–264.

Shumway, R. H., & Stoffer, D. S. (2017). *Time series analysis and its applications: With r examples* (*4th*). Springer.

Siripanthana, S., & Stillman, E. C. (2016). Sufficient reduction methods for multivariate time-dependent health surveillance data. *Communications in Statistics - Theory and Methods*, *45*(4), 950–966.

Testik, M. C. (2005). Model inadequacy and residuals control charts for autocorrelated processes. *Quality and Reliability Engineering International*, *21*(2), 115–130.

Triantafyllopoulos, K. (2021). *Bayesian inference of stats space models: Kalman filtering and beyond*. Springer.

Triantafyllopoulos, K., & Bersimis, S. (2016). Phase II control charts for autocorrelated processes. *Quality Technology and Quantitative Management*, *13*(1), 88–108.

Triantafyllopoulos, K., & Nason, G. P. (2007). A Bayesian analysis of moving average processes with time-varying parameters. *Computational Statistics and Data Analysis*, *52*, 1025–1046.

West, M., & Harrison, P. J. (1997). *Bayesian forecasting and dynamic models* (*2nd*). Springer.

West, M., Prado, R., & Krystal, A. D. (1999). Evaluation and comparison of EEG traces: Latent structure in nonstationary time series. *Journal of the American Statistical Association*, *94*(448), 1083–1095.

Yao, Y., Chakraborti, S., Yang, X., Parton, J., Lewis Jr, D., & Hudnall, M. (2023). Phase i control chart for individual autocorrelated data: Application to prescription opioid monitoring. *Journal of Quality Technology*, *55*(3), 302–317.

3 The Unconditional Performance of Control Charts for Zero-Inflated Processes with Estimated Parameters

Athanasios C. Rakitzis
University of Piraeus, Department of Statistics and Insurance Science
E-Mail:arakitz@unipi.gr. ORCID:0000-0002-3054-4004.

Eftychia Mamzeridou
University of the Aegean, Department of Statistics and Actuarial-Financial Mathematics
E-Mail:emam@aegean.gr

Petros E. Maravelakis
University of Piraeus, Department of Business Administration
E-Mail:maravel@unipi.gr. ORCID:0000-0002-7477-5947.

3.1 INTRODUCTION

Control charts are the main tool of Statistical Process Control (SPC) and can be used for monitoring a process in order to detect changes in it. Walter A. Shewhart (Shewhart, 1926) was the first who proposed the use of control charts, back in the 1920's. The most common and often used are the Shewhart-type charts because of their ease in design and interpretation. Two main categories of control charts are the variables charts and the attributes charts. Control charts for variables are used when the quality characteristic X (random variable, r.v.) follows a continuous distribution while if X follows a discrete distribution then the respective chart is a control chart for attributes. Control charts for attributes are used for process monitoring when the data collected by the process are counts. For example, attributes control charts are used for monitoring the number of non-conforming items in a sample, the number of non-conformities in each inspected unit, the number of new daily infections in a specific region or the daily number of logins in the computers of a workstation.

The *np*- and *c*-chart are two common Shewhart-type attributes charts. The first is used when X is a binomial random variable (such as the number of non-conforming items in the sample) while the second is used when X is a Poisson random variable (such as the number of emergency calls in a hospital during the night shift). See Montgomery, 2009 for the basic properties of the *np*- and *c*-chart.

However, there are several processes where the *np*- and *c*-chart cannot be used for their monitoring because the distributional assumption of the binomial and the Poisson distribution is violated. For example, when there is overdispersion in the data or if there is an excessive number

DOI: 10.1201/9781032664330-3

of zero values (or, sometimes, both). Therefore, practitioners have to select first the model that fits better their data and then develop a control chart based on this model.

Zero-inflated models (see Johnson *et al.*, 2005, pp. 351-356) have been recommended as alternative models that take into account the excessive number of zeros as well as the overdispersion that is caused by them. This excess in zeros is attributed to the technological progress and automation of manufacturing technology which led to high-quality (or high-yield) processes; processes with a very low probability of non-conforming items or with a very low rate in the occurrence of non-conforming items. A zero value occurs when there are no non-conforming items in a sample or no non-conformities in the inspected unit.

In recent years, there has been an increased interest in control charts for monitoring zero-inflated processes. See the recent overview by Mahmood & Xie, 2019. However, the majority of these works is about control charts for Phase II monitoring, where the basic assumption is that the values of the process parameters are known. This is a strong assumption which rarely (if ever) is met in practice. According to Montgomery, 2009 (see also Jensen *et al.*, 2006) there is significant difference between the theoretical and the actual performance of a control chart, which is attributed to the presence of estimation error. With the term theoretical we refer to the case where process parameters are known (or Case K) while the term actual performance refers to the case where the process parameters are unknown and must be estimated (Case U). Albers *et al.*, 2004 discriminated the estimation error into two distinct sources, namely *model error* and *stochastic error*. The first is caused by incorrect model assumptions (e.g. the true distributional model is much different than the selected one) while the latter is due to estimation uncertainty and is related to the selected Phase I sample. In this work the focus is solely on the stochastic (estimation) error. He *et al.*, 2003 were the first who studied the estimation effect on the performance of a Shewhart-type control chart for monitoring a zero-inflated Poisson process. Rakitzis & Castagliola, 2016 extended the work of He *et al.*, 2003 and studied also the performance of a Shewhart-type chart for monitoring a zero-inflated binomial process. Both works assume that only one of the two parameters of these zero-inflated processes is unknown. Recently, Mamzeridou & Rakitzis, 2024 investigated the performance of an upper one-sided Shewhart chart for monitoring a zero-inflated Poisson process, in the case where both process parameters are unknown.

In this work we consider two popular zero-inflated models (namely, the zero-inflated Poisson and the zero-inflated binomial distribution) and study the statistical design and performance of the corresponding Shewhart-type charts in Case U. It should be mentioned that control charts for monitoring a zero-inflated binomial process in Case U have not been considered so far. In addition, we consider two estimation methods, maximum likelihood estimation (MLE) and method of moments (MoM), while the proposed chart is two-sided with Shewhart-type limits, instead of probability limits. The main difference with the work of Mamzeridou & Rakitzis, 2024 is that in this work we evaluate the performance of the charts in Case U from a different perspective. Thus, different performance measures are used. This will be explained later.

The structure of this paper is: In Section 3.2 we provide, in brief, the basic properties of the Shewhart charts for monitoring a zero-inflated Poisson (Section 3.2.1) and a zero-inflated binomial (Section 3.2.2) process, in the case of known parameters (Case K), along with the appropriate measures for their performance (Section 3.2.3). In Section 3.3, we describe the implementation of the charts in Case U, for each of the considered processes, providing also the estimators under each case. Using these estimates we derive the basic properties of the considered Shewhart charts in the case of estimated parameters. Also, we present the steps for designing the charts as well as the appropriate performance measures, under an unconditional perspective. In Section 3.4 we present the findings of a simulation study regarding the in-control (IC) and the out-of-control (OOC) performance of the considered charts in Case U. The results in the case of monitoring a ZIP process are given in Section 3.4.1 while in Section 3.4.2 we provide the results in the case of a ZIB process. A brief discussion on the OOC performance of the charts is given in Section 3.5 while conclusions are summarized in Section 3.6.

3.2 SHEWHART CHARTS FOR MONITORING ZIP AND ZIB PROCESSES

3.2.1 THE SHEWHART CHART FOR ZIP PROCESSES (CASE K)

The zero-inflated Poisson (ZIP) distribution is a generalization of the Poisson distribution that can be used to model count data containing an excessive number of zeros. By definition, if X is a ZIP random variable, it is defined on $\{0,1,\dots\}$ (as for the standard Poisson distribution) and its probability mass function (p.m.f.) is given by

$$f_{ZIP}(x\,|\phi,\lambda) = \begin{cases} \phi + (1-\phi)\, f_P(0\,|\lambda), & x = 0 \\ (1-\phi)\, f_P(x\,|\lambda), & x = 1,2,\dots \end{cases}, \tag{3.1}$$

where $f_P(x\,|\lambda)$ is the p.m.f. of the standard Poisson distribution with parameter $\lambda > 0$ and $\phi \in [0,1]$ is the zero-inflation parameter. If $\phi = 0$, then the ZIP distribution coincides with the standard Poisson distribution. If $\phi = 1$, then the ZIP distribution reduces to the Dirac distribution on $x = 0$. Moreover, it is not difficult to verify that the cumulative distribution function (c.d.f.) of X is given by:

$$F_{ZIP}(x\,|\phi,\lambda) = \phi + (1-\phi)F_P(x\,|\lambda), \tag{3.2}$$

where $F_P(x\,|\lambda)$ is the c.d.f. of the Poisson distribution with parameter λ. The mean and the variance of the ZIP distribution with parameters (ϕ,λ) are, respectively, given by:

$$E(X) = \lambda\,(1-\phi),\ \ V(X) = \lambda\,(1+\lambda\phi)\,(1-\phi). \tag{3.3}$$

Let us now assume that we want to set-up a Shewhart control chart for monitoring a ZIP process. In the case of known parameters (Case K), the upper and the lower control limits (denoted as UCL_{ZIP}, LCL_{ZIP}, respectively) are given by (see, for example, Rakitzis & Castagliola, 2016, page 4196):

$$UCL_{ZIP} = \left\lfloor \lambda_0(1-\phi_0) + L\sqrt{\lambda_0(1+\lambda_0\phi_0)(1-\phi_0)} \right\rfloor, \tag{3.4}$$

$$LCL_{ZIP} = \max\left(0, \left\lceil \lambda_0(1-\phi_0) - L\sqrt{\lambda_0(1+\lambda_0\phi_0)(1-\phi_0)} \right\rceil\right), \tag{3.5}$$

where $\lfloor\dots\rfloor$ and $\lceil\dots\rceil$ denote the rounded down and rounded up integer, ϕ_0 λ_0 are the in-control values of the process parameters ϕ and λ, while $L > 0$ is a constant that plays the role of chart's design parameter. These limits are also known as *Shewhart-type* or $L\sigma$ limits. It should be noted that in Mamzeridou & Rakitzis, 2024 the authors used only one probability limit and not Shewhart-type limits. Also, given ϕ_0 and λ_0, the value of L determines the limits of the interval $[LCL_{ZIP}, UCL_{ZIP}]$. Thus, it is directly related to the false alarm rate (FAR) of the chart. Recall that the FAR is the probability for the chart to give an OOC signal when the process is actually IC. The largest the L the smallest the FAR.

Let $Y_1, Y_2, \dots$ be independent random variables such that $Y_i \sim ZIP(\phi_1,\lambda_1)$, i.e., a ZIP distribution with parameters (ϕ_1,λ_1), where, in general, $\phi_1 \neq \phi_0$, $\lambda_1 \neq \lambda_0$ and $\phi_1 \in [0,1]$, $\lambda_1 > 0$. If $\phi_1 = \phi_0$ and $\lambda_1 = \lambda_0$, then the process is IC, otherwise it is OOC. Then, the probability β which is the probability that the number Y_i of non-conformities is within the interval $[LCL_{ZIP}, UCL_{ZIP}]$ equals:

$$\begin{aligned} \beta &= P(LCL_{ZIP} \le Y_i \le UCL_{ZIP}) \\ &= F_{ZIP}(UCL_{ZIP}|\,\phi_1,\lambda_1) - F_{ZIP}(LCL_{ZIP} - 1|\,\phi_1,\lambda_1) \end{aligned} \tag{3.6}$$

3.2.2 THE SHEWHART CHART FOR ZIB PROCESSES (CASE K)

The zero-inflated binomial (ZIB) distribution is a generalization of the binomial distribution that takes into account the excessive number of zeros, compared to the usual binomial distribution.

By definition, if X is a ZIB random variable, it is defined on $\{0,1,\ldots,n\}$ (same as the binomial distribution) and its p.m.f. is given by:

$$f_{ZIB}(x\,|\phi,n,p) = \begin{cases} \phi + (1-\phi)\, f_B(0\,|n,p), & x = 0 \\ (1-\phi)\, f_B(x\,|n,p), & x = 1,2,\ldots,n \end{cases}, \tag{3.7}$$

where $f_B(x\,|n,p)$ is the p.m.f. of the binomial distribution $B(n,p)$ with parameters $n \in \{1,2,\ldots\}$ and $p \in (0,1)$. Parameter p is the success probability (i.e., the probability of an item being non-conforming) while ϕ is the zero-inflation parameter. If $\phi = 0$, the ZIB distribution coincides with the binomial distribution with parameters n and p while, for $\phi = 1$, the ZIB distribution reduces to the Dirac distribution on $x = 0$. The c.d.f. of X is given by:

$$F_{ZIB}(x\,|\phi,n,p) = \phi + (1-\phi)F_B(x\,|n,p), \tag{3.8}$$

where $F_B(x\,|n,p)$ is the c.d.f. of $B(n,p)$. The mean and the variance of the ZIB distribution with parameters (ϕ,n,p) are, respectively, given by the following two expressions:

$$E(X) = np(1-\phi), \quad V(X) = np(1-p+np\phi)(1-\phi). \tag{3.9}$$

In a similar manner, the control limits of a Shewhart chart for monitoring a ZIB process with known parameters are given by (see, for example, Rakitzis & Castagliola, 2016, page 4196):

$$UCL_{ZIB} = \left\lfloor np_0(1-\phi_0) + L\sqrt{np_0(1-p_0+np_0\phi_0)(1-\phi_0)} \right\rfloor, \tag{3.10}$$

$$LCL_{ZIB} = \max\left(0, \left\lceil np_0(1-\phi_0) - L\sqrt{np_0(1-p_0+np_0\phi_0)(1-\phi_0)} \right\rceil\right), \tag{3.11}$$

where ϕ_0, p_0 are the in-control values of process parameters ϕ and p, while constant $L > 0$ is the chart's design parameter. Let $Y_1, Y_2, \ldots$ be independent random variables with $Y_i \sim ZIB(\phi_1,n,p_1)$, i.e., the number of non-conforming units in a sample of size n is a ZIB random variable with parameters (ϕ_1,n,p_1). Again, in general, $\phi_1 \neq \phi_0$ and $p_1 \neq p_0$ with $\phi_1 \in [0,1]$ and $p_1 \in (0,1)$. If $\phi_1 = \phi_0$ and $p_1 = p_0$ the process is IC, otherwise it is OOC. Then, the probability β which is the probability that the number Y_i of non-conforming items is within the interval $[LCL_{ZIB}, UCL_{ZIB}]$ equals:

$$\begin{aligned} \beta &= P(LCL_{ZIB} \leq Y_i \leq UCL_{ZIB}) \\ &= F_{ZIB}(UCL_{ZIP}|\,\phi_1,n,p_1) - F_{ZIP}(LCL_{ZIP} - 1|\,\phi_1,n,p_1) \end{aligned} \tag{3.12}$$

3.2.3 RUN LENGTH PROPERTIES (CASE K)

The run length of the Shewhart charts for ZIP and ZIB processes with known parameters is a geometric random variable N with parameter $1-\beta$, i.e., the p.m.f. $f_N(\nu)$ and the c.d.f. $F_N(\nu)$ of N are defined for $\nu = 1,2,\ldots$ and they are, respectively, given by:

$$f_N(\nu) = (1-\beta)\beta^{\nu-1}, \quad F_N(\nu) = 1-\beta^{\nu}, \tag{3.13}$$

where β is the probability defined in equations (3.6) and (3.12). Therefore, the Average Run Length (*ARL*) and the Standard Deviation Run Length (*SDRL*) of the Shewhart charts for ZIP and ZIB processes with known parameters equal

$$E(N) = ARL = \frac{1}{1-\beta}, \quad \sqrt{V(N)} = SDRL = \frac{\sqrt{\beta}}{1-\beta}. \tag{3.14}$$

Note also that the moment about the origin of order two of the run length distribution equals

$$\mu_2' = E(N^2) = V(N) + E^2(N) = \frac{\beta}{(1-\beta)^2} + \frac{1}{(1-\beta)^2} = \frac{1+\beta}{(1-\beta)^2}, \tag{3.15}$$

where $\mu_k' = E(N^k)$, $k = 1, 2, \ldots$. It should be mentioned that since β is given by equations (3.6) or (3.12) the run length distribution as well as the *ARL* and *SDRL* refer all to the OOC case. For $\phi_1 = \phi_0$ and $\lambda_1 = \lambda_0$ we get the respective IC measures.

3.3 SHEWHART CHARTS FOR MONITORING ZIP AND ZIB PROCESSES WITH UNKNOWN PARAMETERS

In this section we discuss the properties of the two Shewhart charts, for ZIP and ZIB processes, in Case U. The two processes are treated individually. As already stated, for estimating the process parameters, we consider both the MLE method and the MoM.

3.3.1 THE SHEWHART CHART FOR ZIP PROCESSES (CASE U)

Suppose that we are interested in monitoring a ZIP process but the IC values of the process parameters are unknown. Thus, ϕ_0 and λ_0 must be estimated from a Phase I sample. Let $X_1, X_2, \ldots, X_m$ be a random sample from $ZIP(\phi_0, \lambda_0)$, i.e., $X_i \sim ZIP(\phi_0, \lambda_0)$, $i = 1, 2, \ldots, m$. This is the Phase I sample which is used for estimating ϕ_0 and λ_0.

Using the MoM, the respective moment estimators (ME) for λ_0 and ϕ_0 are given by:

$$\tilde{\lambda} = \frac{\overline{X^2}}{\bar{X}} - 1,\ \tilde{\phi} = 1 - \bar{X}/\tilde{\lambda}_0, \tag{3.16}$$

where

$$\bar{X} = \frac{1}{m}\sum_{i=1}^{m} X_i,\ \overline{X^2} = \frac{1}{m}\sum_{i=1}^{m} X_i^2, \tag{3.17}$$

Also, using the MLE method, the respective estimators are not given in closed form but they are obtained numerically by solving the following non-linear system of equations:

$$\begin{cases} \hat{\lambda} = \bar{X}^{+}(1 - e^{-\hat{\lambda}}) \\ \hat{\phi} = 1 - \bar{X}/\hat{\lambda} \end{cases} \tag{3.18}$$

where $\bar{X}^{+}$ is the average of strictly positive X's.

When ϕ_0, λ_0 are estimated by $\hat{\phi}$, $\hat{\lambda}$, the control limits of the Shewhart chart are denoted as $\widehat{LCL}_{ZIP}$, $\widehat{UCL}_{ZIP}$ and they are simply obtained by substituting ϕ_0, λ_0 with $\hat{\phi}$ and $\hat{\lambda}$ in equations (3.5) and (3.4). In a similar manner, if ϕ_0 and λ_0 are estimated by $\tilde{\phi}$, $\tilde{\lambda}$, the respective limits are denoted as $\widetilde{LCL}_{ZIP}$, $\widetilde{UCL}_{ZIP}$. It is not difficult to see that in Case U the control limits are based on the estimates of process parameters obtained from the available Phase I sample and thus, they are random variables. Given the estimates of ϕ_0 and λ_0 it is possible to calculate the limits and proceed to Phase II process monitoring.

It is not difficult to realize that different practitioners will end up with different estimates for the process parameters, because of the different Phase I samples. Thus, it is possible to use different control limits. In order to reflect the estimation error that it is attributed to the different estimates among all the practitioners, we will use the probability $\hat{\beta} = P(Y_i \in [\widehat{LCL}_{ZIP}, \widehat{UCL}_{ZIP}] \,|\phi_1, \lambda_1)$ (or $\tilde{\beta} = P(Y_i \in [\widetilde{LCL}_{ZIP}, \widetilde{UCL}_{ZIP}] \,|\phi_1, \lambda_1)$, in case of MoM estimators) to investigate the performance of the ZIP-Shewhart chart in Case U. Note also that this probability cannot be calculated in practice because ϕ_1, λ_1 is unknown.

The assessment of the control chart's performance in Case U is made under two perspectives, the "unconditional" or the "conditional". In Case U, the run length distribution of the ZIP-Shewhart chart is no longer a geometric one because the control limits are r.v. However, given the estimates $\hat{\phi}$, $\hat{\lambda}$ (or $\tilde{\phi}$, $\tilde{\lambda}$), the run length of the chart becomes *conditional* on them and thus, it is known as *conditional run length*. Its distribution is now a geometric one. Using the conditional run length distribution it is possible to evaluate the performance of the chart under the "conditional" perspective. This approach was followed by Mamzeridou & Rakitzis, 2024. See also Vakilian *et al.*, 2018; Zhao & Driscoll, 2016 regarding the use of the conditional run length distribution to study the c-chart in Case U.

On the other hand, by averaging the conditional run length distribution over the (joint) distribution of $(\hat{\phi}, \hat{\lambda})$ (or $(\tilde{\phi}, \tilde{\lambda})$) we get the *unconditional* run length distribution. Using the unconditional run length distribution, the performance of the chart is evaluated under the unconditional perspective. This approach was followed by He *et al.*, 2003 and Rakitzis & Castagliola, 2016. The majority of the literature on the performance of attributes control charts in Case U is based on the use of the "unconditional" run length distribution (see, for example, Braun, 1999; Castagliola & Wu, 2012; Castagliola *et al.*, 2014; Johannssen *et al.*, 2022; Testik *et al.*, 2006; Wu *et al.*, 2016 and references therein). An advantage of this perspective is that it can be applied before collecting the Phase I sample. Thus, it is possible to have practical guidelines regarding the size of the Phase I samples that gives the desired IC performance. For further details on the use of the unconditional run length distribution in studying the performance of control charts in Case U; see Jensen *et al.*, 2006. Motivated by the previously mentioned works, we will use the unconditional (or marginal) run length distribution.

In Algorithm 1 we provide the steps for the calculation of the appropriate performance measures of the ZIP-Shewhart chart in Case U. To the best of our knowledge, the joint distribution of $(\hat{\phi}, \hat{\lambda})$ (or $(\tilde{\phi}, \tilde{\lambda})$) is not known. For this reason, we use Monte Carlo simulation. The steps are given in the case of MLE method while in the case of the MoM, the steps are similar. In order to avoid additional complexities in notation we will use the notations *ARL*, μ_2' and *SDRL* in either Case K and Case U. However, in the latter case, these are the *unconditional* measures.

Algorithm 1 Calculation of the unconditional performance measures – ZIP process

---

1: Choose a Phase I sample of size m from a $ZIP(\phi_0, \lambda_0)$ process.
2: Use the Phase I sample and estimate ϕ_0, λ_0 by using the MLE method (and get the $\hat{\phi}$, $\hat{\lambda}$ estimates).
3: Given the values L, $\hat{\phi}$ and $\hat{\lambda}$, determine $\widehat{UCL}_{ZIP}$, $\widehat{LCL}_{ZIP}$ from equations (3.4) and (3.5), respectively.
4: Calculate the IC *ARL* and the IC μ_2' as $1/\left(1-\hat{\beta}\right)$ and $\left(1+\hat{\beta}\right)/\left(1-\hat{\beta}\right)^2$, respectively.
5: Repeat Steps 1-4 T times and calculate (i) the *unconditional* IC *ARL* as the average of the T values $1/\left(1-\hat{\beta}\right)$, (ii) the *unconditional* IC μ_2' as the average of the T values $\left(1+\hat{\beta}\right)/\left(1-\hat{\beta}\right)^2$, (iii) the *unconditional* IC *SDRL* as $\sqrt{\mu_2' - ARL^2}$, where *ARL* and μ_2' are those obtained in (i) and (ii).

3.3.2 THE SHEWHART CHART FOR ZIB PROCESSES (CASE U)

Let $X_1, X_2, \ldots, X_m$ be a random sample from the zero-inflated binomial (ZIB) distribution with in-control parameters n, p_0 and ϕ_0, i.e., $X_i \sim ZIB(\phi_0, n, p_0)$, $i = 1, 2, \ldots, m$. Using the MoM, the respective estimates of the IC process parameters are given by:

$$\tilde{p} = \frac{\overline{X^2} - \bar{X}}{(n-1)\bar{X}}, \; \tilde{\phi} = 1 - \frac{(n-1)\bar{X}^2}{n\left(\overline{X^2} - \bar{X}\right)}, \tag{3.19}$$

where

$$\bar{X} = \frac{1}{m}\sum_{i=1}^{m} X_i, \; \overline{X^2} = \frac{1}{m}\sum_{i=1}^{m} X_i^2, \tag{3.20}$$

Also, similar to the case of ZIP process, the MLEs are not given in closed form but they are obtained numerically by solving the following non-linear system of equations:

$$\begin{cases} \hat{p} = \bar{X}^{+}(1-(1-\hat{p})^{n}) \\ \hat{\phi} = 1 - \bar{X}/(n\hat{p}) \end{cases} \tag{3.21}$$

where $\bar{X}^{+}$ is the average of strictly positive X's. When ϕ_0, p_0 are estimated by $\hat{\phi}$, $\hat{p}$, the control limits of the Shewhart chart are denoted as $\widehat{LCL}_{ZIB}$, $\widehat{UCL}_{ZIB}$. Similarly, if the process parameters are estimated by $\tilde{\phi}$, $\tilde{p}$, the respective limits are denoted as $\widetilde{LCL}_{ZIB}$, $\widetilde{UCL}_{ZIB}$. Therefore, the control limits of the ZIB-Shewhart chart in Case U are calculated by simply replacing the (uknown) values ϕ_0 and p_0 with their respective estimates. Then, practitioners can proceed to the Phase II process monitoring.

The performance of the ZIB-Shewhart chart in Case U is evaluated in terms of the *unconditional ARL* and *SDRL*. The steps for their calculation are identical to those provided in Section 3.3.1 for the ZIP-Shewhart chart, after some straightforward but necessary modifications. Due to space economy, the details are left to the readers.

3.4 NUMERICAL RESULTS

In this section, we present the results of an extensive simulation study, using 50000 simulation runs (i.e., $T = 50000$), regarding the IC performance of the ZIP- and ZIB-Shewhart charts in Case U.

3.4.1 THE UNCONDITIONAL PERFORMANCE OF THE ZIP-SHEWHART CHART

In Tables 3.1 and 3.2 we present the IC *ARL* and *SDRL* values for the ZIP-Shewhart chart in Case U, for $m \in \{100, 200, 500, 1000, 2000, 5000\}$, $\lambda_0 \in \{1, 2, 4, 5, 6, 8\}$ and $\phi \in \{0.9, 0.8, 0.7\}$. The MLE method has been used to get the results in Table 3.1 while the MoM has been used for the results in Table 3.2. The first line in each cell gives the *ARL* while the second line gives the *SDRL*. In the column "Case K" we provide the *ARL* and *SDRL* values in Case K while constant L, given in the respective column, has been determined in Case K (in two decimals accuracy) so as the IC *ARL* is *as close as possible* to the nominal value $ARL_0 = 370.4$, a value that appears very frequently in several SPC textbooks. It is well known (see, for example, Rakitzis & Castagliola, 2016) that due to the discrete nature of the ZIP distribution it is almost impossible to achieve *exactly* the desired ARL_0 value. From the results in column 'Case K' in Table 3.1 (and Table 3.2) we deduce that there are cases where the IC *ARL* in Case K is above 370.4 while for the remaining ones it is below.

It is not difficult to see that for the particular combinations (ϕ_0, λ_0), the IC *ARL* values can be very different between the Case K and Case U. For example, from the results in Table 3.1, when $(\phi_0, \lambda_0) = (0.8, 4)$ the IC *ARL* (*SDRL*) in Case K is $ARL = 234.04$ ($SDRL = 233.54$) while when $m = 200$ we have $ARL = 566.39$ ($SDRL = 1116.81$). Note also that for a relatively small Phase I sample (i.e., for $m = 100$ or 200), the *ARL* and *SDRL* are very large, especially the latter. Actually, for large ϕ_0 values (such as 0.9 or 0.8) and for Phase I samples of size $m = 100$, the *SDRL* values are extremely large. This is an indication of increased uncertainty regarding the actual IC performance of the chart. As m increases, the IC *ARL* values tend to decrease, but not in a monotonic way. This

Table 3.1

IC ARL and $SDRL$ values for the ZIP-Shewhart Chart in Case U when L is from Case K – MLE method

ϕ_0	λ_0	L	100	200	500	1000	2000	5000	Case K
0.9	1	6.66	735.21	426.62	330.69	323.88	325.72	326.11	526.64
			3164.42	1039.20	502.21	434.35	431.99	432.28	526.14
	2	6.41	1268.37	617.42	435.46	395.69	392.04	393.24	189.92
			11282.42	1739.44	694.71	518.63	489.55	489.74	189.42
	4	5.61	1173.64	568.04	390.56	348.66	335.52	335.86	468.09
			14863.58	1527.86	614.13	447.65	392.86	386.74	467.59
	5	5.72	2775.20	1043.62	643.09	555.63	522.05	511.23	314.19
			32197.04	3564.15	1118.23	752.59	624.91	589.38	313.69
	6	5.31	1685.94	714.62	457.20	396.67	374.75	366.50	497.71
			14686.21	2545.71	764.17	522.35	439.39	410.96	497.21
	8	5.15	3201.76	991.22	575.55	486.03	450.12	432.52	292.56
			45373.05	4210.97	1025.86	665.61	538.16	480.22	292.06
0.8	1	6.33	1605.20	1033.08	811.91	797.17	801.98	799.38	263.32
			8203.23	2641.74	1245.20	1115.59	1118.12	1116.19	262.82
	2	5.49	1561.49	943.59	726.60	700.33	696.85	692.86	301.87
			7511.36	2157.01	1075.06	911.12	897.56	894.40	301.37
	4	4.47	850.31	566.39	449.27	424.31	420.75	417.34	234.04
			3305.00	1116.81	620.58	518.11	499.33	496.14	233.54
	5	4.47	1536.53	927.90	707.73	650.36	633.37	634.67	365.09
			6766.62	2016.17	1033.60	808.04	746.39	744.54	364.59
	6	4.09	875.03	568.34	446.35	411.82	403.97	400.85	248.86
			3104.88	1129.01	622.37	498.47	463.78	458.92	248.36
	8	3.89	1092.12	667.40	502.29	459.47	444.99	440.59	289.74
			4205.83	1436.17	717.51	562.35	505.80	494.03	289.24
0.7	1	5.18	785.25	587.16	535.49	538.28	537.27	539.31	175.55
			2352.71	1179.19	757.51	474.96	747.22	748.72	175.05
	2	4.5	693.97	532.31	465.54	462.64	468.50	467.07	201.24
			1799.96	941.45	615.07	597.17	601.32	600.20	200.74
	4	3.66	412.11	328.80	291.87	287.96	290.63	294.41	409.89
			896.25	516.96	362.38	339.81	341.00	344.01	409.39
	5	3.65	658.47	506.09	434.03	420.35	416.23	411.57	243.39
			1584.07	832.45	552.26	497.58	490.16	485.96	242.89
	6	3.34	408.37	327.86	285.55	276.14	276.70	280.81	377.61
			863.19	514.41	353.55	316.93	314.09	317.52	377.11
	8	3.17	475.63	370.34	318.53	304.70	303.99	306.20	404.97
			1114.49	594.23	398.22	347.88	338.69	340.27	404.47

is common for attributes control charts with estimated parameters (see, for example Castagliola & Wu, 2012; Mamzeridou & Rakitzis, 2024; Rakitzis & Castagliola, 2016, and references, therein). Note also that as m increases, it is not necessary that the IC ARL and $SDRL$ obtained in Case U will converge to the respective IC ARL and $SDRL$ values in Case K. Even for very large Phase I samples (such as for $m = 2000$ or 5000), the ARL and $SDRL$ can be significantly different than the respective

Table 3.2

IC *ARL* and *SDRL* values for the ZIP-Shewhart Chart in Case U when *L* is from Case K – MoM

ϕ_0	λ_0	L	100	200	500	1000	2000	5000	Case K
0.9	1	6.66	883.14	427.10	328.68	318.96	322.32	326.28	526.64
			12359.71	1145.97	518.36	432.19	429.41	432.41	526.14
	2	6.41	1281.62	624.01	435.30	398.50	390.07	389.88	189.92
			13991.43	1953.22	708.40	527.32	488.35	486.97	189.42
	4	5.61	1462.50	577.83	391.60	348.63	335.95	335.42	468.09
			58219.25	1993.99	628.37	452.55	394.96	386.37	467.59
	5	5.72	2839.46	1068.15	653.69	558.32	521.99	511.90	314.19
			31349.76	3975.49	1163.99	762.75	628.38	590.31	313.69
	6	5.31	1759.65	721.64	458.78	398.31	373.00	367.80	497.71
			15155.06	2718.01	775.81	533.86	439.46	412.17	497.21
	8	5.15	3049.76	1022.58	578.72	488.46	448.43	433.13	292.56
			36613.36	4224.83	1056.43	674.30	536.22	481.14	292.06
0.8	1	6.33	1920.78	1074.68	814.54	794.96	796.77	802.86	263.32
			13635.01	3116.44	1285.25	1116.00	1114.25	1118.77	262.82
	2	5.49	1647.73	984.21	727.80	696.39	694.18	696.20	301.87
			9176.57	2471.15	1113.19	915.94	895.45	897.04	301.37
	4	4.47	913.25	580.55	453.29	422.83	419.63	417.46	234.04
			3924.83	1208.57	638.30	518.07	498.36	496.24	233.54
	5	4.47	1670.69	948.22	713.63	652.88	637.28	636.91	365.09
			9529.34	2154.13	1047.45	822.13	750.63	746.48	364.59
	6	4.09	941.12	583.22	448.94	412.70	402.90	400.97	248.86
			3628.02	1268.77	632.52	501.68	463.91	459.03	248.36
	8	3.89	1157.29	684.89	509.14	464.15	445.14	441.07	289.74
			4884.43	1528.66	730.69	571.11	506.42	494.43	289.24
0.7	1	5.18	848.42	601.17	532.47	536.05	533.64	539.36	175.55
			3613.84	1268.09	753.96	746.32	744.54	748.75	175.05
	2	4.5	744.48	541.96	468.19	464.60	466.30	465.76	201.24
			2272.43	1000.84	631.23	598.28	599.60	599.18	200.74
	4	3.66	428.65	337.05	291.65	287.69	289.70	293.15	409.89
			1036.65	549.53	367.38	339.81	340.25	343.01	409.39
	5	3.65	698.85	514.16	437.42	419.80	416.28	410.79	243.39
			1788.67	863.89	563.35	497.57	490.20	485.24	242.89
	6	3.34	432.39	333.83	286.85	278.06	278.10	280.96	377.61
			1015.94	532.35	357.67	320.08	315.37	317.64	377.11
	8	3.17	509.23	377.29	322.06	306.22	303.69	305.87	404.97
			1291.78	613.58	407.06	351.84	338.65	339.99	404.47

values in Case K. See, for example, the case $(\phi_0, \lambda_0) = (0.8, 1)$ or $(\phi_0, \lambda_0) = (0.7, 2)$. This fact is also attributed to the discrete nature of the ZIP distribution.

The results so far show that if we want to apply the ZIP-Shewhart chart in the practice, then we must collect a very large Phase I sample, even larger than 5000 preliminary observations, in order to have similar performance between the Case K and Case U. Since this is not always possible in

Table 3.3
"Adjusted" L^* Values for the ZIP-Shewhart Chart – MLE method

		$\phi_0 = 0.9$			$\phi_0 = 0.8$			$\phi_0 = 0.7$		
λ_0	m	L^*	ARL	$SDRL$	L^*	ARL	$SDRL$	L^*	ARL	$SDRL$
1	100	6.54	526.64	1914.88	4.86	263.22	650.62	4.10	179.75	374.91
	200	6.90	528.91	1390.08	5.12	262.39	494.63	4.26	175.45	276.87
	500	7.33	527.17	947.08	5.30	263.03	358.17	4.36	174.57	202.59
	1000	7.51	529.11	739.50	5.38	263.06	288.63	4.41	175.39	177.76
	2000	7.64	527.05	581.89	5.43	263.30	264.45	4.44	175.54	175.05
	5000	7.67	526.33	526.29	5.64	263.34	262.94	4.44	175.55	175.05
2	100	4.98	189.51	526.49	4.53	300.65	806.72	3.80	199.34	400.56
	200	5.30	189.66	360.71	4.73	304.39	543.58	3.92	202.41	309.87
	500	5.52	190.31	265.89	4.86	300.18	406.96	4.00	201.09	243.50
	1000	5.61	190.35	227.59	4.92	301.04	344.89	4.04	201.66	212.20
	2000	5.67	189.86	200.66	4.96	302.01	309.55	4.05	201.21	201.62
	5000	5.74	189.98	189.86	4.98	301.87	301.42	4.05	201.24	200.74
4	100	5.14	473.58	3861.22	3.88	235.33	579.53	3.66	409.93	924.35
	200	5.49	463.95	1431.24	4.02	234.34	390.11	3.75	405.19	654.51
	500	5.74	464.46	753.44	4.12	232.37	302.78	3.82	406.40	520.92
	1000	5.84	466.90	618.94	4.15	234.95	273.28	3.85	408.91	460.79
	2000	5.89	465.12	547.49	4.17	233.91	246.11	3.86	407.91	418.67
	5000	5.94	468.77	487.91	4.19	234.04	234.12	3.88	409.90	409.54
5	100	4.73	323.72	1673.04	3.92	365.10	993.44	3.28	242.37	487.02
	200	5.04	314.12	726.86	4.06	364.24	663.87	3.36	242.89	364.17
	500	5.24	313.38	480.98	4.16	366.44	499.38	3.41	241.53	297.82
	1000	5.32	313.61	400.84	4.20	367.29	442.19	3.44	245.08	272.32
	2000	5.36	311.84	359.94	4.22	365.72	398.56	3.45	243.57	249.88
	5000	5.40	314.64	327.18	4.24	365.54	368.39	3.46	243.46	243.16
6	100	4.81	499.73	2733.80	3.62	250.20	689.91	3.31	381.12	826.60
	200	4.79	492.48	3018.95	3.75	250.10	443.31	3.39	372.96	594.64
	500	5.12	495.15	1425.54	3.83	249.11	326.73	3.45	380.46	482.38
	1000	5.45	493.99	663.90	3.86	248.56	292.40	3.47	378.22	431.74
	2000	5.50	499.64	597.13	3.88	248.15	267.48	3.48	376.75	394.47
	5000	5.53	496.66	530.07	3.89	248.49	250.36	3.49	377.58	377.91
8	100	4.33	299.65	1500.19	3.47	289.30	796.77	3.12	404.77	886.12
	200	4.60	293.34	777.60	3.60	293.82	548.37	3.20	409.21	670.61
	500	4.80	292.26	463.56	3.68	289.34	390.57	3.25	407.76	519.92
	1000	4.87	292.58	378.29	3.71	289.76	344.97	3.27	409.30	473.06
	2000	4.91	293.21	340.75	3.73	292.57	323.01	3.28	406.63	433.95
	5000	4.94	294.13	313.23	3.74	289.99	295.78	3.29	405.99	408.78

practice, we provide in Tables 3.3 (MLE method) and 3.4 (MoM) the "adjusted" L^* values (see column "L^*" in the tables) which have been obtained so as the IC *ARL* in Case U is *as close as possible* to IC *ARL* in Case K, for a given size m for the Phase I sample. Note also that any other value (instead of the IC *ARL* in Case K) can be used instead, according to practitioners needs. Once

Table 3.4
"Adjusted" L^* Values for the ZIP-Shewhart Chart – MoM

		$\phi_0 = 0.9$			$\phi_0 = 0.8$			$\phi_0 = 0.7$		
λ_0	m	L^*	*ARL*	*SDRL*	L^*	*ARL*	*SDRL*	L^*	*ARL*	*SDRL*
1	100	6.25	531.07	3095.24	4.86	267.77	857.32	4.09	175.79	391.94
	200	6.90	543.54	1649.93	5.11	263.45	508.96	4.24	176.52	287.04
	500	7.33	529.79	951.57	5.30	264.72	369.75	4.36	175.91	209.19
	1000	7.51	532.30	756.62	5.38	264.53	296.31	4.42	175.79	180.14
	2000	7.63	527.60	592.97	5.44	263.61	265.94	4.44	175.55	175.19
	5000	7.68	526.70	528.12	5.32	263.32	262.82	4.25	175.55	175.05
2	100	4.98	193.14	792.41	4.52	306.59	855.64	3.80	201.34	419.99
	200	5.30	190.68	384.17	4.72	306.55	578.28	3.92	204.84	325.62
	500	5.53	192.77	273.03	4.86	303.33	422.58	4.00	202.36	250.62
	1000	5.61	190.45	231.20	4.92	304.04	357.06	4.04	201.89	215.39
	2000	5.67	190.04	202.49	4.96	302.50	312.96	4.06	201.38	202.14
	5000	5.71	189.98	189.95	4.97	301.87	301.48	4.04	201.24	400.74
4	100	5.12	468.23	2383.37	3.86	235.30	634.38	3.65	424.18	1115.79
	200	5.48	470.19	3398.92	4.02	237.56	416.71	3.75	412.83	687.30
	500	5.75	471.20	782.49	4.11	234.15	309.95	3.82	412.84	539.78
	1000	5.84	468.29	631.06	4.15	234.46	274.23	3.85	411.61	470.61
	2000	5.90	471.43	561.97	4.18	235.88	250.54	3.87	411.04	425.87
	5000	5.94	468.75	489.51	4.19	234.09	234.39	3.88	409.90	409.62
5	100	4.71	320.47	1739.25	3.90	371.43	1109.44	3.26	237.90	480.98
	200	5.02	314.95	770.66	4.06	373.64	710.68	3.36	248.60	381.95
	500	5.25	320.30	500.50	4.16	366.83	504.38	3.42	248.55	309.66
	1000	5.33	318.47	411.56	4.20	369.94	450.50	3.44	247.02	278.25
	2000	5.37	316.00	367.03	4.22	366.28	401.06	3.45	244.04	251.96
	5000	5.40	314.45	327.80	4.24	365.33	368.72	3.46	243.46	243.30
6	100	4.79	510.34	4858.52	3.61	253.05	691.00	3.30	386.09	883.67
	200	5.13	519.81	1693.26	3.74	249.54	441.93	3.39	382.38	621.35
	500	5.37	509.23	886.18	3.83	251.35	333.00	3.45	384.21	492.48
	1000	5.45	500.16	679.99	3.86	249.25	294.17	3.47	380.40	438.07
	2000	5.50	500.60	601.97	3.88	250.16	272.42	3.49	383.03	405.29
	5000	5.54	502.74	539.86	3.90	249.53	252.24	3.50	378.26	379.39
8	100	4.29	299.06	5744.17	3.46	293.92	923.36	3.10	397.00	952.52
	200	4.60	296.97	784.66	3.60	297.99	576.82	3.18	392.39	645.46
	500	4.80	294.37	468.89	3.68	293.14	400.61	3.24	400.23	513.15
	1000	4.87	293.66	381.99	3.71	291.57	349.50	3.26	400.55	463.44
	2000	4.91	295.21	344.15	3.73	292.42	323.03	3.28	407.42	436.28
	5000	4.94	295.41	315.50	3.74	290.44	296.90	3.29	405.94	408.91

the L^* values has been obtained, we calculate the respective IC *ARL* and *SDRL* values when the constant L^* is used, instead of L.

The L^* values have been determined with a two decimals accuracy, using 50000 simulation runs. The steps of the algorithmic procedure are given at Algorithm 2.

Algorithm 2 Determination of the adjusted constant L^* – ZIP process

1: Choose the nominal IC ARL value, say ARL_0.
2: Collect a Phase I sample of size m from a $ZIP(\phi_0, \lambda_0)$ process.
3: Use the Phase I sample and estimate ϕ_0, λ_0 by using the MLE method (and get the $\hat{\phi}$, $\hat{\lambda}$ estimates).
4: Set the initial value $L \leftarrow 0.01$
5: Given the values L, $\hat{\phi}$ and $\hat{\lambda}$, determine $\widehat{UCL_{ZIP}}$, $\widehat{LCL_{ZIP}}$ from equations (3.4) and (3.5) and calculate the IC ARL.
6: Repeat Steps 2-5 T times and calculate the *unconditional* IC ARL as the average of the T values $1/\left(1-\hat{\beta}\right)$.
7: Compare the ARL obtained in step 6 with the ARL_0. If $|ARL - ARL_0|/ARL_0 > 0.05$ then increase L by 0.01 and go back to Step 1. Otherwise, select L as the "adjusted" L^* value.

For example, in the case $(\phi_0, \lambda_0) = (0.8, 4)$, the IC ARL ($SDRL$) in Case K is equal to $ARL = 234.04$ ($SDRL = 233.54$) with $L = 4.47$. Here $ARL_0 = 234.04$. When the size of the Phase I sample is $m = 200$, then the "adjusted" value for the chart parameter is $L^* = 4.02$ (when the process parameters are estimated via the maximum likelihood estimation method), which gives IC $ARL = 234.34$ and $SDRL = 390.11$. Recall that for $m = 200$ and $L = 4.47$ the $ARL = 566.39$ and $SDRL = 1116.81$. Therefore, the use of the "adjusted" L^* value results in an IC ARL value is much closer to the nominal ARL_0 value while there is also a significant reduction in the $SDRL$ value.

Before closing this section it is worth mentioning that Tables 3.1-3.4 offer a head-to-head comparison between the MLEs and the MEs. Recall that for the MEs there are closed-form expressions while MLEs require the use of a non-linear equation solver and can only be obtained numerically. The results show that for Phase I samples of size $m \geq 500$, the two methods provide (almost) identical results. When $m = 100$ or 200, the respective ARL and $SDRL$ values can be very different. According to Beckett *et al.*, 2014; Wagh & Kamalja, 2018, the MLEs are superior than the estimates obtained by the MoM, in terms of mean square error and bias, especially for small sample sizes. Therefore, we would suggest the use of MLEs when very large Phase I samples are not available. However, for larger Phase I samples (e.g. when $m > 1000$) the MoM estimators of ϕ_0, λ_0 might be more appealing to practitioners due to their simplicity in calculation.

3.4.2 THE UNCONDITIONAL PERFORMANCE OF THE ZIB-SHEWHART CHART

Tables 3.5-3.6 and Tables 3.7-3.8 are counterparts to Tables 3.1-3.2 and Tables 3.3-3.4 but for the ZIB-Shewhart chart. In Tables 3.5 and 3.6 we present the IC ARL and $SDRL$ values for the ZIB-Shewhart chart, for $n \in \{100, 250\}$, $p_0 \in \{0.01, 0.02, 0.030\}$, $m \in \{100, 200, 500, 1000, 2000, 5000\}$ and $\phi \in \{0.9, 0.8, 0.7\}$. The value of L has been determined in order to achieve an IC ARL value in Case K that it is *as close as possible* to the nominal $ARL_0 = 370.4$. Note also that the results in Table 3.5 have been obtained via the MLE method while for those in Table 3.6 we used the MoM. Again, the values of the achieved ARL_0 can be very different than the nominal 370.4 value. As for the case of ZIP-Shewhart chart, this is attributed to the discrete nature of the ZIB distribution and the large value of the zero-inflation parameter ϕ_0.

For example, for $\phi_0 = 0.8$, $p_0 = 0.01$, $n = 100$ and $L = 6.35$, the control limits are obtained from equations (3.10) and (3.11) and they equal $UCL_{ZIB} = 3$, $LCL_{ZIB} = 0$. Therefore, using equation (3.12), the FAR equals $1 - \beta = 1 - F_{ZIB}(3\,|0.8, 100, 0.01) + F_{ZIB}(0 - 1\,|0.8, 100, 0.01) \approx 0.0036748 + 0 = 0.0036748$ and the $ARL_0 = 1/0.0036748 \approx 272.12$.

Conclusions from Tables 3.5 and 3.6 are similar to those from Tables 3.1 and 3.2. Specifically, the IC ARL values can be very different between Case K and Case U. For example, if $(\phi_0, n, p_0) = (0.9, 250, 0.01)$ then the IC ARL and $SDRL$ are 242.82 and 242.32, respectively. But, in Case U

Table 3.5

IC *ARL* and *SDRL* values for the ZIB-Shewhart Chart in Case U when *L* is from Case K – MLE method

ϕ_0	p_0	n	L	100	200	500	1000	2000	5000	Case K
0.9	0.01	100	6.68	905.96	450.09	340.95	332.53	334.73	337.79	544.25
				5067.77	1205.60	518.39	446.73	446.30	448.57	543.75
		250	6.38	1804.08	828.37	551.49	490.08	478.06	475.08	242.82
				16092.61	2962.55	916.65	654.97	590.73	586.15	242.32
0.8	0.01	100	6.35	1959.66	1106.40	870.21	854.52	851.31	858.41	272.12
				11762.09	2957.93	1349.42	1197.39	1193.81	1199.00	271.62
		250	5.31	1845.05	1117.18	836.70	793.15	788.18	783.97	364.92
				9380.28	2617.58	1248.05	1028.49	1002.35	998.90	364.42
0.7	0.01	100	5.19	844.34	617.27	562.32	565.09	561.93	558.17	181.42
				2883.62	1213.13	799.13	793.95	791.61	788.79	180.91
		250	4.35	815.67	604.66	528.87	520.38	513.14	508.91	243.28
				2210.39	1068.48	700.18	664.77	657.86	654.25	242.78
0.9	0.02	100	6.43	1390.59	658.03	463.27	419.49	413.94	413.81	196.73
				9792.72	1864.65	748.52	552.77	522.49	521.12	196.23
		250	5.73	3301.38	1126.98	695.83	596.43	555.99	542.99	329.22
				76501.60	3833.54	1231.11	825.55	668.75	630.13	328.72
0.8	0.02	100	5.51	1752.68	1059.84	800.56	773.11	768.20	766.16	322.92
				8044.89	2697.97	1211.15	1018.73	1000.71	999.11	322.42
		250	4.04	546.84	367.64	300.80	284.05	281.48	284.06	390.77
				2701.38	675.68	402.04	338.08	323.91	325.42	390.27
0.7	0.02	100	4.51	771.87	581.34	509.49	501.80	499.72	489.92	215.28
				2145.68	1034.41	686.15	659.35	657.09	648.97	214.78
		250	3.66	730.52	556.84	475.17	458.28	456.62	451.26	260.51
				1834.69	948.22	613.03	544.71	540.56	535.75	260.01
0.9	0.03	100	6.38	3195.35	1230.92	756.34	657.13	624.07	620.12	320.23
				55772.52	4905.18	1345.95	909.57	774.01	759.32	319.73
		250	5.09	2297.93	808.28	484.48	414.70	388.05	375.15	248.86
				34034.94	2666.92	834.65	557.19	460.84	417.27	248.36
0.8	0.03	100	4.51	570.29	398.58	330.08	319.25	322.23	326.26	470.64
				1891.84	746.76	445.74	390.33	389.40	392.55	470.14
		250	3.86	935.83	576.64	439.76	405.54	391.48	387.97	252.12
				3658.73	1244.32	619.49	494.25	446.75	439.04	251.61
0.7	0.03	100	4.27	1083.34	802.81	680.58	667.35	668.83	668.83	313.76
				3109.93	1467.40	910.28	842.23	841.27	840.69	313.26
		250	3.4	1035.11	760.02	621.56	589.34	578.33	567.52	363.24
				2753.87	1373.60	816.73	697.60	667.16	656.36	362.74

and for $m = 200$, the IC *ARL* and *SDRL* are 828.37 and 2962.55, respectively. Clearly, these values are much larger than the respective ones in Case K, especially the IC *SDRL*. Therefore, very large Phase I samples are needed (e.g. $m \geq 1000$) while there are cases where even $m = 5000$ preliminary observations might not be enough in order to have similar performance between Case K and Case U. See, for example, the cases $(\phi_0, n, p_0) = (0.8, 100, 0.01)$ or $(\phi_0, n, p_0) = (0.9, 100, 0.02)$. However, as the size m increases, the IC *SDRL* reduces while the IC *ARL* converges (not monotonically) to a specific value, which is not necessarily the same with the IC *ARL* under Case K.

Table 3.6

IC *ARL* and *SDRL* values for the ZIB-Shewhart Chart in Case U when *L* is from Case K – MoM

ϕ_0	p_0	n	L	100	200	500	1000	2000	5000	Case K
0.9	0.01	100	6.68	819.19	442.89	337.97	329.77	330.31	335.92	544.25
				5757.19	1528.21	537.91	449.36	442.94	447.19	543.75
		250	6.38	1964.63	813.32	553.84	490.51	476.76	475.55	242.82
				38479.06	2549.29	952.35	663.62	590.62	586.55	242.32
0.8	0.01	100	6.35	2066.18	1128.33	856.90	844.78	849.41	853.56	272.12
				14747.90	3252.57	1376.88	1196.22	1192.42	1195.46	271.62
		250	6	1907.18	1126.10	839.75	789.24	783.95	779.37	364.92
				11480.70	2864.85	1286.38	1029.17	998.91	995.10	364.42
0.7	0.01	100	5.19	859.75	620.89	557.01	555.93	557.25	553.53	181.42
				3177.49	1270.84	798.42	787.10	788.10	785.27	180.91
		250	4.35	814.89	609.44	525.10	517.14	512.98	503.76	243.28
				2371.18	1135.87	710.05	662.48	657.73	649.79	242.78
0.9	0.02	100	6.43	1411.89	664.56	460.10	417.44	411.19	410.55	196.73
				19130.12	2215.13	772.55	558.28	520.09	518.42	196.23
		250	5.73	3139.49	1149.55	691.13	594.82	554.54	542.81	329.22
				36337.64	4337.88	1249.93	826.06	673.75	629.65	328.72
0.8	0.02	100	5.51	1805.57	1050.28	804.90	763.27	764.15	770.48	322.92
				8924.58	2581.10	1249.33	1010.73	997.52	1002.50	322.42
		250	4.04	524.27	364.56	301.95	283.03	281.73	284.55	390.77
				1687.15	693.27	412.39	337.16	324.26	325.83	390.27
0.7	0.02	100	4.51	781.04	586.60	505.38	500.09	498.89	490.52	215.28
				2441.09	1146.70	684.72	657.40	656.41	649.48	214.78
		250	3.66	728.28	556.02	473.93	457.86	455.61	451.69	260.51
				1916.02	959.71	618.42	546.08	539.76	536.13	260.01
0.9	0.03	100	6.38	3265.02	1217.89	749.80	653.31	624.05	618.16	320.23
				47087.82	4698.07	1359.02	923.07	776.75	757.65	319.73
		250	5.09	2174.34	796.58	483.49	416.57	386.64	373.77	248.86
				22699.92	3104.88	854.60	570.90	460.04	416.15	248.36
0.8	0.03	100	4.51	571.83	398.43	329.15	318.58	319.97	324.40	470.64
				2249.32	805.98	454.06	392.87	387.60	391.10	470.14
		250	3.86	904.74	574.09	441.43	404.73	391.78	387.30	252.12
				3503.86	1234.12	632.01	495.61	446.67	483.42	251.61
0.7	0.03	100	4.27	1111.46	805.86	680.79	668.27	666.85	668.61	313.76
				3585.85	1501.58	922.51	845.54	839.90	841.10	313.26
		250	3.4	1029.06	754.55	622.78	586.09	578.36	567.83	363.24
				2848.15	1366.98	825.29	695.60	667.58	656.65	362.74

In Tables 3.7 and 3.8 we provide the "adjusted" values L^* of the chart's design parameter L, so as for a pre-specified size m of the Phase I sample, the IC *ARL* under Case K is the closest possible to the IC *ARL* under Case K. The L^* has been determined with a two decimals accuracy, using Monte Carlo simulation. The steps of the related algorithmic procedure are similar to those presented in Section 3.3.1, for the case of the ZIP-Shewhart chart. Some direct but necessary modifications are needed. The results in both Tables 3.7 and 3.8 show that the use of the "adjusted" L^* results in IC *ARL* values much closer to the respective value in Case K while the IC *SDRL* is much lower, compared to the case where the unadjusted L value is used.

Table 3.7
"Adjusted" L^* Values for the ZIB-Shewhart Chart – MLE method

			$\phi_0 = 0.9$			$\phi_0 = 0.8$			$\phi_0 = 0.7$		
p_0	n	m	L^*	*ARL*	*SDRL*	L^*	*ARL*	*SDRL*	L^*	*ARL*	*SDRL*
0.01	100	100	6.22	547.62	2898.06	4.83	275.82	734.92	4.10	187.96	416.23
		200	6.86	553.26	1464.83	5.13	277.34	536.52	4.27	182.05	293.11
		500	7.36	554.95	984.83	5.30	272.52	379.78	4.37	181.63	215.06
		1000	7.53	550.44	783.88	5.40	272.71	300.60	4.43	181.48	184.54
		2000	7.66	544.80	603.72	5.46	272.36	274.22	4.44	181.43	181.08
		5000	7.73	544.28	544.97	5.34	272.12	271.62	4.21	181.42	180.91
0.02	100	100	4.99	196.97	718.36	4.55	326.95	914.88	3.83	218.21	455.28
		200	5.30	197.17	380.91	4.74	324.82	606.08	3.94	216.92	338.19
		500	5.54	197.61	277.31	4.88	324.36	450.06	4.03	217.99	267.40
		1000	5.63	196.87	237.15	4.94	322.97	373.78	4.06	216.04	228.13
		2000	5.70	197.22	208.97	4.98	323.02	331.20	4.08	215.36	215.65
		5000	5.74	196.78	196.75	5.00	322.93	322.55	4.01	215.28	214.78
0.03	100	100	5.09	323.34	1293.61	4.42	477.90	1513.62	3.73	317.90	673.94
		200	5.44	323.53	747.21	4.60	473.39	965.78	3.83	315.70	503.62
		500	5.68	320.37	483.84	4.73	472.69	677.53	3.90	314.05	400.09
		1000	5.77	320.28	415.16	4.78	471.26	576.32	3.94	316.69	349.47
		2000	5.83	320.38	362.67	4.82	472.82	504.36	3.95	313.84	317.76
		5000	5.88	320.43	324.91	4.85	470.93	471.65	3.96	313.76	313.26
0.01	250	100	5.05	248.87	943.69	4.47	375.30	1005.00	3.76	246.02	511.11
		200	5.38	245.81	525.73	4.65	368.71	689.54	3.86	244.62	379.11
		500	5.62	245.20	356.34	4.79	370.87	515.43	3.94	244.92	304.14
		1000	5.70	243.80	304.10	4.83	365.06	432.10	3.97	243.45	260.65
		2000	5.76	243.18	265.94	4.87	365.57	381.55	3.99	243.39	244.93
		5000	5.81	243.10	244.36	4.90	364.99	364.83	3.99	243.28	242.78
0.02	250	100	4.74	337.48	1522.58	3.92	396.38	1240.43	3.29	262.28	539.24
		200	5.04	332.37	875.33	4.07	393.34	748.01	3.37	262.76	397.89
		500	5.25	331.23	511.43	4.17	393.15	537.20	3.43	266.14	331.43
		1000	5.33	330.02	426.54	4.21	395.09	480.39	3.45	263.90	295.47
		2000	5.38	331.36	385.70	4.23	392.28	427.76	3.46	261.44	268.93
		5000	5.41	329.94	343.60	4.25	390.37	394.67	3.48	260.72	260.59
0.03	250	100	4.28	252.68	1350.49	3.44	252.94	674.34	3.10	359.07	821.73
		200	4.54	250.29	646.02	3.57	260.90	480.26	3.17	351.75	562.90
		500	4.73	250.57	390.05	3.65	256.79	344.24	3.23	362.81	461.89
		1000	4.80	251.69	321.47	3.68	257.94	307.20	3.24	354.55	406.47
		2000	4.84	251.71	290.24	3.69	252.70	276.63	3.26	362.99	384.36
		5000	4.87	252.18	267.24	3.71	253.99	258.71	3.26	362.34	363.33

3.5 OUT-OF-CONTROL PERFORMANCE OF THE CHARTS IN CASE U

In this section, we investigate the OOC performance of the ZIP- and ZIB-Shewhart charts. It is expected that when the "adjusted" constant L^* is used, there will be differences in the detection ability of the chart. These differences also depend on the size m of the Phase I sample as well as on the magnitude of shift(s) in process parameters. Again, we use Monte Carlo simulation and the *ARL*, *SDRL* values are provided for several shifts in process parameters. Due to space economy, only the results in the case of MLE method are presented.

Table 3.8
"Adjusted" L^* Values for the ZIB-Shewhart Chart – MoM

			$\phi_0 = 0.9$			$\phi_0 = 0.8$			$\phi_0 = 0.7$		
p_0	n	m	L^*	ARL	$SDRL$	L^*	ARL	$SDRL$	L^*	ARL	$SDRL$
0.01	100	100	6.22	549.55	8769.46	4.88	272.38	892.13	4.12	182.62	413.66
		200	6.92	544.46	1711.07	5.12	272.35	542.85	4.26	182.66	307.14
		500	7.35	544.48	970.99	5.31	273.55	386.41	4.38	182.00	218.65
		1000	7.53	549.34	794.59	5.40	273.23	305.94	4.43	181.43	185.40
		2000	7.66	545.77	613.90	5.46	272.41	274.95	4.45	181.46	181.25
		5000	7.73	544.25	545.20	5.37	272.12	271.62	4.26	181.42	180.91
0.02	100	100	5.01	198.13	728.96	4.54	325.00	979.55	3.83	215.85	476.67
		200	5.33	196.99	395.82	4.75	325.78	619.55	3.94	216.50	345.00
		500	5.55	199.01	284.76	4.88	325.51	461.30	4.02	215.86	268.80
		1000	5.63	197.01	240.24	4.94	324.76	382.56	4.06	216.52	231.72
		2000	5.70	197.76	211.85	4.97	323.00	333.42	4.08	215.41	216.14
		5000	5.73	196.74	196.76	5.00	322.95	322.62	4.01	215.28	214.78
0.03	100	100	5.10	323.72	1962.17	4.42	477.24	1577.82	3.73	320.25	747.63
		200	5.45	323.21	829.87	4.60	472.48	1011.32	3.83	315.39	514.54
		500	5.69	320.64	497.95	4.73	474.02	692.41	3.90	314.86	408.38
		1000	5.77	320.75	420.67	4.78	471.66	584.45	3.94	317.75	356.21
		2000	5.83	320.69	366.18	4.82	473.45	510.22	3.96	314.94	321.24
		5000	5.87	320.24	325.46	4.83	470.68	471.69	3.96	313.78	313.31
0.01	250	100	5.07	244.13	891.46	4.46	364.97	1082.59	3.76	244.34	555.38
		200	5.40	248.42	543.63	4.66	368.34	699.62	3.87	248.16	398.63
		500	5.62	244.77	360.81	4.78	366.04	516.36	3.94	244.81	308.94
		1000	5.70	243.84	307.25	4.84	367.57	442.50	3.97	243.55	264.50
		2000	5.76	243.47	268.96	4.87	366.00	385.89	3.99	243.30	245.08
		5000	5.81	243.20	244.94	4.90	365.03	365.05	3.96	243.28	242.78
0.02	250	100	4.75	344.06	1986.21	3.93	402.80	1373.22	3.29	262.04	565.00
		200	5.04	332.03	840.28	4.07	392.60	756.94	3.37	262.55	410.71
		500	5.26	332.81	522.39	4.17	393.90	550.32	3.43	265.68	333.49
		1000	5.34	332.74	432.10	4.21	395.62	482.49	3.45	264.35	298.61
		2000	5.38	330.73	386.37	4.23	392.27	430.20	3.46	261.68	270.50
		5000	5.41	330.15	345.08	4.25	391.27	394.91	3.47	260.55	260.38
0.03	250	100	4.30	259.09	1453.56	3.45	258.11	842.59	3.10	358.13	868.61
		200	4.55	250.38	628.92	3.57	258.28	484.50	3.18	362.44	592.21
		500	4.73	249.27	389.73	3.65	256.53	345.74	3.22	351.99	449.17
		1000	4.80	251.08	321.89	3.68	257.24	307.31	3.24	354.47	407.83
		2000	4.84	251.44	290.47	3.69	253.20	278.18	3.25	356.52	377.35
		5000	4.87	252.08	267.51	3.71	254.03	259.04	3.26	362.22	363.50

3.5.1 OUT-OF-CONTROL PERFORMANCE OF THE ZIP-CHARTS IN CASE U

In Table 3.9, we give the *ARL* and *SDRL* values for two ZIP processes: $ZIP(0.8,2)$ and $ZIP(0.7,1)$. The shifts in process parameters are $\phi_1 = \tau\phi_0$ with $\tau \in \{0.8, 0.6\}$ and $\lambda_1 = \delta\lambda_0$ with $\delta \in \{1.2, 1.5\}$.

Table 3.9
OOC Performance of the ZIP-Shewhart Chart – MLE

		Case K, $L = 5.49$		Case U, $L = 5.49$, $m = 200$		Case U, $L^* = 4.73$, $m = 200$	
ϕ_1	λ_1	*ARL*	*SDRL*	*ARL*	*SDRL*	*ARL*	*SDRL*
0.80	2.0	301.87	301.37	936.24	2109.12	303.87	549.85
0.80	2.4	140.16	139.66	353.55	681.22	135.87	213.47
0.80	3.0	59.58	59.08	121.06	196.63	55.77	75.56
0.64	2.0	167.70	167.20	525.50	1226.86	168.79	301.97
0.64	2.4	77.87	77.37	197.93	382.76	75.39	119.99
0.64	3.0	33.10	32.60	67.44	107.50	31.17	42.15
0.48	2.0	116.10	115.60	362.21	820.71	116.05	207.98
0.48	2.4	53.91	53.41	138.30	283.77	52.26	81.71
0.48	3.0	22.92	22.41	45.54	74.05	21.58	29.25
		Case K, $L = 5.18$		Case U, $L = 5.18$, $m = 500$		Case U, $L^* = 4.36$, $m = 500$	
ϕ_1	λ_1	*ARL*	*SDRL*	*ARL*	*SDRL*	*ARL*	*SDRL*
0.70	1.0	175.55	175.05	537.67	757.06	174.84	202.52
0.70	1.2	98.71	98.21	260.15	351.78	97.92	109.00
0.70	1.5	50.78	50.27	113.41	145.43	50.17	53.25
0.56	1.0	119.69	119.19	366.41	516.63	119.55	138.96
0.56	1.2	67.30	66.80	177.81	241.10	66.81	73.99
0.56	1.5	34.62	34.12	77.84	99.37	34.22	36.19
0.42	1.0	90.80	90.30	277.36	391.92	90.46	104.06
0.42	1.2	51.06	50.55	134.04	180.94	50.72	56.12
0.42	1.5	26.27	25.76	59.01	75.39	25.93	27.24

For the first process, we assume that the size of the Phase I sample is $m = 200$ while in the second $m = 500$.

The OOC *ARL* values are compared for both ZIP-Shewhart charts with either "adjusted" or unadjusted L. Also, we provide the OOC *ARL* values in Case K. The results show that the use of the "adjusted" L^* value results in *ARL* values that are very close to the theoretical values (i.e., those under Case K) while there is a reduction in the *SDRL* values. Note also, that even for large shifts in process parameters (e.g. for $\delta = 1.5$ and $\tau = 0.6$) the respective IC *ARL* and *SDRL* in Case U when the unadjusted L is used can be much different than the corresponding values under Case K.

3.5.2 OUT-OF-CONTROL PERFORMANCE OF THE ZIB-CHARTS IN CASE U

In Table 3.10, we provide the *ARL* and *SDRL* values for two ZIB processes: $ZIB(0.9, 0.03, 250)$ and $ZIB(0.8, 0.01, 100)$. Similar to the case of a ZIP process, the shifts in process parameters are $\phi_1 = \tau\phi_0$ with $\tau \in \{0.8, 0.6\}$ and $p_1 = \delta p_0$ with $\delta \in \{1.2, 1.5\}$. For the first process, we assume that the size of the Phase I sample is $m = 500$ while in the second $m = 1000$.

The results in Table 3.10 are similar to those in Table 3.9. Specifically, they show that the use of the "adjusted" L^* values gives *ARL* values that are very close to the theoretical ones (i.e., those under Case K), while the *SDRL* are much lower than the respective ones in the case of "unadjusted" L. Like the case of ZIP processes, the OOC *ARL* and *SDRL* can be much larger, compared to the respective values in Case K, even for large shifts in process parameters (such as $\delta = 1.5$ and $\tau = 0.6$) if the "unadjusted" L is used.

Table 3.10
OOC Performance of the ZIB-Shewhart Chart – MLE

			Case K, $L = 5.09$		Case U, $L = 5.09$, $m = 500$		Case U, $L^* = 4.73$, $m = 500$	
ϕ_1	p_1	n	*ARL*	*SDRL*	*ARL*	*SDRL*	*ARL*	*SDRL*
0.90	0.030	250	248.86	248.36	483.73	833.67	250.84	382.82
0.90	0.036	250	83.19	82.69	132.07	187.96	80.43	104.01
0.90	0.045	250	24.68	29.17	38.55	44.62	28.41	31.08
0.72	0.030	250	88.88	88.38	173.84	306.16	89.08	135.42
0.72	0.036	250	29.71	29.21	47.17	67.06	28.62	36.72
0.72	0.045	250	10.60	10.09	13.78	15.67	10.12	10.76
0.54	0.030	250	54.10	53.60	105.20	180.56	54.52	83.48
0.54	0.036	250	18.08	17.58	28.51	39.97	17.46	22.24
0.54	0.045	250	6.45	5.93	8.36	9.35	6.17	6.38
			Case K, $L = 6.35$		Case U, $L = 6.35$, $m = 1000$		Case U, $L^* = 5.40$, $m = 1000$	
ϕ_1	p_1	n	*ARL*	*SDRL*	*ARL*	*SDRL*	*ARL*	*SDRL*
0.80	0.010	100	272.12	271.62	848.45	1195.33	272.55	299.14
0.80	0.012	100	152.31	151.81	410.53	556.14	153.99	161.53
0.80	0.015	100	77.86	77.36	178.83	229.69	77.64	80.66
0.64	0.010	100	151.18	150.68	472.46	662.73	151.42	165.80
0.64	0.012	100	84.62	84.12	229.06	309.98	89.65	90.28
0.64	0.015	100	43.26	42.75	98.62	127.01	43.08	44.46
0.48	0.010	100	104.66	104.16	325.56	458.49	105.11	115.89
0.48	0.012	100	58.58	58.08	158.08	213.82	58.63	62.54
0.48	0.015	100	29.95	29.44	68.74	88.10	29.86	30.80

3.6 CONCLUSIONS

In this work we studied the design and performance of Shewhart-type charts for zero-inflated processes, when the process parameters are unknown and must be estimated. Using Monte Carlo simulation and two estimation methods (maximum likelihood and method of moments) we evaluated the performance of the considered charts for several Phase I sample sizes, in terms of the unconditional *ARL* and *SDRL* performance measures. The numerical results showed that in the case of zero-inflated processes, very large Phase I samples are needed in order to obtain a reliable design of the chart, whose performance in the case of the estimated parameters is similar to its theoretical performance. In almost all of the examined cases, the suggested size m of the Phase I sample is larger than 1000 preliminary observations while there are cases where even more than 5000 preliminary observations do not guarantee that the actual performance will be close to the theoretical one.

For relatively small preliminary samples (e.g. for $m = 100$ or 200), the IC *ARL* and *SDRL* values can be very large. Therefore, we do not advise practitioners to set-up their chart when the process parameters are estimated from a Phase I sample of this size. Note also that in case of large ϕ_0 values, a small Phase I sample might end up to a sample with only zero values, making the estimate of process parameters impossible. As a general guideline, we suggest that Phase I samples with at least 1000 observations are necessary to obtain reliable estimates of process parameters. As m increases, the IC *ARL* and *SDRL* reduce, but not necessarily in a monotonic way (the former).

In an attempt to assist practitioners in ending up with reliable designs for their charts, we provided "adjusted" values for the design parameter L in order to obtain an IC *ARL* very close to the IC *ARL* in Case K, for a given size m of the Phase I sample. Note also that the proposed methodology works for any other nominal ARL_0 (instead of the IC *ARL* value in Case K) that meets practitioners' needs.

Finally, we used R (see R Core Team (2023)) for all the necessary computations in this work. The related programs are available from the authors upon request.

ACKNOWLEDGEMENTS

The authors are grateful to the two anonymous reviewers for their helpful and constructive comments on an earlier draft of this article.

REFERENCES

Albers, W., Kallenberg, W. C. M., & Nurdiati, S. (2004). Parametric control charts. *Journal of Statistical Planning and Inference*, *124*(1), 159–184.

Beckett, S., Jee, J., Ncube, T., Pompilus, S., Washington, Q., Singh, A., & Pal, N. (2014). Zero-inflated Poisson (ZIP) distribution: Parameter estimation and applications to model data from natural calamities. *Involve, a Journal of Mathematics*, *7*(6), 751–767.

Braun, W. J. (1999). Run length distributions for estimated attributes charts. *Metrika*, *50*, 121–129.

Castagliola, P., & Wu, S. (2012). Design of the *c* and *np* charts when the parameters are estimated. *International Journal of Reliability, Quality and Safety Engineering*, *19*(02), 1250010.

Castagliola, P., Wu, S., Khoo, M. B. C., & Chakraborti, S. (2014). Synthetic Phase II Shewhart-type attributes control charts when process parameters are estimated. *Quality and Reliability Engineering International*, *30*(3), 315–335.

He, B., Xie, M., Goh, T. N., & Ranjan, P. (2003). On the estimation error in zero-inflated Poisson model for process control. *International Journal of Reliability, Quality and Safety Engineering*, *10*(2), 159–169.

Jensen, W. A., Jones-Farmer, L. A., Champ, C. W., & Woodall, W. H. (2006). Effects of parameter estimation on control chart properties: A literature review. *Journal of Quality Technology*, *38*(4), 349–364.

Johannssen, A., Chukhrova, N., & Castagliola, P. (2022). The performance of the hypergeometric *np* chart with estimated parameter. *European Journal of Operational Research*, *296*(3), 873–899.

Johnson, N. L., Kemp, A. W., & Kotz, S. (2005). *Univariate discrete distributions*. John Wiley & Sons, Inc. New York, NY, USA.

Mahmood, T., & Xie, M. (2019). Models and monitoring of zero-inflated processes: The past and current trends. *Quality and Reliability Engineering International*, *35*(8), 2540–2557.

Mamzeridou, E., & Rakitzis, A. (2024). On the design and performance of upper one-sided Shewhart charts for zero-inflated Poisson processes with estimated parameters. *Quality Technology & Quantitative Management*, *21*(5), 611–632.

Montgomery, D. C. (2009). *Introduction to Statistical Quality Control* (*6th*). John Wiley & Sons.

Rakitzis, A. C., & Castagliola, P. (2016). The effect of parameter estimation on the performance of one-sided Shewhart control charts for zero-inflated processes. *Communications in Statistics-Theory and Methods*, *45*(14), 4194–4214.

R Core Team (2023). *R: A language and environment for statistical computing*. R Foundation for Statistical Computing. Vienna, Austria. https://www.R-project.org/.

Shewhart, W. A. (1926). Quality control charts. *The Bell System Technical Journal*, *5*(4), 593–603.

Testik, M. C., McCullough, B. D., & Borror, C. M. (2006). The effect of estimated parameters on Poisson EWMA control charts. *Quality Technology & Quantitative Management*, *3*(4), 513–527.

Vakilian, F., Amiri, A., & Faraz, A. (2018). The guaranteed adaptive *c*-charts with estimated parameter. *Quality and Reliability Engineering International*, *34*(8), 1575–1589.

Wagh, Y. S., & Kamalja, K. K. (2018). Zero-inflated models and estimation in zero-inflated poisson distribution. *Communications in Statistics-Simulation and Computation*, *47*(8), 2248–2265.

Wu, S., Castagliola, P., & Khoo, M. B. (2016). Run rules based phase II *c* and *np* charts when process parameters are unknown. *Communications in Statistics-Theory and Methods*, *45*(4), 1182–1197.

Zhao, M. J., & Driscoll, A. R. (2016). The *c*-chart with bootstrap adjusted control limits to improve conditional performance. *Quality and Reliability Engineering International*, *32*(8), 2871–2881.

4 A Comparative Study of Self-starting CUSUM Control Charts for Location Shifts

Konstantinos Bourazas
Department of Mathematics and Statistics & KIOS Research and Innovation Center of Excellence, University of Cyprus, Nicosia, Cyprus
E-Mail: kbourazas@aueb.gr. ORCID: 0000-0002-2691-4869.

4.1 INTRODUCTION

Statistical Process Control and Monitoring (SPC/M) is an area of Statistics whose primary objective is the real-time detection of anomalies in an ongoing process. Specifically, the goal is to identify when a process transitions from its In Control (IC) state, where it operates only under natural causes of variation, to the Out Of Control (OOC) state, where assignable causes of variation are present Montgomery, 2019. The implementation of SPC/M methods is mainly performed via control charts, while most of these methods involve two phases. Phase I, which is the training phase and involves the collection of independent IC data to calibrate the monitoring scheme and drive reliable estimates of the distribution and unknown parameters. Then, in Phase II, i.e., the testing phase, we collect new observations for comparison against the IC standards established in Phase I. The limitations of the Phase I/II separation are well-documented, mainly tied to assumptions and procedures during Phase I. For instance, the estimation error of Phase I estimates, which negatively affects performance during Phase II, is typically not taken into account (see Qiu, 2013). To eliminate this impact, the Phase II data could be used to update the Phase I estimates. However, this available information is usually wasted since the learning stops at the end of Phase I. Additionally, if a shift in a parameter occurs during Phase I, it can deteriorate the performance of both phases. An extensive discussion of Phase I methods is provided in Jones-Farmer *et al.*, 2014, highlighting major issues and advancements. Self-starting methods in SPC/M aim to mitigate issues related to Phase I/II separation. These methods do not require a calibration phase (Phase I) to estimate the unknown parameters, but they can provide testing from the start of a process while estimating the unknown parameters.

Self-starting control charts have been developed for both of the two major parametric approaches in Statistics, the frequentist and the Bayesian. The concept of self-starting control charts was introduced in Hawkins, 1987, where a frequentist CUSUM-type control chart for location and scale shifts was developed. Staying within the frequentist framework, notable are the self-starting CUSUMs (SSCs, Hawkins & Olwell, 1998) that are widely used. Additional noteworthy frequentist self-starting methods for monitoring the process mean are presented in works such as Hawkins *et al.*, 2003, Hawkins & Zamba, 2005, or Tercero-Gómez *et al.*, 2014. From the Bayesian perspective, a family of Bayesian self-starting control charts based on the posterior predictive distribution

DOI: 10.1201/9781032664330-4

introduced in Bourazas *et al.*, 2023b, while Bayesian self-starting methods for monitoring the mean of a process can be found in Tsiamyrtzis & Hawkins, 2005, Tsiamyrtzis & Hawkins, 2008, and Tsiamyrtzis & Hawkins, 2010. Furthermore, we should highlight the online change point in Adams & MacKay, 2007, who employed a machine learning algorithm to estimate a change point, utilizing the posterior distribution of the length of an IC sequence. Concerning the outlier detection, numerous self-starting methods have been proposed, with the frequentist Q charts in Quesenberry, 1991c , Quesenberry, 1991a, and Quesenberry, 1991b for Normal, Binomial, and Poisson data, and the Bayesian Predictive Control Charts Bourazas *et al.*, 2022 to be representative.

In this work, we focus on the comparison of two well-established self-starting parametric methods for univariate data, the frequentist Self-Starting CUSUM Hawkins & Olwell, 1998 and the Bayesian Predictive Ratio CUSUM Bourazas *et al.*, 2023b for persistent location shifts in Normal data. In Section 4.2, we briefly present the assumptions and the testing statistics of the methods. Section 4.3 follows with an extensive simulation study, while Section 4.4 concludes this work.

4.2 COMPETING METHODS

In this section, we will provide the assumptions, and the testing procedures of the two competing methods; the frequentist self-starting CUSUM (SSC) (Hawkins & Olwell, 1998) and the Bayesian Predictive Ratio CUSUM (PRC) (Bourazas *et al.*, 2023b). The under-study control charts are focused on detecting shifts in the mean of Normal data, assuming both the mean and the variance of the process are unknown.

4.2.1 SELF STARTING CUSUM (SSC)

We start presenting competing methods with SSC. Assume $\boldsymbol{X}_n = (x_1, ..., x_n)$ to be a random sample from a Normal distribution with unknown mean and variance. Let $\bar{x}_n = \sum_{j=1}^{n} x_j$ and $s_n = \sum_{j=1}^{n}(x_j - \bar{x}_n)^2/(n-1)$ be the sample mean and the sample variance of $\boldsymbol{X}_n$ respectively. For the next observable X_{n+1}, we define its standardized form as

$$T_{n+1} = \frac{X_{n+1} - \bar{x}_n}{s_n}. \tag{4.1}$$

It is known that $\sqrt{\frac{n}{n+1}} T_{n+1} \sim t_{n-1}$, i.e., a Student-$t$ distribution with $n-1$ degrees of freedom. Finally, we transform the T_{n+1} into a random variable U_{n+1} that follows an exact standard Normal distribution and it is

$$U_{n+1} = \Phi^{-1} F_{n-1}\left(\sqrt{\frac{n}{n+1}} T_{n+1}\right), \tag{4.2}$$

where $\Phi^{-1}(\cdot)$ is the inverse cumulative function of a standard Normal distribution and F_{n-1} is the cumulative function of a Student-t distribution with $n-1$ degrees of freedom. The CUSUM statistic for x_{n+1} will be

$$\begin{aligned} C_{n+1}^{+} &= \max\{0, C_n + U_{n+1} - k_{SSC}\}, \ or \\ C_{n+1}^{-} &= \min\{0, S_n + U_{n+1} + k_{SSC}\}, \end{aligned} \tag{4.3}$$

when we are interested in detecting upward or downward shifts respectively and setting the CUSUM statistics for $n = 1$ or 2 equal to zero. The design parameter k_{SSC} is associated with the magnitude of the shift targeted for detection, and it is conventionally set to half the size of the shift we want to detect. We ring an alarm for an upward or downward shift if $C_{n+1}^{+} > h_{SSC}^{+}$ or $C_{n+1}^{+} < h_{SSC}^{-}$, respectively. For a two-sided SSC (and assuming $h_{SSC}^{+} = -h_{SSC}^{-} = h_{SSC}$) we can plot $C_{n+1} = \max\{C_{n+1}^{+}, |C_{n+1}^{-}|\}$ and an alarm is ringed whenever the statistic C_{n+1} exceeds h_{SSC}.

4.2.2 PREDICTIVE RATIO CUSUM (PRC)

Now, we continue with PRC, which appears to be the Bayesian analog of SSC. PRC is based on the ratio of two posterior predictive distributions, one that results directly from the data and represents the IC state, and a shifted version of the IC predictive that represents the OOC state. Let $X_i|\left(\theta_1,\theta_2^2\right) \sim N\left(\theta_1,\theta_2^2\right)$, with $i = 1,...,n$ and both parameters being unknown. As the joint prior distribution of the mean and the variance, we assume a Normal Inverse Gamma, and specifically $\left(\theta_1,\theta_2^2\right) \sim NIG(\mu_0,\lambda,a,b)$. The choice of NIG is convenient as it is conjugate to the Normal likelihood leading to closed-form posterior predictive distributions. NIG can be considered as a hierarchical prior set-up in the mean θ_1 and the variance θ_2^2, where $\theta_1|\theta_2^2 \sim N(\mu_0,\theta_2^2/\lambda)$ and $\theta_2^2 \sim IG(a,b)$. The hyperparameters μ_0, λ, a, and b are considered to be known with fixed values. Regarding the prior hyperparameters determination, a standard way to set is by the desired values for the mean and variance of the priors. However, the prior hyperparameters can also have further interpretation through imaginary data use. In that case, we could estimate θ_1 from λ observations with a mean μ_0 and θ_2^2 from 2α observations with a sum of the squared deviations of 2β. These relationships arise directly from the form of the posterior of the imaginary data, which will be again a NIG due to the conjugacy. We present this rationale for deriving the hyperparameters of the weakly informative prior used in Section 4.3. For more information on the interpretation of prior parameters, see Gelman *et al.*, 2013. Additionally, the power priors by Ibrahim & Chen, 2000 are advantageous and flexible as they allow the direct use of historical data from similar processes and their combination with other sources of information. Some guidelines and suggestions regarding the determination of the prior hyperparameters for control charts are provided by Bourazas *et al.*, 2022. In case of prior ignorance, the non-informative reference prior Berger *et al.*, 2009 utilization is suggested. The reference prior maximizes the expected Kullback–Leibler divergence of the posterior distribution relative to the prior. For this model the reference prior is $\pi\left(\theta_1,\theta_2^2\right) \propto 1/\theta_2^2 \equiv NIG(0,0,-1/2,0)$. It is worth mentioning that, even if this prior is improper (it does not integrate to 1), it leads to a proper posterior.

If we denote $\boldsymbol{X}_n = (x_1,...,x_n)$, then the IC posterior and predictive distributions will be $\left(\theta_1,\theta_2^2\right)|\boldsymbol{X}_n \sim NIG\left(\hat{\mu}_n,\hat{\lambda}_n,\hat{a}_n,\hat{b}_n\right)$ and $f(X_{n+1}|\boldsymbol{X}_n) = t_{2\hat{a}_n}\left(\hat{\mu}_n,(\lambda_n+1)\cdot\hat{b}_n/(\lambda_n\cdot\hat{a}_n)\right)$ respectively, where t denotes the Student-t distribution and

$$\hat{\mu}_n = \left(\lambda\mu_0 + \sum_{j=1}^{n} x_j\right) \Big/ (\lambda+n), \hat{\lambda}_n = \lambda+n, \hat{a}_n = a+n/2$$

$$\hat{b}_n = b + \left(\lambda\mu_0^2 + \sum_{j=1}^{n} d_j^2\right) \Big/ 2 - \left(\lambda\mu_0 + \sum_{j=1}^{n} d_j\right)^2 \Big/ (2(\lambda+n)). \tag{4.4}$$

To express the OOC shift we introduce a step change of size $k_{PRC}\cdot\hat{\theta}_2$ to the location parameter of the posterior marginal distribution of the mean, where $\hat{\theta}_2 = \sqrt{\hat{b}_n/\hat{a}_n}$. The shift will be upward or downward if $k_{PRC} > 0$ or $k_{PRC} < 0$ respectively. Thus, the location parameter of the OOC (intervened) posterior $\hat{\mu}_n + k_{PRC}\cdot\hat{\theta}_2$ and the corresponding OOC predictive will be $f'(X_{n+1}|\boldsymbol{X}_n) = t_{2\hat{a}_n}\left(\hat{\mu}_n + k_{PRC}\cdot\hat{\theta}_2, \left(\hat{\lambda}_n+1\right)\cdot\hat{b}_n/\left(\hat{\lambda}_n\cdot\hat{a}_n\right)\right)$. The logarithmic predictive ratio $log(L_{n+1})$ for the next observable x_{n+1} will be

$$log(L_{n+1}) = \frac{f'(X_{n+1}|\boldsymbol{X}_n)}{f(X_{n+1}|\boldsymbol{X}_n)} = (\hat{a}_n+1/2)\cdot log\frac{2\hat{a}_n + Z_{n+1}^2}{2\hat{a}_n + \left(Z_{n+1} - k_{PRC}\cdot\hat{\lambda}_n/(\hat{\lambda}_n+1)\right)^2}, \tag{4.5}$$

where $Z_{n+1} = (X_{n+1}-\hat{\mu}_n)/\sqrt{\left(\hat{\lambda}_n+1\right)\cdot\hat{b}_n/\left(\hat{\lambda}_n\cdot\hat{a}_n\right)}$. Note that the design parameter k_{PRC} has a different interpretation from k_{SSC}; we typically set it to be equal to the size of the shift we want to detect, e.g., for a change of 1 standard deviation we set $k_{PRC} = 1$. The CUSUM statistic for the next data point at time $n+1$ will be

$$
\begin{aligned}
S_{n+1}^{+} &= \max\{0, S_n + log\,(L_{n+1})\}, \;\; or \\
S_{n+1}^{-} &= \min\{0, S_n - log\,(L_{n+1})\},
\end{aligned}
\tag{4.6}
$$

when we are interested in detecting upward or downward shifts respectively, and setting the first CUSUM statistics equal to 0, or the first two CUSUM statistics when using the reference prior. Like SSC, a alarm is raised when $C_{n+1}^{+} > h_{PRC}^{+}$ or $C_{n+1}^{+} < h_{PRC}^{-}$, upwards and downwards respectively. For two-sided procedures (and assuming $h_{PRC}^{+} = -h_{PRC}^{-} = h_{PRC}$), we raise an alarm if $S_{n+1} = \max\{S_{n+1}^{+}, |S_{n+1}^{-}|\} > h$.

4.3 SIMULATIONS

In this section, we will evaluate and compare the performance of the methods described in Section 4.2; namely the PRC and SSC. The two competing methods had been compared in Bourazas *et al.*, 2023b for short runs. However, in this study, we will evaluate them for long runs using the appropriate performance metric. Starting from the false alarm tolerance, we configure the control charts for two-sided shifts, and we set the decision limits h_{SSC} and h_{PRC} to achieve an IC Average Run Length (ARL_0) of $ARL_0 = 370$ for both methods. Algorithmically, we determine the decision limits using the Regula Falsi method for 10,000 simulated sequences from the standard Normal, similar to the approach described in Algorithm 2 in Bourazas *et al.*, 2023a. For the IC state, we assume that the data follow a standard Normal distribution, while for the OOC sequences, the post-change data come again from a Normal distribution with a shift in the mean of size δ. To assess the performance of the methods for different magnitudes of jumps, we consider four values for the step change: $\delta = 0.5$, 1, 1.5, or 2. Continuing with the location of change point τ, we assume different scenarios, where τ appears from very early to later stages in the data sequence, examining in this way its impact on the detection perfomance. Specifically, we set 10 values for τ with $\tau = 11$, 21,..., 101. This translates that for the smallest value of τ, the change point occurs after only 10 IC data points available, while for the last value of τ, we have an IC history of 100 data points. As OOC metric, we employ the Conditional Expected Delay (CED), which is given by the formula:

$$
CED(\tau) = E_{\tau}(T - \tau + 1 | T \geq \tau) = \frac{E\,(T - \tau + 1)^{+}}{P(T \geq \tau)}, \tag{4.7}
$$

where T is the stopping time and $x^{+} = max(0,x)$. T is defined for each method as the first time that its cumulative statistic plots beyond the corresponding decision threshold, and the $CED(\tau)$ is the average delay of an alarm given that the change point has occurred. For every OOC scenario, the $CED(\tau)$ estimate is based on 10,000 iterations.

To investigate the performance of the methods under varying values of the design parameter k, three different values are specified for each method. Following the suggestions in Zantek, 2006, we set for SSC the values $k_{SSC} = 0.25$, 0.375, or 0.5, while for PRC we set the corresponding values $k_{PRC} = 0.5$, 0.75, or 1. This means that we have $4 \times 10 \times 3 = 120$ scenarios in total for each method. Regarding the prior for PRC, we conduct a small sensitivity analysis by setting two priors to evaluate the performance of PRC under the absence or the presence of prior information. Therefore, we have two processes for PRC:

- PRC_n for the non-informative reference prior $NIG(0,0,-1/2,0)$
- PRC_i for the weakly informative prior $NIG(0,4,2,1.5)$.

Regarding the weakly informative prior, it is positioned at the correct values of the parameters and assigns equal information weight as four observations from the IC distribution. Specifically, if

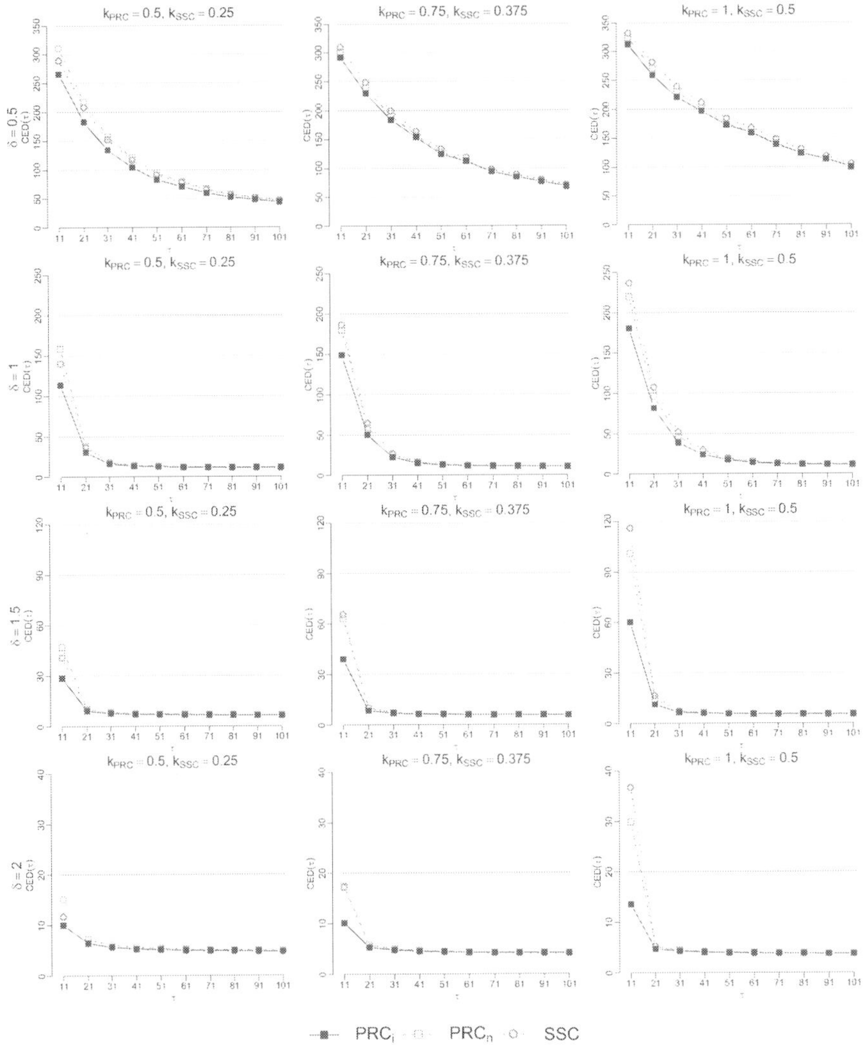

Figure 4.1 The conditional expected delay $CED(\tau)$ at locations $\tau = 11, 21, ..., 101$, for shifts size of $\delta = 0.5$, 1, 1.5, or 2 of SSC, and PRC, under the non-informative reference (PRC$_n$) or a moderately informative (PRC$_i$) prior. The design parameters are set $k_{PRC} = 0.5$, 0.75, or 1, and $k_{SSC} = k_{PRC}/2$.

we assume four imaginary values y_i, where $i = 1, ..., 4$, from the standard Normal and denoting their sample mean as $\bar{y}$, then from properties of Normal distribution we have $E(\bar{y}) = 0$ and $E\left(\sum_{i=1}^{4}(y_i - \bar{y})^2\right) = 3$, i.e., the expected sample mean is zero and the expected sum of squared deviations equal to three. Thus, we set $\mu_0 = 0$, $\lambda = 4$, $\alpha = 4/2$ and $\beta = 3/2$ to illustrate the effect of the available prior information on the detection performance. Note that this is the prior setting used in the simulation study in Bourazas *et al.*, 2023b. The results are presented graphically in Figure 4.1, and they are tabulated in Table 4.1 in Appendix.

First of all, we observe that the larger the jump, the more rapidly all methods detect it. Additionally, both methods enhance their detection performance when the change point τ occurs later, i.e., when more IC data points are available to estimate the unknown parameters. For jumps of size 1, 1.5, and 2, the performance seems to stabilize beyond a certain point, implying that the methods do not require many data points to achieve their "best" performance for jumps of these magnitudes. However, this is not the case for the smaller jump of 0.5, where (for various values of k) it takes time for $CED(\tau)$ to stabilize (or has not stabilized even with 100 IC data points). This indicates that a larger number of IC data points is needed for the efficient detection of small shifts. Notably, for $\delta = 0.5$ when $\tau = 11$, the detection ability appears to be very limited, as, in most cases, the CED values are greater than 300, i.e., close to the nominal value of $ARL_0 = 370$. Comparing the methods, we notice the positive impact of the prior in decreasing CED for PRC_i, particularly when τ occurs early where only a limited number of data points are available. On the other hand, the performance of PRC_i converges with PRC_n and SSC when τ occurs later, and the volume of IC data increases, washing out the impact of the prior. Comparing PRC_n and SSC, we observe that, generally, they demonstrate similar performances with small differences in CED. SSC has a slight advantage for small values of ks and τ occurs early in the process. However, in those cases, the CED values are exceptionally high for both methods, rendering them ineffective for settings like short runs. On the other hand, PRC_n shows slightly better performance for larger values of ks.

4.4 CONCLUSIONS

In this study, the focus was on online testing for a change point in the mean of univariate Normal data without the presence of Phase I. Specifically, we evaluated and compared the CED performance for two parametric self-starting methods, the frequentist SSC, and the Bayesian PRC, under various magnitudes of shifts for the mean, and for a wide range of different change point positions. For PRC, we also conducted a small prior sensitivity analysis, assuming two priors: the non-informative reference prior and a weakly informative prior.

As expected, both methods detect large changes more quickly compared to small changes, and improve their performance as the IC horizon increases, or in other words, when τ appears at later stages. Moreover, large values of the design parameters k enhance performance for significant changes but simultaneously increase the risk of absorbing smaller changes. Comparing the methods, we observe the beneficial effect of the prior in reducing CED, especially at the beginning of the process when only a few data points are available. Regarding the comparison of the PRC with the non-informative prior (PRC_n) against SSC, we see that they exhibit similar performances with very small differences in CED overall. SSC has a slight advantage by setting a small value for the design parameter k, while PRC performs better for larger values of k.

REFERENCES

Adams, R., & MacKay, D. (2007). Bayesian online changepoint detection. *arXiv preprint arXiv:0710.3742*.

Berger, J., Bernardo, J., & Sun, D. (2009). The formal definition of reference priors. *Annals of Statistics*, *37*(2), 905–938.

Bourazas, K., Kiagias, D., & Tsiamyrtzis, P. (2022). Predictive control charts (pcc): A bayesian approach in online monitoring of short runs. *Journal of Quality Technology*, *54*(4), 367–391.

Bourazas, K., Sobas, F., & Tsiamyrtzis, P. (2023a). Design and properties of the predictive ratio cusum (PRC) control charts. *Journal of Quality Technology*, *55*(4), 404–421.

Bourazas, K., Sobas, F., & Tsiamyrtzis, P. (2023b). Predictive ratio cusum (PRC): A bayesian approach in online change point detection of short runs. *Journal of Quality Technology*, *55*(4), 391–403.

Gelman, A., Carlin, J. B., Stern, H. S., & Rubin, D. B. (2013). *Bayesian data analysis*. Chapman; Hall/CRC.

Hawkins, D. (1987). Self-starting cusum charts for location and scale. *Journal of the Royal Statistical Society Series D: The Statistician*, *36*(4), 299–316.

Hawkins, D., & Olwell, D. (1998). *Cumulative sum charts and charting for quality improvement.* Springer Science & Business Media.

Hawkins, D., Qiu, P., & Kang, C. W. (2003). The changepoint model for statistical process control. *Journal of Quality Technology*, *35*(4), 355–366.

Hawkins, D., & Zamba, K. D. (2005). Statistical process control for shifts in mean or variance using a changepoint formulation. *Technometrics*, *47*(2), 164–173.

Ibrahim, J. G., & Chen, M.-H. (2000). Power prior distributions for regression models. *Statistical Science*, 46–60.

Jones-Farmer, A., Woodall, W., Steiner, S., & Champ, C. (2014). An overview of phase i analysis for process improvement and monitoring. *Journal of Quality Technology*, *46*(3), 265–280.

Montgomery, D. (2019). *Introduction to statistical quality control.* John wiley & sons.

Qiu, P. (2013). *Introduction to statistical process control.* CRC press.

Quesenberry, C. (1991a). Spc Q charts for a binomial parameter p: Short or long runs. *Journal of Quality Technology*, *23*(3), 239–246.

Quesenberry, C. (1991b). Spc Q charts for a poisson parameter λ: Short or long runs. *Journal of Quality Technology*, *23*(4), 296–303.

Quesenberry, C. (1991c). Spc Q charts for start-up processes and short or long runs. *Journal of Quality Technology*, *23*(3), 213–224.

Tercero-Gómez, V., Cordero-Franco, A., Pérez-Blanco, A., & Hernández-Luna, A. (2014). A self-starting cusum chart combined with a maximum likelihood estimator for the time of a detected shift in the process mean. *Quality and Reliability Engineering International*, *30*(4), 591–599.

Tsiamyrtzis, P., & Hawkins, D. (2005). A bayesian scheme to detect changes in the mean of a short-run process. *Technometrics*, *47*(4), 446–456.

Tsiamyrtzis, P., & Hawkins, D. (2008). A bayesian ewma method to detect jumps at the start-up phase of a process. *Quality and Reliability Engineering International*, *24*(6), 721–735.

Tsiamyrtzis, P., & Hawkins, D. (2010). Bayesian startup phase mean monitoring of an autocorrelated process that is subject to random sized jumps. *Technometrics*, *52*(4), 438–452.

Zantek, P. (2006). Design of cumulative sum schemes for start-up processes and short runs. *Journal of Quality Technology*, *38*(4), 365–375.

APPENDIX

Table 4.1

The conditional expected delay $CED(\tau)$ at locations $\tau = 11, 21, ..., 101$, for shifts size of $\delta = 0.5, 1, 1.5$, or 2 of SSC, and PRC, under the non-informative reference (PRC_n) or a moderately informative (PRC_i) prior. The design parameters are set $k_{PRC} = 0.5$, 0.75, or 1, and $k_{SSC} = k_{PRC}/2$.

		$k_{PRC} = 0.5, k_{SSC} = 0.25$			$k_{PRC} = 0.75, k_{SSC} = 0.375$			$k_{PRC} = 1, k_{SSC} = 0.5$		
	τ	PRC_i	PRC_n	SSC	PRC_i	PRC_n	SSC	PRC_i	PRC_n	SSC
$\delta = 0.5$	11	265.557	310.294	289.331	291.529	301.762	309.652	312.098	323.628	332.204
	21	182.606	217.874	207.950	229.683	238.441	248.640	258.898	270.766	281.146
	31	134.312	158.108	152.737	183.739	190.724	199.355	220.929	229.446	239.264
	41	104.387	120.656	117.141	153.817	157.457	163.413	196.752	201.528	211.785
	51	82.235	94.272	91.157	124.477	127.504	133.084	172.396	175.305	183.147
	61	70.252	79.018	77.140	112.104	113.172	118.558	159.620	159.514	167.806
	71	59.161	65.934	64.483	94.000	94.591	97.963	138.930	140.639	147.576
	81	52.126	56.685	55.674	84.490	84.767	88.688	123.703	123.719	130.485
	91	47.692	51.401	50.791	76.172	76.633	79.237	113.584	113.323	118.368
	101	43.430	46.278	45.934	68.223	67.173	70.161	99.971	99.039	105.028
$\delta = 1$	11	113.563	158.367	140.094	148.950	179.849	186.042	180.157	219.954	236.668
	21	30.118	38.499	36.310	50.252	59.526	64.747	81.708	96.471	107.350
	31	15.902	18.064	17.547	22.315	25.397	27.012	38.858	46.752	52.245
	41	13.015	14.119	13.647	14.825	15.886	16.256	23.787	27.662	30.198
	51	12.251	12.968	12.652	12.543	13.039	13.258	16.910	17.771	19.111
	61	11.634	12.213	11.910	11.700	12.058	12.189	13.787	14.687	15.241
	71	11.342	11.829	11.557	11.011	11.201	11.259	12.131	12.694	12.969
	81	11.115	11.566	11.317	10.904	11.046	11.078	11.564	11.808	12.159
	91	11.039	11.411	11.213	10.604	10.677	10.690	11.338	11.497	11.602
	101	10.776	11.145	10.937	10.335	10.396	10.415	10.856	11.054	11.090
$\delta = 1.5$	11	28.762	47.111	40.836	39.077	63.196	65.549	60.090	101.115	116.028
	21	9.124	10.325	9.463	8.508	9.787	10.018	11.327	14.633	16.613
	31	7.783	8.418	7.934	6.907	7.280	7.131	6.705	7.205	7.358
	41	7.250	7.697	7.341	6.437	6.632	6.517	6.070	6.308	6.292
	51	7.062	7.408	7.120	6.264	6.396	6.302	5.880	6.081	6.058
	61	6.847	7.139	6.894	6.091	6.177	6.114	5.699	5.792	5.774
	71	6.727	6.975	6.765	5.959	6.023	5.966	5.619	5.689	5.660
	81	6.678	6.911	6.723	5.964	6.012	5.963	5.612	5.662	5.652
	91	6.612	6.825	6.656	5.906	5.931	5.884	5.517	5.555	5.544
	101	6.533	6.727	6.564	5.853	5.868	5.829	5.475	5.525	5.507
$\delta = 2$	11	10.018	15.146	11.752	10.152	17.401	17.170	13.562	29.935	36.761
	21	6.405	7.179	6.380	5.302	5.781	5.390	4.739	5.189	5.222
	31	5.670	6.099	5.631	4.799	5.020	4.794	4.309	4.523	4.392
	41	5.320	5.636	5.273	4.554	4.679	4.502	4.095	4.220	4.101
	51	5.158	5.409	5.120	4.425	4.514	4.371	3.993	4.079	3.989
	61	5.027	5.233	4.986	4.320	4.373	4.274	3.904	3.959	3.892
	71	4.929	5.118	4.902	4.261	4.301	4.215	3.854	3.901	3.834
	81	4.906	5.071	4.881	4.234	4.262	4.182	3.849	3.877	3.826
	91	4.841	4.996	4.824	4.191	4.205	4.138	3.801	3.818	3.771
	101	4.806	4.945	4.785	4.176	4.180	4.126	3.791	3.811	3.770

Section II

Statistical Methods in Healthcare Systems Assurance

5 Capturing Responder Effects in Randomized Clinical Trials with a Mixture Model

Daniel R. Jeske
University of California, Riverside, Department of Statistics.
E-Mail: daniel.jeske@ucr.edu.

5.1 INTRODUCTION

The two-group randomized clinical trial (RCT) is known to be the gold standard design for comparing a control group of experimental units to a treated group of experimental units. In this design, control group observations $X_1, X_2, \ldots, X_m$ are assumed to be independent and identically distributed observations from a cumulative distribution function,F, and treatment group observations $Y_1, Y_2, \ldots, Y_n$ are assumed to be independent and identically distributed observations from a cumulative distribution function G. The independent and identically distributed assumptions are met by randomly selecting experimental units from a homogenous population of experimental units. Defining μ_F and μ_G to be the means of F and G, respectively, the treatment effect is typically defined as the difference between the means, $\mu_G - \mu_F$.

Statistical inference such as hypothesis testing, point estimation, and interval estimation about the treatment effect depends on what type of assumptions are made about F and G. If parametric models for F and G are assumed, classical frequentist or Bayesian inference methods might be used. The most common parametric model for F and G assumes they are normal distributions. Alternatively, a nonparametric analysis might be used that requires little in the way of assumptions about F and G. For example, it might only be assumed that G is a shift of F. The shift model specifies $G(u) = F(u - \delta)$ or $G(u) = F(u + \delta)$, for $\delta > 0$, depending on if the quantity being measured is expected to be larger or smaller within the treatment group, respectively.

A key design issue in the RCT is determining the sample sizes, m and n. Assumptions made about F and G are also central to the various approaches taken for sample size calculations. Typically the RCT is powered so that if the treatment effect was as large as the smallest effect of practical importance a size α test of the null hypothesis $H_0: \ F = G$ will get rejected with a high probability, $1 - \beta$. Here, α and β denote the type-1 and type-2 errors of the hypothesis test, respectively. Sample size calculations may also incorporate the relative cost of treating experimental units.

While the discussion and results in this paper apply to any application of a RCT, including for example, where the experimental units are animals, plants, or inanimate objects, the primary focus will be the context of clinical trials. Therefore, from here on the experimental units will be referred to as patients. Furthermore, it will be assumed that the quantity being measured is expected to be larger for treated patients. It would be known from the context of the experiment whether larger or smaller response values correspond to an efficacious treatment. In the case where smaller values

DOI: 10.1201/9781032664330-5

correspond to an efficacious treatment effect, the amount of decrease can be recorded as the response variable. The application in section 5 is exactly this situation.

The recent notion of personalized medicine is accompanied by the recognition that patients can vary considerably in how they respond to a treatment. Specifically, there may be some patients who do not respond to a treatment that is effective for others. In these cases, there will be some patients in the treatment group of an RCT that have responses that look like they came from F rather than G. Such patients are described as being non-responders to the treatment. See, for example, Manegold *et al.*, 2016 and Spear *et al.*, 2001.

To capture the reality of non-responders, a mixture model can be used to represent G. Let θ denote the fraction of the treated population that are responders. Then the observations from the treatment group can be taken to come from $G(u) = (1-\theta)F(u) + \theta F(u-\delta)$. As both θ and δ are necessary to understand the full effect of the treatment, the treatment effect is now generalized to (θ, δ). The parameter space for the model is the set of continuous distribution functions for F together with $\Omega = \{(\theta, \delta) : \delta > 0, 0 < \theta \leq 1\} \cup (0,0)$, where the point $(0,0)$ represents the null hypothesis $H_0 : F = G$. The special case $\theta = 1$ gives the standard pure shift model for the treatment group. Because observations from the control group provide information about F, the mixture model for G is identifiable.

The rest of this paper discusses aspects of the design and analysis of an RCT with the mixture model representation for G. Section 5.2 discusses sample size calculations that power the RCT to a desired level. Section 5.3 discusses estimation and confidence interval procedures for the generalized treatment effect (θ, δ). Section 5.4 discusses the context of group sequential RCTs. A real-life example illustrating the inference methodologies is presented in section 5.5 and the paper concludes with a summary and future work directions in section 5.6.

5.2 SAMPLE SIZE CONSIDERATIONS

5.2.1 GENERAL RESULTS

This section summarizes results in Jeske & Yao, 2020, where a sample size formula was derived to test $H_0 : F = G$ using the Wilcoxon rank sum test statistic, defined as $W = \sum_{i=1}^{n} R_i$, where $R = (R_1, \ldots, R_n)$ is the vector of ranks of the $Y_1, \ldots, Y_n$ observations amongst all $N = m + n$ observations. Let $\lambda = \lim_{N \to \infty} (m/N) < 1$ denote the asymptotic ratio of sample sizes. It is well known (e.g., Randles and Wolfe, 1979) that

$$\frac{W - n(m+n+1)/2}{\sqrt{\frac{mn(m+n+1)}{12}}}$$

has a limiting (as $N \to \infty, \lambda$ fixed) standard normal distribution under H_0. When larger responses are anticipated from the treatment group a one-sided size α test would reject when $W > n(m+n+1)/2 + z_\alpha \sqrt{mn(m+n+1)/12}$.

The limiting distribution of W under general alternatives is also known (e.g., Randles and Wolfe, 1979). Considering the mixture model alternative, define

$$\begin{aligned}
\gamma(\theta, \delta, F) &= P(X_1 < Y_1) \\
\xi_1(\theta, \delta, F) &= P(X_1 < Y_1, X_1 < Y_2) - \gamma^2(\theta, \delta, F) \\
\xi_2(\theta, \delta, F) &= P(X_1 < Y_1, X_2 < Y_1) - \gamma^2(\theta, \delta, F).
\end{aligned}$$

Then $\sqrt{N}(W - \mu_W)/(mn)$ has a limiting normal distribution with mean zero and variance $\xi_1(\theta, \delta, F)/\lambda + \xi_2(\theta, \delta, F)/(1-\lambda)$, where $\mu_W = n(m\,\gamma(\theta, \delta, F) + (n+1)/2)$. It follows that the asymptotic power of the Wilcoxon test is

$$\pi(G) = \Phi\left(\frac{\gamma(\theta,\delta,F) - \frac{1}{2} - z_\alpha\sqrt{\frac{m+n+1}{12mn}}}{\sqrt{\frac{\xi_1(\theta.\delta.F)}{m} + \frac{\xi_2(\theta.\delta.F)}{n}}}\right).$$

With $\rho = n/m$, asymptotic group sizes that detect the mixture model alternative with power $1-\beta$ are (Proposition 2.1 in Jeske & Yao, 2020)

$$m = \left(\frac{z_\alpha\sqrt{\frac{\rho+1}{12\rho}} + z_\beta\sqrt{\xi_1(\theta,\delta,F) + \frac{\xi_2(\theta.\delta.F)}{\rho}}}{\gamma(\theta,\delta,F) - \frac{1}{2}}\right)^2, \quad n = \rho m. \tag{5.1}$$

For balanced designs, $\rho = 1$ and the sample size formula becomes

$$m = n = \left(\frac{z_\alpha/\sqrt{6} + z_\beta\sqrt{\xi_1(\theta,\delta,F) + \xi_2(\theta,\delta,F)}}{\gamma(\theta,\delta,F) - 1/2}\right)^2 \tag{5.2}$$

5.2.2 LOCATION-SCALE FAMILIES

The sample size formula in 5.2 simplifies when F belongs to a location-scale family. That is, $F(u) = \Psi\left(\frac{u-\mu}{\sigma}\right)$, where μ is the location parameter, σ is the scale parameter, and $\Psi(\cdot)$ is a known cumulative distribution function involving no parameters and having probability density function $\psi(\cdot)$. Within the location-scale family, $X \sim F$ if and only if $X \sim \mu + \sigma Z$, where $Z \sim \Psi$. No generality is lost by assuming that $\Psi(\cdot)$ has zero mean and unit variance so that μ and σ can be interpreted as the mean and standard deviation of F. Under the mixture alternative $Y \sim \mu + \sigma Z$ with probability $(1-\theta)$ and $Y \sim \delta + \mu + \sigma Z$ with probability θ. If F is a location-scale distribution, then Proposition 2.3 in Jeske & Yao, 2020 shows that the power of the Wilcoxon rank sum test for mixture alternatives depends on Ψ and (θ, K), where $K = \delta/\sigma$.

To illustrate the sample size calculation, suppose a researcher designs the study under the assumption that Ψ is the standard normal distribution, selects the shift size K based on what is the smallest shift size that would have any practical consequence, and selects θ based on the smallest fraction of responders in the treatment group that would make the treatment clinically meaningful. Under the assumption $\rho = 1$, Table 5.1 shows the required control group and treatment group sizes that are necessary to detect the shift alternative for various choices of K and θ.

The last row of Table 5.1 shows the case when $\theta = 1$, and corresponds to the assumption that all patients in the treatment group are responders. The much larger group sizes needed when $\theta < 1$ highlights the fact that designing the RCT under the all responders assumption can lead to a significantly underpowered design when there really are non-responders in the treatment group. Jeske & Yao, 2020 discuss similar findings when $\Psi$is the standardized kernel of the logistic, Laplace, and the t distribution with 3 degrees of freedom. It was also shown that the difference in the required group sizes when using the Wilcoxon rank sum test versus the optimal linear rank sum test was minimal.

5.3 INFERENCE FOR (θ, δ)

In the mixture model setting for G, the treatment effect is the pair (θ, δ). Lubich *et al.*, 2022 investigated a method of moments estimator for (θ, δ). Let $(\mu_F,\ \sigma_F^2)$ denote the mean and variance

Table 5.1

Sample sizes $m = n$ that are necessary for a 5% test based on W to have 80% power for the mixture alternative when Ψ is the standardized normal distribution.

	$K = \delta/\sigma$			
θ	.4	.6	.8	1
.5	331	152	89	60
.6	230	105	61	41
.7	169	77	45	30
.8	129	59	34	23
.9	102	46	27	18
1.0	82	37	21	14

of the control group response, and (μ_G, σ_G^2) be the same for the treatment group response. With $\Delta = \theta\,\delta$, it can be shown that (θ, δ) satisfy

$$\theta = \left\{1 + \frac{\sigma_G^2 - \sigma_F^2}{\Delta^2}\right\}^{-1}, \delta = \Delta\left\{1 + \frac{\sigma_G^2 - \sigma_F^2}{\Delta^2}\right\}. \tag{5.3}$$

Note that the mixture model for the treatment group implies $\sigma_G^2 \geq \sigma_F^2$, with equality if and only if $\theta = 1$.

The relationships in 5.3 motivate the following method of moment estimators,

$$\hat{\theta} = \left\{1 + \frac{(S_Y^2 - S_X^2)_+}{[(\bar{Y} - \bar{X})_+]^2 + \varepsilon_N}\right\}^{-1}, \hat{\delta} = (\bar{Y} - \bar{X})_+\left\{1 + \frac{(S_Y^2 - S_X^2)_+}{[(\bar{Y} - \bar{X})_+]^2 + \varepsilon_N}\right\}, \tag{5.4}$$

where $t_+ = t$ if $t > 0$ and 0 otherwise, $(\bar{X}, S_X^2)$ are the mean and variance of the control group observations, $(\bar{Y}, S_Y^2)$ are the same for the treatment group observations, and $\varepsilon_N = a_N S_X^2$, with $a_N = \log N^2/N$. The purpose of ε_N is to bound the denominators away from zero in a way that preserves the consistency and asymptotic normality of the estimators that are established by the following proposition given in Lubich *et al.*, 2022 and stated for the case where $m = n$.

Proposition 5.1. *(Consistency) For continuous F, any $0 < \theta \leq 1$ and any $\delta > 0$, $\hat{\theta} \xrightarrow{P} \theta$ and $\hat{\delta} \xrightarrow{P} \delta$.*

(Asymptotic Normality) Let $\mu_X = E(X)$, $\mu_Y = E(Y)$, $\sigma_X^2 = Var(X) < \infty$, and $\sigma_Y^2 = Var(Y) < \infty$ denote the means and variances of the control group and treatment group observations, respectively. For continuous F, any $0 < \theta < 1$ and any $\delta > 0$, $\sqrt{n}\,(\hat{\theta} - \theta) \to N(0\,, \sigma_\theta^2)$ and $\sqrt{n}\,(\hat{\delta} - \delta) \to N(0\,, \sigma_\delta^2)$, where

$$\sigma_\theta^2 = \left(1 + \frac{\sigma_Y^2 - \sigma_X^2}{(\mu_Y - \mu_X)^2}\right)^{-4}\left\{\frac{4\,(\sigma_Y^2 - \sigma_X^2)^2}{(\mu_Y - \mu_X)^6}\,(\sigma_Y^2 + \sigma_X^2) + \text{two more terms}\right\}$$

$$\sigma_\delta^2 = \left(1 - \frac{\sigma_Y^2 - \sigma_X^2}{(\mu_Y - \mu_X)^2}\right)^2 (\sigma_Y^2 + \sigma_X^2) + \text{two more terms}.$$

The additional terms in each variance expression are detailed in Lubich *et al.*, 2022. It follows that asymptotic confidence intervals for the case $\rho = 1$ are

$$\left[\hat{\theta} - z_{\alpha/2}\left(\frac{\hat{\sigma}_\theta}{\sqrt{n}}\right), \hat{\theta} + z_{\alpha/2}\left(\frac{\hat{\sigma}_\theta}{\sqrt{n}}\right)\right], \left[\hat{\delta} - z_{\alpha/2}\left(\frac{\hat{\sigma}_\delta}{\sqrt{n}}\right), \hat{\delta} + z_{\alpha/2}\left(\frac{\hat{\sigma}_\delta}{\sqrt{n}}\right)\right] \tag{5}$$

Table 5.2
Coverage probabilities of asymptotic 95% confidence interval for θ.

			Sample Size			
Standardized F	θ	δ	25	50	100	500
Normal	.5	1	.88	.87	.91	.93
		3	.90	.93	.94	.95
	.8	1	.94	.95	.95	.96
		3	.90	.93	.94	.94
Logistic	.5	1	.95	.95	.95	.97
		3	.91	.92	.93	.95
	.8	1	.94	.94	.94	.95
		3	.88	.90	.93	.94
Laplace	.5	1	.88	.87	.89	.93
		3	.92	.94	.94	.96
	.8	1	.92	.92	.94	.95
		3	.92	.93	.93	.95

where $\hat{\sigma}_\theta$ and $\hat{\sigma}_\delta$ are the asymptotic standard deviations computed using sample moments $\bar{X}$, $\bar{Y}$, S_x^2, and S_Y^2 in place of μ_X, μ_Y, σ_X^2, and σ_Y^2, respectively. However, positivity adjustments analogous to those of the estimators are made. That is, $\sigma_Y^2 - \sigma_X^2$ is estimated with $(S_Y^2 - S_X^2)_+$ and denominators of the form $(\mu_Y - \mu_X)^k$ are estimated with $[(\bar{Y} - \bar{X})_+]^k + a_N S_X^k$. Also, when necessary, the boundaries of the asymptotic confidence intervals are truncated at the edge of the parameter spaces.

Table 5.2 and Table 5.3 show the result of a simulation study to assess the coverage probabilities for nominal 95% asymptotic confidence intervals defined by (5). The estimated coverage probabilities are based upon 1000 simulated data sets that had equal sample sizes, $m = n$. Note that $\delta = 1$ represents a small shift in the component distributions which often results in an unimodal distribution, while $\delta = 3$ results in a bimodal distribution.

Table 5.2 shows the coverage probability of the 95% asymptotic interval for θ is well calibrated except for the case where the sample size is 25. As the sample size increases the coverage probabilities converge to 0.95 quickly. Table 3 shows the coverage probability for the 95% asymptotic interval for δtends to be conservative when the component distributions are not well separated and are fairly well calibrated otherwise, even for small sample sizes. As the sample size increases, the coverage probabilities converge to 0.95 rather slowly when δ is small.

5.4 GROUP SEQUENTIAL CLINICAL TRIALS

While RCTs are popular and effective designs used for testing the efficacy of a treatment, group sequential clinical trials extend the idea by incorporating the possibility of terminating the trial early by rejecting the null hypothesis due to strong evidence of efficacy, or alternatively, strong evidence of futility. Instead of a one-time test of the null hypothesis, group sequential clinical trials conduct the hypothesis test at each of a predetermined number of stages. At each stage, $i = 1, 2, \ldots, S$, new control group, and new treatment group patients are added to the study and a test of the null hypothesis is carried out on the accumulation of all patients in the study up to that point.

Friel & Jeske, 2023 considered the design of group sequential clinical trials under the mixture model alternative for G. Accept and reject boundaries at each stage were determined by the error spending functions of Gordon Lan & DeMets, 1983. It was assumed that at each stage the size of the new cohorts of control group and treatment group patients was constant, though not necessarily

Table 5.3
Coverage probabilities of asymptotic 95% confidence interval for δ

			Sample Size			
Standardized F	θ	δ	25	50	100	500
Normal	.5	1	.99	.99	.99	.96
		3	.96	.95	.95	.95
	.8	1	.96	.97	.97	.96
		3	.94	.96	.94	.95
Logistic	.5	1	.98	.99	.99	.97
		3	.94	.95	.95	.94
	.8	1	.95	.96	.96	.96
		3	.94	.94	.94	.96
Laplace	.5	1	.98	.99	.99	.98
		3	.95	.96	.94	.96
	.8	1	.97	.97	.98	.97
		3	.94	.96	.95	.94

equal. Two options for the test statistic at each stage were considered. Here, only the sequential average rank (SAR) statistics are discussed.

Let W_i denote the Wilcoxon rank sum test at stage i. The mean and variance of W_i under the null are $\mu = n\,(m+n+1\,)/2$ and $\sigma^2 = m\,n\,(m+n+1\,)/12$, respectively. The definition of the SAR statistic at stage s is

$$Z_s = \frac{\sum_{i=1}^{s} W_i/s - \mu}{\sigma/\sqrt{s}}.$$

Critical values to either accept or reject the null at each stage are $\{a_i\}_{i=1}^{S}$ and $\{r_i\}_{i=1}^{S}$, respectively, with the constraint $a_S = r_S$ imposed to ensure a decision is made by stage S. The critical values are calculated to ensure the overall size of the sequential procedure is α and the overall power of the sequential procedure under a specified alternative is $1-\beta$. At the first $S-1$ stages, the null is rejected if the SAR exceeds the associated critical value to reject, the null is accepted if the SAR is smaller than the associated critical value to accept, and otherwise, the trial continues to the next stage. At the final stage, if it is reached, the null is rejected if the SAR exceeds the common value $a_S = r_S$, and otherwise accepts the null.

The critical values and the arm sizes of the sequential procedure depend on a specified alternative. The specified alternative encompasses a choice of F, θ, and K. However, due to the null distribution-free property of the Wilcoxon rank sum test, the size of the sequential procedure is guaranteed even F is misspecified. Friel & Jeske, 2023 provide R code for finding the critical values and arm sizes. Inputs to the R code are: 1) the maximum number of stages, S, 2) the exponent, a, of the Lan and Demets error spending functions, 3) the arm size ratio, 4) the overall type-1 and type-2 error rates, and 5) the specified alternative, expressed as $(\,F,\theta,K\,)$, for which power of $1-\beta$ is desired. Choices for F include normal, logistic, Laplace, and the t distribution with three degrees of freedom. Table 4 illustrates the required arm sizes.

The comparison of arm sizes in Table 5.4 for the alternative sequential design with the fixed sample design ($S = 1$) illustrates the potential savings of overall sample size that early stopping allows. However, because the number of stages in the sequential designs is random, it is more appropriate to compare the expected total sample size of the sequential designs with the fixed sample design. Table 5.5 offers that comparison for the same setting illustrated in Table 5.4 for the arm sizes and it can be seen the sequential designs have significantly smaller expected total sample sizes.

Table 5.4

Arm sizes per group per stage necessary to detect the mixture alternative with the SAR test statistic, equal arm sizes, $\alpha = .05$, $\beta = .2$, and $a = 2$.

		$K = .5$			$K = 1.0$		
θ	S	F Normal	F Logistic	F Laplace	F Normal	F Logistic	F Laplace
.4	1	335	298	231	95	86	75
	2	175	155	121	50	45	40
	3	121	106	83	35	31	27
	4	93	82	64	27	24	21
.6	1	149	131	103	41	38	33
	2	79	69	54	22	20	18
	3	54	47	37	16	14	13
	4	41	37	29	12	11	10
.8	1	84	73	58	24	21	18
	2	44	39	31	13	12	10
	3	30	27	22	9	8	7
	4	24	21	17	7	7	6
1.0	1	53	47	37	15	14	12
	2	29	25	20	8	8	7
	3	20	18	14	6	6	5
	4	16	14	11	5	5	4

Table 5.5

Expected total sample sizes to detect the mixture alternative with the SAR test statistic, equal arm sizes, $\alpha = .05$, $\beta = .2$, and $a = 2$.

		$K = .5$			$K = 1.0$		
θ	S	F Normal	F Logistic	F Laplace	F Normal	F Logistic	F Laplace
.4	1	670	596	462	190	172	150
	2	572	503	396	163	147	130
	3	538	475	373	152	139	121
	4	518	456	360	147	135	118
.6	1	298	262	206	82	76	66
	2	254	225	176	72	66	59
	3	240	213	165	68	63	54
	4	231	204	163	68	62	56
.8	1	168	146	116	48	42	36
	2	144	127	101	40	37	33
	3	134	120	94	40	36	32
	4	131	118	91	40	35	33
1.0	1	106	94	74	30	28	24
	2	91	82	65	27	24	21
	3	89	77	62	27	23	22
	4	85	74	62	28	23	22

Friel & Jeske, 2023 considered the estimation of the treatment effect (θ, δ) based on pooling the arm cohorts across all of the stages that were needed for the trial. Letting s' denote the number of stages that were needed, the realized control group and treatment group sizes are $m' = s'm$ and $n' = s'n$, respectively. Method of moments estimators (4), using m' and n' as the group sizes can then be calculated. A comparison with a clustering based estimation method showed the method of moments estimator was preferable. The adequacy of using the asymptotic confidence intervals in (5) with the pooled cohort data is an open question.

5.5 APPLICATION

A blood pressure study by Green *et al.*, 2008 investigated the impact of online counseling and care support on patients trying to manage their blood pressure. In a single stage RCT, $m = 246$ patients on a blood pressure reduction plan did not receive the intervention and represent the control group. The treatment group is $n = 235$ additional patients on a blood pressure reduction plan who did receive the intervention. The dependent variable is the reduction in diastolic blood pressure (DPB). Summary statistics for the two groups are $\bar{x} = 3.793$ mmHg, $s_X^2 = 71.78$ mmHg^2, $\bar{y} = 6.354$ mmHgand $s_Y^2 = 89.73$ mmHg^2.

The standard analysis of an RCT would assume $\theta = 1$, leaving only δ as the treatment effect, which would be estimated by $\bar{y} - \bar{x} = 2.56$ mmHg, a very modest impact since Pater, 2005 has suggested anything less than 3 mmHg is not clinically meaningful. On the other hand, calculating the method of moments estimators of the mixture model in (4) gives $(\hat{\theta}, \hat{\delta}) = (.319, 8.04)$. Lubich *et al.*, 2022 use these estimates along with an empirical probability integral transformation to demonstrate the adequacy of the mixture model for the treatment group observations.

Analysis using the mixture model changes the conclusion of the RCT to a clinically relevant finding of the form that approximately 1/3 of patients will respond to the pharmacist intervention and see an average drop in DBP of about 8 mmHg. A larger range of possibilities for the treatment effect is captured by asymptotic 90% confidence intervals for θ and δ, calculated from (5), which are $(.16, .47)$ and $(1.3, 14.8)$, respectively.

5.6 SUMMARY AND FUTURE WORK

The mixture of the control response distribution and a shift of that distribution can be used as a model for the treatment response distribution when trying to capture the existence of responders and non-responders within the treatment group. In this paper, the method of moment estimators developed in previous research has been presented to estimate the treatment effect that is characterized by both the proportion of responders in the treated population as well as the size of the treatment effect for responders. Confidence interval procedures were presented based on the asymptotic normality of the method of moment estimators. Contexts considered were randomized clinical trials and group sequential trials.

Current research efforts are focused on alternative pseudo likelihood based inference procedures for (θ, δ). Specifically, Lubich & Jeske, 2024 are working under the assumption that F is log-concave and use a smoothed log-concave estimator of the cumulative distribution function $\hat{F}$ and the corresponding smoothed estimate of the probability density function $\hat{f}(u) = \partial \hat{F}(u)/\partial u$. The proposed pseudo-likelihood estimator is based on substituting $\hat{f}(u)$in the true likelihood in place of $f(u)$, to give

$$\begin{aligned} \mathrm{PSL}(\tilde{\theta}, \tilde{\delta}) &= \left\{ \prod_{i=1}^{m} \hat{f}(X_i) \right\} \left\{ \prod_{i=1}^{n} \left[(1 - \tilde{\theta}) \hat{f}(Y_i) + \tilde{\theta} \hat{f}(Y_i - \tilde{\delta}) \right] \right\} \\ &\propto \prod_{i=1}^{n} \left[\hat{f}(Y_i) + \tilde{\theta} \left(\hat{f}(Y_i - \tilde{\delta}) - \hat{f}(Y_i) \right) \right]. \end{aligned}$$

The proposed maximum pseudo-likelihood estimators are then,

$$(\hat{\theta}_{PSL}, \hat{\delta}_{PSL}) = \underset{0<\tilde{\theta}<1, \tilde{\delta}>0}{\arg\max} \; \mathrm{PSL}(\tilde{\theta}, \tilde{\delta}) .$$

Confidence interval procedures are being proposed based on likelihood ratio test statistics with Satterthwaite, 1946 type correction factors.

REFERENCES

Friel, D. C., & Jeske, D. R. (2023). Wilcoxon rank-sum tests to detect one-sided mixture alternatives in group sequential clinical trials. *Statistical Methods in Medical Research*, *32*(9), 1711–1727.

Gordon Lan, K., & DeMets, D. L. (1983). Discrete sequential boundaries for clinical trials. *Biometrika*, *70*(3), 659–663.

Green, B. B., Ralston, J. D., Fishman, P. A., Catz, S. L., Cook, A., Carlson, J., Tyll, L., Carrell, D., & Thompson, R. S. (2008). Electronic communications and home blood pressure monitoring (e-bp) study: Design, delivery, and evaluation framework. *Contemporary clinical trials*, *29*(3), 376–395.

Jeske, D. R., & Yao, W. (2020). Sample size calculations for mixture alternatives in a control group vs. treatment group design. *Statistics*, *54*(1), 97–113.

Lubich, B., Jeske, D., & Yao, W. (2022). Statistical inference for method of moments estimators of a semi-supervised two-component mixture model. *The American Statistician*, *76*(4), 376–383.

Lubich, B., & Jeske, D. R. (2024). Improved semi-parametric inference for a mixture model of responses from a control versus treatment group trial. *Statistical Methods in Medical Research*, *33*(3), 515–531.

Manegold, C., Adjei, A., Bussolino, F., Cappuzzo, F., Crino, L., Dziadziuszko, R., Ettinger, D., Fennell, D., Kerr, K., Le Chevalier, T., *et al.* (2016). Novel active agents in patients with advanced nsclc without driver mutations who have progressed after first-line chemotherapy. *ESMO open*, *1*(6), e000118.

Pater, C. (2005). Beyond the evidence of the new hypertension guidelines. blood pressure measurement–is it good enough for accurate diagnosis of hypertension? time might be in, for a paradigm shift (i). *Current controlled trials in cardiovascular medicine*, *6*, 1–14.

Satterthwaite, F. E. (1946). An approximate distribution of estimates of variance components. *Biometrics bulletin*, *2*(6), 110–114.

Spear, B. B., Heath-Chiozzi, M., & Huff, J. (2001). Clinical application of pharmacogenetics. *Trends in molecular medicine*, *7*(5), 201–204.

6 Polygenic Risk Scores and Applications in Clinical Practice

Fragiskos Bersimis
Department of Statistics and Insurance Science, University of Piraeus
E-Mail: fbersimis@unipi.gr. ORCID: 0000-0002-8450-0453.

Pantelis Bagos
Department of Computer Science and Biomedical Informatics, University of Thessaly
E-Mail: pbagos@compgen.org. ORCID: 0000-0003-4935-2325.

6.1 INTRODUCTION

6.1.1 OVERVIEW OF GENETIC INFLUENCES ON TRAITS AND DISEASES

Various human phenotypic characteristics (Height, Body Mass Index, etc.), including common complex disorders and non-communicable diseases with a considerable impact on public health (Cancer types, Cardiovascular diseases, Neurological disease, etc.), are tied to the information hidden in a person's genetic material, thus heritable to some extent (Kafkas *et al.*, 2021; Kunkle *et al.*, 2019; Locke *et al.*, 2015; Mavaddat *et al.*, 2015; Patel *et al.*, 2023; Yang *et al.*, 2010). Moreover, genetics has shown that individual variants affect traits corresponding to phenotypic characteristics, as well as traits related to human health (Wray *et al.*, 2021). Traits affected by one or a very low number of genes are often referred to as monogenic disorders (Mendelian disorders) and strongly affect the chance that an individual will develop a certain trait whether disease-related or not (Robinson & Katsanis, 2010; Sugrue & Desikan, 2019; Vinuesa *et al.*, 2023). Examples of monogenic disorders include phenylketonuria and sickle cell disease (Chial, 2008; Hardouin *et al.*, 2023). However, most traits, corresponding to phenotypic characteristics, including complex health disorders, are polygenic, meaning they are affected by a great number of variants with small individual effects each, and when they are influenced by non-genetic factors (e.g., environmental factors, lifestyle habits, etc.) are often called multifactorial disorders (Jung *et al.*, 2024; Sugrue & Desikan, 2019).

6.1.2 INTRODUCTION TO GWAS METHODOLOGY AND HISTORICAL FINDINGS

In a polygenically-derived trait, causal variants are identified as those that seem to be strongly associated with a specific phenotype, being present in higher frequencies to a significant degree in individuals showing the trait of interest when compared to a sample set of persons not having the specific trait (Cano-Gamez & Trynka, 2020; Dima & Breen, 2015; Wang, Namba, *et al.*, 2023). The above-mentioned variants are usually identified by association with a trait (phenotypic characteristics including health disorders) via Genomewide Association Studies (GWAS) utilizing genomes of multiple individuals displaying variation in the trait of interest (Abdellaoui *et al.*, 2023;

DOI: 10.1201/9781032664330-6

Kontou & Bagos, 2024; Uffelmann *et al.*, 2021; Wolc & Dekkers, 2022). The observed relationship between human phenotypic characteristics or complex diseases and their genetic information, via variants, has been strengthened in the last two decades or so due to the many GWAS that have been carried out lately and continue to take place nowadays (Uffelmann *et al.*, 2021). GWASs aim to detect and identify mostly variations of single nucleotide polymorphisms (SNPs[1]), strongly associated with a variety of traits, i.e., phenotypic characteristics, such as hair colour and habitual physical activity resistance, as well as, health disorders, such as type 2 diabetes mellitus (T2D) and rheumatoid arthritis, etc. (Hardy & Singleton, 2009; Klimentidis *et al.*, 2018; Morgan *et al.*, 2018). More specifically, GWAS is a study design used to pinpoint genetic variations, in the genome that are associated to specific traits and, methodologically, includes the DNA extraction, as well as, genotyping from people participating in a collected sample of a large number (usually thousands) of participants. This sample in most of the cases is divided into two clusters regarding the trait (e.g., a phenotypic characteristic such as green eyes, dark hair, etc., or a disease such as diabetes mellitus, hypertension, a cancer type, etc.), which is under investigation. The first cluster includes participants having the trait of interest (cases) and the second cluster includes those participants who do not have the trait of interest (controls). The process of statistical analysis involves the detection of specific genetic variants that occur with high frequency in individuals with a specific trait compared to those without the trait via suitable association tests (Hardy & Singleton, 2009; Klimentidis *et al.*, 2018). Since the number of SNPs is extremely large, the number of individual statistical tests is also extremely large, and thus a particularly low threshold of statistical significance is used to decrease the number of false positives (Yang *et al.*, 2010). It is noted that the logic that governs GWASs relates to ideas resulting from Ronald Fisher's infinitesimal model, and specifically, if a trait is polygenic, "random sampling of alleles at each gene generates a continuous normally-distributed phenotype in a population" (Boyle *et al.*, 2017; Visscher & Goddard, 2019). Historically, very important findings have emerged through GWAS, such as the recognition of the correlation between specific SNPs within the BRCA1 and the BRCA2 genes with the increased risk of breast cancer (Milne & Antoniou, 2011), the finding of the relationship between specific SNPs within the FTO gene with the increased risk of obesity (Huang *et al.*, 2023), as well as, the influence of specific SNPs within the APOE gene on increased risk of Alzheimer disease appearance (Aslam *et al.*, 2023). From the previous, it is perceived that every SNP related to a health disorder or a disease, could be moderately useful in estimating a person's risk of developing this health disorder or disease, providing this way a potential valuable clinical tool for predicting a disease's risk.

But a single SNP is unlikely to be a determinant in the formation of a hereditary disease when there is a very large number of SNPs associated with the specific disease in human DNA (Gibbs *et al.*, 2003; Lewis & Vassos, 2020). Furthermore, as it is already mentioned, most diseases are complex conditions influenced by a combination of genetic, environmental, and lifestyle factors, therefore, a single SNP may not contain enough information as it constitutes proportionately, a very small part of the whole genome, but a specific combination of many SNPs is necessary to obtain sufficient information (Lewis & Vassos, 2020).

For example, in the Japanese population, SNP rs200787930 was found to be associated with Coronary Artery Disease (CAD), however, those who have the corresponding risk allele [2] will not develop this disease necessarily (Yamada *et al.*, 2018). However, the human genome has been found to include a very large number of SNPs, and recent applications of PRS have been found to include information arising from millions of them, such as e.g., in the case of CAD where even 6,630,150 SNPs have been used in the South Asian population and in the case of Breast cancer where even 6,494,889 SNPs have been used in European population (Mars, Kerminen, *et al.*, 2022; Wang *et al.*, 2020). As mentioned, GWAS give out summary statistics on trait-associated variants which however

---

[1] Single nucleotide polymorphisms (SNPs) are the most common type of genetic variation among human populations and corresponds to a population frequency of at least 1% (Wright, 2001)

[2] An allele is an alternative form at a specific location in the genome that may differ through SNPs and risk allele is the one increasing the probability of a health disorder or a disease (Choi *et al.*, 2020)

are not causal and thus their effect mechanism is generally unknown (Crouch & Bodmer, 2020; Sun *et al.*, 2022). It is important to consider that GWAS operate on populations and thus are not fully informative at an individual level, however, their results do provide biological insights towards traits of interest (Lewis & Vassos, 2020; Ruotsalainen *et al.*, 2021).

6.1.3 DEFINITION OF POLYGENIC RISK SCORES AND CURRENT CHALLENGES

In the recent literature, various methods have been proposed for retrieving suitable information from different SNPs in order to highlight heritability by analyzing GWAS data such as Linkage Disequilibrium Score Regression (LDSR) and Polygenic Risk Score calculation (Choi *et al.*, 2020; Duncan *et al.*, 2018). The latter, i.e., Polygenic Risk Score (PRS) constitutes one of the most widespread and useful method of utilizing GWAS data and represents a fairly popular way of assessing the genetic risk regarding a disease (Ala-Korpela & Holmes, 2020; Choi *et al.*, 2020; Dima & Breen, 2015; Lewis & Vassos, 2020; Wang, Namba, *et al.*, 2023). PRS is a mathematical expression that can be attributed to the association between a disease occurrence and the genetic underpinning, which could be defined as a single-value numerical estimation of the genetic predisposition to the appearance of a specific disease, distinct for each individual of a general population (Arnold & Koenig, 2021; Choi *et al.*, 2020). Therefore, since each selected SNP is related to a phenotypic characteristic or a disease, then PRS, which represents the sum of many such small, considered independent, effects, will also be related to this disease, i.e., PRS express essentially a cumulative combination of SNPs' risk alleles that appear to be associated with the phenotype of interest in GWASs (Ala-Korpela & Holmes, 2020; Lambert *et al.*, 2019). The result produced by PRS usually differs for different persons in a population, providing an indication of their likelihood of developing the trait of interest (i.e., the risk of developing the related disease) and PRSs' accuracy is greater when the SNPs used in the calculation are derived from GWAS with large samples and strong correlations of variant to trait (Crouch & Bodmer, 2020; Lewis & Vassos, 2020; Xiang *et al.*, 2024). It may seem at first a simple approach as combining different variants in a single sum without considering interactions between genes is used, but PRS provides an estimate of the available genetic information for each discrete member of various populations (Lewis & Vassos, 2020). Moreover, the methodology of combining various (non-genetic) features into a single numerical value that expresses the probability of the occurrence of a disease, (e.g., hypertension), has been used extensively in the past with great success, as well as, many scales nowadays are used as auxiliary medical tools in the clinical diagnosis of diseases such as psychiatric, cardiology, etc (Bersimis & Varlamis, 2019; Bersimis *et al.*, 2020). The development and application of PRSs raise several ethical concerns regarding genetic discrimination and privacy, equity and accessibility, as well as the psychological impact regarding a variety of populations. Specifically, PRS clinical testing may lead to genetic discrimination, particularly in the insurance industry and healthcare provision (Lewis *et al.*, 2021; Slunecka *et al.*, 2021; Yanes *et al.*, 2024; Zhao *et al.*, 2023). Furthermore, PRSs rely on genetic data, which could be misused if not adequately protected, meaning that, unauthorized access or sharing of genetic information may lead to breaches of confidentiality (Knoppers, 2014). In addition, access to PRS-based tools may be limited by socioeconomic factors, potentially widening health disparities, especially because they are considered expensive (Xia *et al.*, 2024). Finally, the psychological impact of learning that an individual is at risk for a certain phenotype, and specifically a disease in a clinical setting is still under investigation and specifically long-term impact is unknown (Yanes *et al.*, 2024).

Despite the fact that many PRSs have been developed in the recent literature for many diseases, they represent only a trait's hereditary factor and do not take into account the causal role of environmental aspects, as well as, lifestyle behaviour, therefore, the concern that exists in the scientific community is whether PRS can be used as prognostic clinical tools, that is, whether there is a real and measurable benefit for patients in the health care system at the level of diagnosis or treatment (Koch *et al.*, 2023). In this work, it is investigated whether PRSs could be taken into account in clinical practice as informative predictive tools.

6.2 OVERVIEW OF POLYGENIC RISK SCORES

Polygenic risk scores (PRSs), which are similarly referred to as polygenic scores (PGSs) or genetic risk scores (GRSs), are numerical metrics expressing a person's probability of developing a phenotypic characteristic or a health disorder based on the effect that many genetic variants (SNPs) linked to the trait have (Collister *et al.*, 2022). PRS prediction has been applied to various traits, such as male pattern baldness, vision disorders such as myopia, mental disorders such as schizophrenia, cardiovascular diseases such as hypertension, cancer types such as breast cancer, etc (Henne *et al.*, 2023; Kurniansyah *et al.*, 2022; Mavaddat *et al.*, 2019; Mojarrad *et al.*, 2020; Taniguchi *et al.*, 2020; Toulopoulou *et al.*, 2019). Recent work highlighted an additional application of PRSs and related methods since they can be utilized to predict the occurrence of traits in ancient genomes (Carlson *et al.*, 2022).

6.2.1 HISTORICAL REVIEW OF PRS DEVELOPMENT & CHALLENGES

Historically, until the current application of PRSs have been reached, genomics research has gone through several stages, some of the main ones being the initiation of the Human Genome Project in 1990, the completion of the first GWASs in 2005 and the publication of the PRS for categorizing people according their likelihood of developing a mental disease by the Schizophrenia Consortium in 2009 (Collins *et al.*, 2003; Haines *et al.*, 2005; The International Schizophrenia Consortium, 2009). Researchers' current approach to the development of Polygenic Risk Scores (PRSs) has been greatly impacted by past difficulties in their development and implementation. Some of those earlier challenges shaped current PRS development approaches. An important challenge is that early GWAS were being carried out in small sample sizes datasets identifying a limited number of SNPs associated with a variety of diseases, leading to constrained predictive power of PRSs (Abdellaoui *et al.*, 2023; Visscher *et al.*, 2017). In addition, lack of diverse populations in early GWAS is another challenge that still exists, to a lesser extent, leading to PRSs that were less accurate and less transferable to non-European populations, as until 2021, 86% of GWAS participants were of European ancestry (Abdellaoui *et al.*, 2023; Fatumo *et al.*, 2022; Martin *et al.*, 2019). Furthermore, early PRS models sometimes overfitted to specific datasets (e.g., European ancestry samples), leading to reduced reproducibility in independent cohorts, as well as, lacked clear clinical utility, making their adoption by healthcare systems challenging (Duncan *et al.*, 2019). Finally, PRS calculations required significant computational resources and earlier, compared to today, there were computational and infrastructure barriers (Wang *et al.*, 2022).

6.2.2 CURRENT APPROACHES OF PRS DEVELOPMENT METHODOLOGY

According to the general PRS development methodology, PRS is usually produced by the sum of considered as independent common SNPs' risk alleles, with minor allele frequency greater than 1%, weighted by the corresponding effect size estimates derived from GWAS summary statistics data (Choi *et al.*, 2020; Dudbridge, 2013; Ma & Zhou, 2021). This cumulative model appears to be applicable to many traits as a large meta-analysis of twin heritability has shown (Polderman *et al.*, 2015). Therefore, a PRS for a specific trait is calculated via a weighted sum of as many product factors as there are SNPs that the particular method deems suitable to be included in the PRS (Choi *et al.*, 2020).

For the aforementioned set of independent SNPs, the numbers of risk alleles for each SNP that receive values of 0, 1 or 2 (for each person is different) are summed up, weighted by the effect size of each SNP (Choi *et al.*, 2020). The general mathematical formula for calculating PRS is listed below in equation 6.1:

$$PRS = \sum_{i=1}^{n} (SE_i) \cdot (RA_i) \tag{6.1}$$

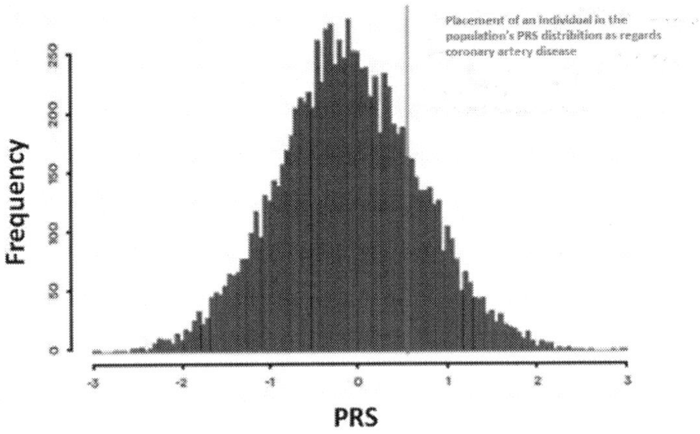

Figure 6.1 PRS distribution for Coronary Artery Disease (Chatziandreou, 2022).

where:

- SE_i each SNP's size effect
- RA_i the number of risk alleles for each SNP
- n the number of SNPs utilized in PRS

The result produced by equation 6.1 is an one-dimensional numerical estimate of the genetic effect on the occurrence of a disease or a phenotypic characteristic for an individual. This happens by placing this individual within a normal distribution of risk towards developing certain diseases or expressing a phenotype (Lewis *et al.*, 2021). For example, figure 6.1 shows the placement of an individual in the population PRS distribution for CAD expressing a high risk for developing the disease (Chatziandreou, 2022). Of course, such estimates have inherent degrees of uncertainty and are probabilistic by nature (Ding *et al.*, 2022). In addition, it is possible to calculate more than one PRS and then select that which is deemed to be most applicable (or performs the best) for the case at hand (Collister *et al.*, 2022). There are many PRS development variations, such as P+T, LDpred, PRS-CS, AnnoPred, etc., as well as, other methods including the use of machine learning algorithms and artificial intelligence prediction models are utilized that could identify individuals at risk for specific diseases (e.g., atrial fibrillation) (Raben *et al.*, 2023; Wang, Khurshid, *et al.*, 2023; Ye *et al.*, 2021). It should be noted that, generally, there is no rigorous methodological framework for PRS development that systematically outperforms others clearly in terms of performance (Monti *et al.*, 2024).

The SNPs, participating in the calculation of PRS, are selected based on the importance of their association with the appearance of the trait being investigated, using as a criterion usually the corresponding p-value threshold from training GWAS, as well as, filtered based on the Linkage Disequilibrium (LD) correlation degree among them (Euesden *et al.*, 2015). More specifically, LD, p-value thresholding, SNPs' effect sizes, training and testing (target) data set are all inseparable parts of the PRS construction process (Choi *et al.*, 2020; Duncan *et al.*, 2019). Effect size estimates are derived from GWAS summary statistics and are essentially an expression of the association of the variant in question with the phenotype being investigated (Choi *et al.*, 2020; Lambert *et al.*, 2019). Generally, PRS tend to increase in power with a higher number of SNPs integrated into their construction but there is an implicit inverse correlation between SNP count and effect size estimation accuracy with higher SNP counts usually leading to more noisy estimates and usually, many PRSs are constructed utilizing different parameterization before selecting that assumed to have the highest performance (Choi *et al.*, 2020; Collister *et al.*, 2022; Lambert *et al.*, 2019). Furthermore, effect sizes are measured differently depending on the type of study and nature of the trait distribution, in some cases represented as Odds ratios and Hazard ratios (OR / HR) or as beta coefficients (Lambert *et al.*, 2019; Lewis & Vassos, 2020). In addition, LD refers to the non-stochastic correlation between SNPs' alleles in different loci (Qanbari, 2020; Slatkin, 2008). Two methods that utilize LD and are pertinent to constructing a PRS, namely "clumping and thresholding (C+T)" and "pruning and thresholding" (P+T) are used to increase PRS prediction accuracy (Chen, Gatev, *et al.*, 2023; Choi *et al.*, 2020; Jiang *et al.*, 2024; Sally *et al.*, 2024). The discussed criteria used to select the SNPs participating in the development of a PRS are decisive for their accuracy since either a small number of SNPs may be available which may constitute an insufficient list of genetic variants, or a very large number of SNPs which may constitute a list with a lot of noise.

In addition to the different SNPs selection methodologies, there are further specializations when it comes to the construction of PRS, such as taking into account demographic factors, such as gender and age, as well as, focusing on specific populations. For example, PRS aiming to assess mostly the risk of cardiovascular diseases, are developed where gender and age have been considered by research design, to show their true effect on discrimination of individuals with low and high probability of a disease's appearance (Wong *et al.*, 2022). In another publication, evidence of interaction between PRS and gender was found in predicting Coronary Heart Disease (CHD) risk in special surveys (Feitosa *et al.*, 2021). Moreover, it was revealed that PRS has a notable and discrete impact in a variety of diseases according to sex and age (Jermy *et al.*, 2024). In the case of special populations, such as women who have overcome breast cancer, and have a high genetic risk for cardiovascular diseases, they can benefit from a healthy lifestyle, to reduce the risk of developing any cardiovascular disease (Peng *et al.*, 2023). In another publication, it is highlighted that prevalence of Atrial Fibrillation was higher in a special population of individuals with heart failure when considering the relative genetic risk score (Kloosterman *et al.*, 2020). From the above it appears that when developing a PRS, factors such as gender, age and potential special populations should be taken into account.

6.2.3 PRS' PREDICTIVE ABILITY EVALUATION

As mentioned previously, the use of PRS is a question in daily clinical practice that has not been well established, even though it is suggested by many scientists (Roberts *et al.*, 2022). Thus, the correct and unbiased evaluation of the clinical utility of PRS is imperative. However, there is evidence of the PRSs' added value in clinical assessment in specific diseases, nevertheless more research is needed to identify stratified monitoring and prompt medical treatment (Mars, Koskela, *et al.*, 2020). Thus, in the case that an individual is clinically tested for a disease or a disorder, PRS may provide the genetic information about that individual's risk of developing that disease. In this case, PRS's predictive clinical utility may be evaluated by prospectively monitoring the individual in order to determine, whether suffers or not, at the present time or in the future, from the disease that the specific PRS aims to assess. More generally, in the case of populations with common genetic characteristics, PRSs' predictive ability for a disease is evaluated via case-control studies, where area under the receiver operating characteristic curve (AUC), odds ratio (OR – e.g., comparing highest quintile versus lowest quintile), relative ratio (RR) and the pseudo-R2 are calculated (Aragam & Natarajan, 2020; Damask *et al.*, 2020; Lee *et al.*, 2012; Morieri *et al.*, 2018).

6.3 POLYGENIC RISK SCORES COMBINED WITH ENVIRONMENTAL AND CLINICAL FACTORS

There are several indications that traditional clinical tools (non-genetic) for a variety of diseases may not estimate the real risk to the onset of the disease (O Sullivan *et al.*, 2022; Panagiotou *et al.*, 2023). This assessment for the risk of a disease could be optimized using PRS with potential effect the reclassification of individuals to a different risk category, as well as reduce the corresponding assessing mean cost (Mujwara *et al.*, 2022; Torkamani *et al.*, 2018).

6.3.1 USE CASES OF PRS COMBINED WITH NON GENETIC FACTORS

Furthermore, in various cancer types, the combination of genetic and non-genetic risk factors has been previously exploited. For example, in the study of sixteen different cancer types (CA), the integration of PRS with traditional risk factors strengthens the total predictive value (Kachuri *et al.*, 2020), as well as, in the case of hepatocellular carcinoma (HCC), PRS was combined with non-genetic factors and the discrimination of people at risk was also improved (Zou *et al.*, 2024). Furthermore, PRSs have been used in conjunction with non-genetic well known risk factors (mammographic density (MD), hormone levels, etc.) for Breast Cancer (BC) and seem to have better results than using BC risk prediction model alone to better distinguish individuals at risk (Zhang *et al.*, 2018). Additionally, BC risk stratification is enhanced by a specific PRS, calculated using a substantial amount of SNPs, partnered with traditional risk indicators and MD (Brentnall *et al.*, 2020). For example, in a recent publication, it was reported that C statistic from a 10-year risk BC model (Tyrer–Cuzick model) was increased from 0.57 to 0.67 when suitable BC PRS was integrated, highlighting the magnitude of improvement from combining PRS with other factors (Collister *et al.*, 2024). Thus, BC incidence may have a lifestyle /environmental-related component, and specifically, nutrition, smoking, alcohol consumption, physical inactivity, pollution, hormone levels are among the non-genetic variables that could cooperate with developed PRSs for improving BC risk stratification (Daniele *et al.*, 2021; Demarest & Shah, 2024; Gao *et al.*, 2024; Hiatt & Brody, 2018; Neagu *et al.*, 2024). The combination of genetic and non-genetic risk factors has already been used in cardiovascular diseases. For example, in the case of CAD, detecting patients more likely to have a negative prognosis may be easier by combining traditional risk factors and a suitable metaPRS, as well as, by combining PRS with conventional risk factors, the accuracy of identifying people who are predisposed to CAD is increased (Agbaedeng *et al.*, 2021; Qin *et al.*, 2024). For example, recent publications reported that C index of meta GRS for coronary artery disease is about 0.67 and with the inclusion of conventional risk factors, C index increase to almost 0.70, while, by using a PRS combined with established risk tools, the corresponding net reclassification increase was 5.9%, illustrating the improvement from combining CAD PRS with conventional factors (O Sullivan *et al.*, 2022; Riveros-Mckay *et al.*, 2021). In addition, for Atrial Fibrillation (AF), it was demonstrated that classic risk factors in combination with a suitable PRS, are associated with the 10-year risk of this disease, leading to the conclusion that individuals at risk may be able to be distinguished (Zhang *et al.*, 2023). Furthermore, it was noticed that healthy lifestyle adherence reduced the absolute risk for type 2 diabetes, especially in individuals belonging to upper PRS classes, meaning that the interaction between PRSs and lifestyle behaviour is significant (Ye *et al.*, 2021). In addition to CA types and cardiovascular diseases, the contribution of PRS to traditional risk factors has also been highlighted in other disorders, such in the case of fracture risk, when suitable PRS is incorporated into currently used fracture prediction models, it can assist patients and physicians in more accurately determining a patient's risk (Nguyen & Eisman, 2020).

6.3.2 CHALLENGES IN COMBINING PRS WITH NON-GENETIC FACTORS

Although the combination of genetic and non genetic factors seems to improve PRSs' predictive ability for various diseases, the integration of non genetic factors and PRSs present a series of challenges (O Sullivan *et al.*, 2022; Shi *et al.*, 2020). Some of these challenges arise from the nature of genetic and non genetic data that are high dimensional, heterogeneous and often correlated (Wang, 2024). Specifically, both genetic (usually thousands of SNPs or more) and non genetic (numerous variables corresponding to lifestyle attitudes and environmental factors over time) data are high dimensional (Wang, 2024). In addition, non genetic factors are mostly not stable over time in a person's lifetime and there are measurement inconsistencies (different units of measurement, etc), as well as, these factors are often correlated (smoking habit and alcohol consumption, etc). The abovementioned lead to the necessity of using advanced models for integrating PRS with multifaceted lifestyle environmental data such as machine learning and AI approaches, Bayesian and probabilistic models and interaction models. These kinds of models are able to handle high dimensional data, identify non linear relationships and complex patterns between genetic and non genetic data (Hrytsenko *et al.*, 2024). The aforementioned models could manage more effectively population bias due to disparities in GWAS representation (Jin *et al.*, 2024). Furthermore, Bayesian approaches and related methods are useful for understanding probabilistic relationships due to PRS's probabilistic nature (Timmins *et al.*, 2024). Combining PRSs with other factors in disease prediction and risk assessment may lead to the risk of overestimating the impact of PRSs due to the correlations between non genetic factors that could lead to multicollinearity.

6.4 POLYGENIC RISK SCORES DEVELOPED FOR VARIOUS DISEASES & CLINICAL APPLICATION

There is a clear trend towards developing PRSs for various diseases such as cardiological diseases and cancer types, even though there are questions about whether they really measure the pure genetic risk for developing these diseases (Janssens, 2019). Apart from cardiological diseases and various cancer types that are particularly important given their prevalence and societal impact, PRSs have been developed for a variety of diseases such as type 2 diabetes, ophthalmological diseases, depression disorder, psychosis-schizophrenia, etc. (Bray *et al.*, 2024; Guffanti *et al.*, 2019; Mandla *et al.*, 2024; Perkins *et al.*, 2020; Qassim *et al.*, 2021; Ralapanawa & Sivakanesan, 2021; Stark *et al.*, 2024). PRSs have also been developed for rarer diseases such as neurodegenerative diseases (Parkinson's disease, etc) and autoimmune conditions (systemic lupus erythematosus, etc) presenting satisfactory performance in identifying patients at high risk (Ibanez *et al.*, 2019; Khunsriraksakul *et al.*, 2022).

Recent literature states that PRSs may be used in clinical care to identify individuals at higher risk of developing a disease, guide early intervention and assist in directing treatment decisions (Khera *et al.*, 2018; Lewis & Vassos, 2020; Torkamani *et al.*, 2018). The trend of developing PRSs (and Polygenic Scores (PGSs) in general) has become so widespread that there are approximately 4.850 polygenic scores for more than 650 traits in many publications according to PGS Catalog (Lambert *et al.*, 2021). Specifically for a variety of cancer types (e.g., breast cancer, prostate cancer, colorectal cancer, melanoma, etc.) and cardiovascular diseases (e.g., coronary artery diseases, hypertension, stroke, atrial fibrillation, etc.), a multitude of PRSs have been developed that aim to assess the risk of developing the above-mentioned diseases and to stratify individuals from general population according to their risk (Chen, Madduri, *et al.*, 2023; Kim *et al.*, 2023; Ma *et al.*, 2023; Maj *et al.*, 2022; Mega *et al.*, 2015; Mishra *et al.*, 2022; Shieh *et al.*, 2016; Tian *et al.*, 2024). Furthermore, evidence suggests that cancer types and cardiovascular diseases are at least partially hereditary and are amongst the leading causes of death according to multiple studies, including statistics provided by such organizations such as the International Agency for Research on Cancer, among others (Bray *et al.*, 2024; Mars, Lindbohm, *et al.*, 2022; Ralapanawa & Sivakanesan, 2021; Truong *et al.*, 2024).

6.4.1 CANCER TYPES

Using the PRS methodology, it has been shown that common risk variants influence not only whether a cancer will manifest or not, but also the age at which this cancer type will first appear (Namba *et al.*, 2023). Breast cancer (BC) is reported to be the most diagnosed cancer in women and one of the leading causes of death by cancer globally (Bray *et al.*, 2024). The GLOBOCAN 2022 estimates showed it to be the second most diagnosed cancer type globally and to constitute about 7% of all cancer-related deaths (Bray *et al.*, 2024). Although BC is most common among females, BC can also rarely affect males, and additionally, there is evidence of increasing incidence of BC in both males and females in recent years (Łukasiewicz *et al.*, 2021; Peng *et al.*, 2024). There is a strong genetic component to BC with multiple genes (e.g., mutations at BRCA1 and BRCA2 are strongly linked to an increased risk of BC corresponding to ORs of 7.62 and 5.23 respectively) having been identified as conferring susceptibility to BC over a female's lifetime (Hu *et al.*, 2021; Mars, Widén, *et al.*, 2020). For this reason, there have been many attempts to create BC disease-targeted PRSs (Jee *et al.*, 2023; Sun *et al.*, 2024). In addition, BC leads to several consequences of varying severity, culminating in loss of life, but also to less serious ones that can be caused by treatment such as dry eye disease and mild cognitive impairment (Albers *et al.*, 2024; Machairoudia *et al.*, 2024). Thus, the detection of the individual predisposition risk for the occurrence of BC, would be extremely useful for organizing appropriate health screening, intervention and prevention strategies that lead to increased survival rates (Roberts *et al.*, 2023). Moreover, PRSs have recently been recognized as a promising method for BC risk prediction and it has been suggested that combing them with traditional clinical factors could lead to a more accurate personalized BC risk assessment by avoiding overscreening (Gao *et al.*, 2021; Mabey *et al.*, 2024; Roberts *et al.*, 2023; Yang *et al.*, 2022). Additionally, PRS methods are highlighted as an effective and trustworthy indicator of BC risk that could enhance prevention strategies since they are useful in risk stratification (e.g., female individuals in the top 1% of the PRS distribution have multiple times the risk of developing breast cancer compared to those falling around the middle of the PRS distribution) (Ho *et al.*, 2020; Mavaddat *et al.*, 2019). In addition, certain PRSs have been developed that appear robust to confounding caused by individual characteristics or other factors that could be considered as a separate component linked to contralateral BC risk thereby enhancing the corresponding stratification (Kramer *et al.*, 2020). Furthermore, many types of PRSs have been developed for BC, being variable both in regards to the number of the SNPs utilized, as well as, the ancestry distribution of the training sample. In another work it emerges that a PRS, based on 313 risk variants, was successful in classifying women as possessing higher risk than the control group, even when they did not appear to carry the known pathogenic variants in the classic genes associated with breast cancer such as BRCA, showing possible clinical utility of the specific PRS in genetic analysis of females with hereditary breast cancer (Baliakas *et al.*, 2024). Regarding ancestral variation, PRSs have been developed for the evaluation of BC across ancestral gradients ranging from continental-scale populations such as for women of African ancestry to local groups such as Cypriot women (Du *et al.*, 2021; Yiangou *et al.*, 2021). Specifically, it is noted that PRS may be useful for regular healthcare decision-making and in alleviating some of the issues that women of African ancestry face in regards to breast cancer, where the mortality rates for them can be 40% higher than those of European ancestry (Zirpoli *et al.*, 2024). Moreover, it is also noted that Asian females' breast cancer risk could be predicted using a PRS designed for European females, which could assist in health programs (Ho *et al.*, 2020). Additionally, in another work, it is highlighted that implementing a specific PRSs that incorporate variants relating to both BC and ancestry fare better than other tested PRSs, suggesting that multi-ancestry BC PRS studies have potential for the clinical assessment and prevention of BC (Simmons *et al.*, 2024). Integration of PRSs in clinical practice appears promising and particularly so when combined with traditional testing methods (such as the Tyrer-Cuzick test) and family history (Kapoor *et al.*, 2021; Mabey *et al.*, 2024), as well as enabling more personalized risk prediction and stratification strategies (Gao *et al.*, 2021). It is however important to communicate breast PRS results clearly and accurately when it is utilized in a clinical setting (Gupta *et al.*, 2021).

Recent studies have indicated that prostate cancer (PC) ranks in the 2nd place as regards to prevalence globally amongst men and is also the 5th most common cause of death by cancer type with evidence suggesting PC to be the most commonly diagnosed cancer in more than 110 out of 185 countries globally (Bray *et al.*, 2024). Research shows that a multitude of factors contribute to the appearance of prostate cancer in an individual, with one's genetics having a considerable influence but often in conjunction with environmental, lifestyle and psychological factors (Csizmadi & Freedland, 2024; Gong *et al.*, 2024; Wang *et al.*, 2018; Xu *et al.*, 2013). In addition, PRS-based studies have been conducted investigating PC incidence, using SNPs inferred from relevant GWASs (Oh *et al.*, 2020). Such studies have shown that individuals with statistically significant higher risk scores compared to control groups did in fact have PC at higher rate than average (Sipeky *et al.*, 2020). Another study noted significant associations between higher PRS values and a younger age at which PC is diagnosed and suggested that close relatives (of the first degree) are also at risk of PC diagnosis providing additional evidence for PRS utility in clinical practice (Spears *et al.*, 2024). PRS has also been shown to predict higher risk for PC in elderly individuals of European descent (cohort age greater than 70 years) that could assist in decreasing over diagnosis, despite not being informative on clinical significance (Bakshi *et al.*, 2021). In addition, integration of multiple known pathogenic SNPs including BRCA2 and BRCA1 variants, show that PC risk estimation may be age dependent, which could direct next research on PC risk assessment (Nyberg *et al.*, 2023).

6.4.2 CARDIOVASCULAR DISEASES

As regards cardiovascular diseases, coronary artery disease (CAD) is one of the leading causes of death worldwide and is expected to affect more than 24 million globally by the next decade, while, its prevalence is age, gender and region dependent (Dimitriou *et al.*, 2024; Ralapanawa & Sivakanesan, 2021; Stark *et al.*, 2024; Zeinali-Nezhad *et al.*, 2024). There are several factors that contribute to CAD; most of them are believed to be controllable through lifestyle modifications and related interventions (e.g. obesity, low physical activity, smoking, etc) (Zeinali-Nezhad *et al.*, 2024). The genetic background of CAD is well established and innovative PRS frameworks have been proposed improving prediction accuracy for CAD recently and are tested in a variety of populations (Dimitriou *et al.*, 2024; Truong *et al.*, 2024). More specifically, evidence suggest a substantial genetic component to CAD, with some studies reporting that genetics account for half of CAD predisposition (Roberts *et al.*, 2022). PRSs have been proposed as a useful avenue of research for CAD risk prediction and patients' stratifications as regards their death risk since high PRSs are highly correlated with elevated CAD risk (Agbaedeng *et al.*, 2021; Qin *et al.*, 2024). There are also, beyond traditional clinical risk factors for CAD, health diseases, such as diabetes mellitus or hypertension which if identified can lead to therapeutically interventions such as statin therapy (Roberts *et al.*, 2022; Zeinali-Nezhad *et al.*, 2024). As regards the PRS utilization, it was noted that individuals on the top 20% of the PRS distribution were three times more likely to develop CAD compared to those in the bottom 20% indicating huge potential to pinpoint CAD high-risk persons for tailored clinical utility interventions (Lu *et al.*, 2022). Moreover, it is noted that PRS seems to be a useful supplementary tool particularly in identifying risk-prone young adults (risk stratification) and helping design personalized preventive interventions and targeted therapy (Damask *et al.*, 2020; Marston *et al.*, 2023). In another publication, it is highlighted that even before the development of risk factors, CAD PRSs may detect individuals who are at high risk to suffer from CAD and direct therapy intervention, early in life, meaning that CAD PRSs appear to be more useful in youngers while older persons may benefit more from focusing on classical risk factors (Fahed & Natarajan, 2023; Tada & Takamura, 2024). Stroke is one of the most common causes of death globally, accounting for around 11-12% of all fatalities and its prevalence is rising, especially in low-income nations (Collaborators *et al.*, 2021). A variety of PRSs have been developed for stroke due to its hereditary mechanisms, besides related environmental factors and it is highlighted that a special form of PRS (metaPRS) appeared to be significantly promising for stroke risk stratification and prediction in general populations (Lu *et al.*, 2021). Compared to earlier research on stroke, as opposed to stroke traits, it was found that a specific PRS approach could detect additional relevant risk factors to include, such as common walking pace, arthropathies, and lipoprotein(a) (Truong *et al.*, 2024). Furthermore, it is highlighted that stroke PRS is developed with low effect sizes, compared to PRSs for other cardiovascular diseases, which is probably due to the heterogeneity of the causal mechanisms that cause the various types of stroke (Mars, Lindbohm, *et al.*, 2022). In addition, recent findings offer vital information to guide next scientific studies on the etiology of strokes, identify promising therapeutic targets for intervention, and develop instruments for predicting genetic risk in European, East Asian and African people via PRS (Mishra *et al.*, 2022). Atrial fibrillation (AF) is another cardiovascular disease with a high prevalence between 20% and 33% after 45 years (Linz *et al.*, 2024). It is highlighted that AF's appearance is based to a significant extent on genetic factors, in addition to non-genetic (e.g., environmental – polluted atmosphere) ones and, moreover, it was discovered that AF PRS enhanced the corresponding classification models using age and gender (Ma *et al.*, 2023; Mars, Koskela, *et al.*, 2020).

6.4.3 DIABETES MELITUS

Type 2 Diabetes (T2D) is the most common form of diabetes and affects millions of people globally, with an increase most apparent among younger individuals, including children (Ahmad *et al.*, 2022; Chatterjee *et al.*, 2017). The relationship of T2D is close to the occurrence of cardiovascular disease (CVD) as people suffering from T2D are more likely to develop CVD than those without T2D as a result of vascular complications (Ahmad *et al.*, 2024). T2D has been shown to be caused by both genetic and non-genetic (environmental, lifestyle, etc) factors with the former being modulated via the latter (Murea *et al.*, 2012). A polygenic risk score (PRS) approach may be beneficial in risk stratification and early intervention as to avoid the incidence of T2D and associated adverse side-effects aiming to be introduced in every day clinical practice (Ge *et al.*, 2022; Lee *et al.*, 2024). Research so far, while confirming that both genetic and environmental factors lead to T2D, supports the utility of PRS in T2D prevention moderately, optimally in conjunction with lifestyle-based interventions (e.g., healthy diet) and other risk predictors such as an Leicester Risk Assessment (LRA score) (Liu *et al.*, 2024; Merino *et al.*, 2022). The actual performance of PRS in T2D prediction is still debated, with some studies only finding modest additional affects over existing risk predictors while others emphasizing the clinical applications of T2D-related PRS (Kim *et al.*, 2024; Liu *et al.*, 2024). However, even if, until recently, a large proportion of published PRSs for T2D had not been evaluated, vast majority of these PRSs had a substantial association with T2D, even if their accuracy was relatively less than what was stated (Briviba *et al.*, 2024). Also, it is mentioned that a PRS for a Korean population showed high accuracy in discriminating cases and controls assisting in the detection of high-risk persons of T2D (Song *et al.*, 2024), while there are indications of PRS utility for early diagnosis of T2D in a Taiwanese population (Wang *et al.*, 2024). While physical exercise and other lifestyle choices may decrease the frequency of T2D incidence and prevent CVD issues, there is evidence to suggest that the positive impact of physical exercise and associated interventions is greatly reduced for individuals with a high genetic risk score for T2D (Lee *et al.*, 2024).

In this section, the potential of integrating PRSs in clinical practice was highlighted with more extensive reference made for important common diseases such a BC, PC, CAD and T2D. While research has indicated the prospective benefits of applying PRSs in the clinic, there are still some issues that if left unresolved may hinder the successful implementation of PRSs in practice. Some of the most important ones are listed in the following paragraph.

6.5 CLINICAL IMPLEMENTATION ISSUES OF POLYGENIC RISK SCORES

PRSs are numerical metrics for estimating the probability of a variety of traits occurrence and so far, have shown great promise to predict the risk of specific diseases' occurrence, as well as, to be included in everyday clinical care, complementary to traditional risk factors (Khera *et al.*, 2018; Torkamani *et al.*, 2018). Despite this, PRS-based methods do of course have some limitations in their clinical use related to a variety of issues (Privé *et al.*, 2022). One of the major issues preventing PRSs application in clinical practice is that PRSs do not perform equally in different populations, and in particular their predictive ability decreases as the genetic gap widens between the test and training groups, and also, it is a fact that the European populations have been used more than others (Privé *et al.*, 2022). Therefore, the issue of low transferability, even between populations with little genetic divergence, ought to be sufficiently managed prior to employing PRSs in daily clinical practice (Kachuri *et al.*, 2024; Privé *et al.*, 2022). For example, a specific PRS developed in a training sample of Europeans demonstrated poor portability when tested on non-Europeans which highlights the need for extensive GWAS investigations in underrepresented populations, i.e., non-European populations (Tanigawa *et al.*, 2022). Therefore, the importance of expanding GWAS representation beyond European populations is high and there are ongoing global initiatives aiming at improving diversity in PRS development (e.g., H3Africa, All of Us Research Program) (All of Us Research Program Investigators, 2019; Mulder *et al.*, 2018). Moreover, CAD PRS's forecasting ability was found to be quite improved in all ancestry groups with the exception of individuals of African origin (Smith *et al.*, 2024). Clearly, there is also the approach of creating many different ones PRS (e.g. for CAD), one for each ancestry so that the diversity in the genetic background is properly considered (Smith *et al.*, 2024). This may be due to a variety of factors such as LD differences between populations and differences in allele frequencies (Grinde *et al.*, 2019) but it is worth mentioning that new methods are being developed and worked on which could make it possible to apply PRS across populations (Amariuta *et al.*, 2020; Ge *et al.*, 2022; Kachuri *et al.*, 2024). Another issue, is that there are many variants with a risk allele frequency lower than that investigated to date, therefore, there might exist a large number of low frequency variants that may need to be thoroughly considered (Kunkle *et al.*, 2019). The investigation of rare or very rare variants, compared to what has already been studied to date, may offer further information on the reasons that lead to traits corresponding to phenotypic characteristics and complex diseases. Another important issue is that for most traits (phenotypic characteristics or diseases), there is both a genetic and an environmental influence on their occurrence, so the measured genetic signal calculated by the PRSs is not complete and thus usually accounts for a small fraction of the variance that a trait exhibits (Slunecka *et al.*, 2021). In addition, the integration of PRSs into the health care system will bring about substantial challenges in terms of the relevant training, of the clinicians, as well as, of the patients (Slunecka *et al.*, 2021). Finally, it should be noted that there is limited research on PRSs' validation, as well as, there is no standardized PRS construction methodology that is universally accepted to be optimal, that is, there is no one PRS methodology that has been tested in many different surveys and consistently produces the most accurate results (Collister *et al.*, 2022; Slunecka *et al.*, 2021). Of course, emerging methods for standardizing PRS development are growing, such as fine mapping techniques, Bayesian approaches, or machine learning models (Cui *et al.*, 2024; Hrytsenko *et al.*, 2024; Jin *et al.*, 2024). The above mentioned problems, among others, pertain directly to Polygenic Risk Scores, and their use in clinical practice or healthcare at large (Slunecka *et al.*, 2021). It should also be mentioned that PRS implementation in a wide clinical scale for risk stratification is considered expensive and the corresponding cost should be reduced (Xia *et al.*, 2024). In addition, there have also been analyses indicating that polygenic scores do not substantially add to a disease's prediction (e.g., coronary atherosclerosis or obstructive CAD) over classical risk factors and as such PRSs are not yet fit for clinical application (Kujala *et al.*, 2024). For example, in cases where individuals previously have experienced a cardiovascular incident or have high cardiovascular risk, a CAD PRS provides limited information for treatment stratification (Kavousi & Schunkert, 2022).

Despite the caveats mentioned in this paragraph, recent literature make note of PRSs' utility and for some of the aforementioned issues, many efforts are being made to correct them, such as the use of multi-ancestry cohorts that is proving effective at improving CAD risk prediction (Patel *et al.*, 2023; Xiang *et al.*, 2024).

6.6 DISCUSSION

Recent literature has shown that PRS prediction could have clinical utility and, while in many cases not yet integrated into clinical practice, holds great promise for personalized disease risk assessment and individuals' stratification according their risk, as well as, for early diseases diagnosis and drug layering (Corpas & Fatumo, 2023; Cross *et al.*, 2022; Ho-Le *et al.*, 2023; Nguyen & Eisman, 2020). In previous paragraphs, examples of PRSs aimed at estimating the risk of a variety of diseases with a significant prevalence in general populations, such as cardiovascular diseases, cancer types, diabetes mellitus type 2, etc., were given. In many cases there were recommendations for clinical use of PRSs as there is evidence that they may assist in risk assessment and individuals' risk classification (Baliakas *et al.*, 2024). For example, it was recently highlighted that T2D PRS may assist to detect individuals in high risk for a disease even if traditional clinical factors characterize them as being in low risk (Mandla *et al.*, 2024). In addition, PRS derived from cross-ancestry meta-analysis could forecast an elevated risk of mortality from cardiovascular causes (Miyazawa *et al.*, 2023). It is mentioned that BC PRS could enhance monitoring and prevention programs in public health by layering properly the specific populations according to their risk (Mavaddat *et al.*, 2019), as well as, PRS-based approaches could offer valuable insights in BC disease's risk prediction and prevention (Baliakas *et al.*, 2024). Furthermore, it is pointed out that a verified risk model providing individualized PC risk could support males as well as, enable upcoming methods for risk-stratified population assessment (Nyberg *et al.*, 2023). Moreover, it is worth mentioning that PRS has been proven to yield valid predictions in both "population-based" and "electronic records-based studies", as well as PRS methods were considered that could be broadly applicable in clinical practice to predict the risk of common diseases (Lewis & Vassos, 2020; Torkamani *et al.*, 2018). Finally, it has been clearly suggested that PRS should be incorporated into clinical practice for common diseases (Khera *et al.*, 2018).

From all of the above it could be assumed, according to the large multitude of PRSs that have been developed recently for various diseases, that scientists of the respective fields believe in PRSs' properties and that finally, PRSs could potentially have a significant role in clinical application in every day medical practice, both at the individual level using the risk estimated by the PRS for a disease, as well as at the population level for making decisions concerning public health. In any case, it is necessary to continue the research on the PRSs' methodologies in order to evaluate their clinical utility for many diseases (e.g., coronary artery disease, atrial fibrillation, type 2 diabetes, etc.) and to enhance predicting the risk of disease, as well as, how their utilization could guide individualized management of specific patients (McElligott *et al.*, 2023; Wong *et al.*, 2022). In conclusion, PRSs show promise to estimate diseases' risk and guide in early interventions, as well as, to improve the personalized healthcare, but require further validation and refinement for widespread clinical adoption.

REFERENCES

Abdellaoui, A., Yengo, L., Verweij, K. J., & Visscher, P. M. (2023). 15 years of gwas discovery: Realizing the promise. *The American Journal of Human Genetics*, *110*(2), 179–194.

Agbaedeng, T. A., Noubiap, J. J., Mato, E. P. M., Chew, D. P., Figtree, G. A., Said, M. A., & van der Harst, P. (2021). Polygenic risk score and coronary artery disease: A meta-analysis of 979,286 participant data. *Atherosclerosis*, *333*, 48–55.

Ahmad, A., Lim, L.-L., Morieri, M. L., Tam, C. H.-t., Cheng, F., Chikowore, T., Dudenhöffer-Pfeifer, M., Fitipaldi, H., Huang, C., Kanbour, S., *et al.* (2024). Precision prognostics for cardiovascular disease in type 2 diabetes: A systematic review and meta-analysis. *Communications medicine*, *4*(1), 11.

Ahmad, E., Lim, S., Lamptey, R., Webb, D. R., & Davies, M. J. (2022). Type 2 diabetes. *The Lancet*, *400*(10365), 1803–1820.

Ala-Korpela, M., & Holmes, M. V. (2020). Polygenic risk scores and the prediction of common diseases.

Albers, E., Kieffer, J., & Schagen, S. (2024). The prevalence of mild cognitive impairment in breast cancer patients receiving chemotherapy according to the criteria of the national institute on aging-alzheimer's association (nia-aa). *Journal of Cancer Survivorship*, 1–8.

All of Us Research Program Investigators. (2019). The "All of Us" research program. *New England Journal of Medicine*, *381*(7), 668–676.

Amariuta, T., Ishigaki, K., Sugishita, H., Ohta, T., Koido, M., Dey, K. K., Matsuda, K., Murakami, Y., Price, A. L., Kawakami, E., *et al.* (2020). Improving the trans-ancestry portability of polygenic risk scores by prioritizing variants in predicted cell-type-specific regulatory elements. *Nature genetics*, *52*(12), 1346–1354.

Aragam, K. G., & Natarajan, P. (2020). Polygenic scores to assess atherosclerotic cardiovascular disease risk: Clinical perspectives and basic implications. *Circulation research*, *126*(9), 1159–1177.

Arnold, N., & Koenig, W. (2021). Polygenic risk score: Clinically useful tool for prediction of cardiovascular disease and benefit from lipid-lowering therapy? *Cardiovascular Drugs and Therapy*, *35*(3), 627–635.

Aslam, M. M., Fan, K.-H., Lawrence, E., Bedison, M. A., Snitz, B. E., DeKosky, S. T., Lopez, O. L., Feingold, E., & Kamboh, M. I. (2023). Genome-wide analysis identifies novel loci influencing plasma apolipoprotein e concentration and alzheimer s disease risk. *Molecular psychiatry*, *28*(10), 4451–4462.

Bakshi, A., Riaz, M., Orchard, S. G., Carr, P. R., Joshi, A. D., Cao, Y., Rebello, R., Nguyen-Dumont, T., Southey, M. C., Millar, J. L., *et al.* (2021). A polygenic risk score predicts incident prostate cancer risk in older men but does not select for clinically significant disease. *Cancers*, *13*(22), 5815.

Baliakas, P., Munters, A. R., Kämpe, A., Tesi, B., Bondeson, M.-L., Ladenvall, C., & Eriksson, D. (2024). Integrating a polygenic risk score into a clinical setting would impact risk predictions in familial breast cancer. *Journal of Medical Genetics*, *61*(2), 150–154.

Bersimis, F. G., & Varlamis, I. (2019). *Use of health-related indices and classification methods in medical data*. In *Classification techniques for medical image analysis and computer aided diagnosis* (31–66). Elsevier.

Bersimis, F. G., Varlamis, I., Vamvakari, M., & Panagiotakos, D. B. (2020). Binary classification techniques: An application on simulated and real bio-medical data. *Data Analysis and Applications 3: Computational, Classification, Financial, Statistical and Stochastic Methods*, *5*, 145–175.

Boyle, E. A., Li, Y. I., & Pritchard, J. K. (2017). An expanded view of complex traits: From polygenic to omnigenic. *Cell*, *169*(7), 1177–1186.

Bray, F., Laversanne, M., Sung, H., Ferlay, J., Siegel, R. L., Soerjomataram, I., & Jemal, A. (2024). Global cancer statistics 2022: Globocan estimates of incidence and mortality worldwide for 36 cancers in 185 countries. *CA: a cancer journal for clinicians*, *74*(3), 229–263.

Brentnall, A. R., van Veen, E. M., Harkness, E. F., Rafiq, S., Byers, H., Astley, S. M., Sampson, S., Howell, A., Newman, W. G., Cuzick, J., *et al.* (2020). A case–control evaluation of 143 single nucleotide polymorphisms for breast cancer risk stratification with classical factors and mammographic density. *International journal of cancer*, *146*(8), 2122–2129.

Briviba, M., Atava, I., Pečulis, R., Elbere, I., Ansone, L., Rozenberga, M., Silamiķelis, I., & Kloviņš, J. (2024). Evaluating the efficacy of type 2 diabetes polygenic risk scores in an independent european population. *International Journal of Molecular Sciences*, *25*(2), 1151.

Cano-Gamez, E., & Trynka, G. (2020). From gwas to function: Using functional genomics to identify the mechanisms underlying complex diseases. *Frontiers in genetics*, *11*, 424.

Carlson, M. O., Rice, D. P., Berg, J. J., & Steinrücken, M. (2022). Polygenic score accuracy in ancient samples: Quantifying the effects of allelic turnover. *PLoS Genetics*, *18*(5), e1010170.

Chatterjee, S., Khunti, K., & Davies, M. J. (2017). Type 2 diabetes. *The lancet*, *389*(10085), 2239–2251.

Chatziandreou, E. (2022). Mt5 adjusted polygenic risk score enables personalized cardiovascular disease prevention and clinical management. *Value in Health*, *25*(12), S378–S379.

Chen, F., Madduri, R. K., Rodriguez, A. A., Darst, B. F., Chou, A., Sheng, X., Wang, A., Shen, J., Saunders, E. J., Rhie, S. K., *et al.* (2023). Evidence of novel susceptibility variants for prostate cancer and a multiancestry polygenic risk score associated with aggressive disease in men of african ancestry. *European urology*, *84*(1), 13–21.

Chen, J., Gatev, E., Everson, T., Conneely, K. N., Koen, N., Epstein, M. P., Kobor, M. S., Zar, H. J., Stein, D. J., & Huels, A. (2023). Pruning and thresholding approach for methylation risk scores in multi-ancestry populations. *Epigenetics*, *18*(1), 2187172.

Chial, H. (2008). Mendelian genetics: Patterns of inheritance and single-gene disorders. *Nature Education*, *1*(1), 63.

Choi, S. W., Mak, T. S.-H., & O'Reilly, P. F. (2020). Tutorial: A guide to performing polygenic risk score analyses. *Nature protocols*, *15*(9), 2759–2772.

Collaborators, G. 2. S., *et al.* (2021). Global, regional, and national burden of stroke and its risk factors, 1990–2019: A systematic analysis for the global burden of disease study 2019. *The Lancet. Neurology*, *20*(10), 795.

Collins, F. S., Morgan, M., & Patrinos, A. (2003). The human genome project: Lessons from large-scale biology. *Science*, *300*(5617), 286–290.

Collister, J. A., Liu, X., & Clifton, L. (2022). Calculating polygenic risk scores (prs) in uk biobank: A practical guide for epidemiologists. *Frontiers in genetics*, *13*, 818574.

Collister, J. A., Liu, X., Littlejohns, T. J., Cuzick, J., Clifton, L., & Hunter, D. J. (2024). Assessing the value of incorporating a polygenic risk score with nongenetic factors for predicting breast cancer diagnosis in the uk biobank. *Cancer Epidemiology, Biomarkers & Prevention*, *33*(6), 812–820.

Corpas, M., & Fatumo, S. (2023). Generalisation of genomic findings and applications of polygenic risk scores. *BMC Medical Genomics*, *16*(1), 175.

Cross, B., Turner, R., & Pirmohamed, M. (2022). Polygenic risk scores: An overview from bench to bedside for personalised medicine. *Frontiers in Genetics*, *13*, 1000667.

Crouch, D. J., & Bodmer, W. F. (2020). Polygenic inheritance, gwas, polygenic risk scores, and the search for functional variants. *Proceedings of the National Academy of Sciences*, *117*(32), 18924–18933.

Csizmadi, I., & Freedland, S. J. (2024). Adherence to the world cancer research fund lifestyle recommendations and incidence of prostate cancer in uk biobank. *International journal of epidemiology*, *53*(1), dyae008.

Cui, R., Elzur, R. A., Kanai, M., Ulirsch, J. C., Weissbrod, O., Daly, M. J., Neale, B. M., Fan, Z., & Finucane, H. K. (2024). Improving fine-mapping by modeling infinitesimal effects. *Nature genetics*, *56*(1), 162–169.

Damask, A., Steg, P. G., Schwartz, G. G., Szarek, M., Hagström, E., Badimon, L., Chapman, M. J., Boileau, C., Tsimikas, S., Ginsberg, H. N., *et al.* (2020). Patients with high genome-wide polygenic risk scores for coronary artery disease may receive greater clinical benefit from alirocumab treatment in the odyssey outcomes trial. *Circulation*, *141*(8), 624–636.

Daniele, A., Divella, R., Pilato, B., Tommasi, S., Pasanisi, P., Patruno, M., Digennaro, M., Minoia, C., Dellino, M., Pisconti, S., *et al.* (2021). Can harmful lifestyle, obesity and weight changes increase the risk of breast cancer in brca 1 and brca 2 mutation carriers? a mini review. *Hereditary Cancer in Clinical Practice*, *19*, 1–8.

Demarest, K., & Shah, P. D. (2024). Polygenic risk scores for breast cancer. *Current Breast Cancer Reports*, 1–9.

Dima, D., & Breen, G. (2015). Polygenic risk scores in imaging genetics: Usefulness and applications. *Journal of Psychopharmacology*, *29*(8), 867–871.

Dimitriou, M., Moulos, P., Kalafati, I. P., Saranti, G., Rallidis, L. S., & Dedoussis, G. V. (2024). Evaluation of polygenic risk scores for prediction of coronary artery disease in a greek case-control study. *Journal of Personalized Medicine*, *14*(6), 565.

Ding, Y., Hou, K., Burch, K. S., Lapinska, S., Privé, F., Vilhjálmsson, B., Sankararaman, S., & Pasaniuc, B. (2022). Large uncertainty in individual polygenic risk score estimation impacts prs-based risk stratification. *Nature genetics*, *54*(1), 30–39.

Du, Z., Gao, G., Adedokun, B., Ahearn, T., Lunetta, K. L., Zirpoli, G., Troester, M. A., Ruiz-Narváez, E. A., Haddad, S. A., PalChoudhury, P., *et al.* (2021). Evaluating polygenic risk scores for breast cancer in women of african ancestry. *JNCI: Journal of the National Cancer Institute*, *113*(9), 1168–1176.

Dudbridge, F. (2013). Power and predictive accuracy of polygenic risk scores. *PLoS genetics*, *9*(3), e1003348.

Duncan, L., Shen, H., Gelaye, B., Meijsen, J., Ressler, K., Feldman, M., Peterson, R., & Domingue, B. (2019). Analysis of polygenic risk score usage and performance in diverse human populations. *Nature communications*, *10*(1), 3328.

Duncan, L. E., Shen, H., Ballon, J. S., Hardy, K. V., Noordsy, D. L., & Levinson, D. F. (2018). Genetic correlation profile of schizophrenia mirrors epidemiological results and suggests link between polygenic and rare variant (22q11. 2) cases of schizophrenia. *Schizophrenia Bulletin*, *44*(6), 1350–1361.

Euesden, J., Lewis, C. M., & O reilly, P. F. (2015). Prsice: Polygenic risk score software. *Bioinformatics*, *31*(9), 1466–1468.

Fahed, A. C., & Natarajan, P. (2023). Clinical applications of polygenic risk score for coronary artery disease through the life course. *Atherosclerosis*, 117356.

Fatumo, S., Chikowore, T., Choudhury, A., Ayub, M., Martin, A. R., & Kuchenbaecker, K. (2022). A roadmap to increase diversity in genomic studies. *Nature medicine*, *28*(2), 243–250.

Feitosa, M. F., Kuipers, A. L., Wojczynski, M. K., Wang, L., Barinas-Mitchell, E., Kulminski, A. M., Thyagarajan, B., Lee, J. H., Perls, T., Christensen, K., *et al.* (2021). Heterogeneity of the predictive polygenic risk scores for coronary heart disease age-at-onset in three different coronary heart disease family-based ascertainments. *Circulation: Genomic and Precision Medicine*, *14*(3), e003201.

Gao, C., Polley, E. C., Hart, S. N., Huang, H., Hu, C., Gnanaolivu, R., Lilyquist, J., Boddicker, N. J., Na, J., Ambrosone, C. B., *et al.* (2021). Risk of breast cancer among carriers of pathogenic variants in breast cancer predisposition genes varies by polygenic risk score. *Journal of Clinical Oncology*, *39*(23), 2564–2573.

Gao, X., Qi, J., Du, B., Weng, X., Lai, J., & Wu, R. (2024). Combined influence of nutritional and inflammatory status and breast cancer: Findings from the nhanes. *BMC Public Health*, *24*(1), 2245.

Ge, T., Irvin, M. R., Patki, A., Srinivasasainagendra, V., Lin, Y.-F., Tiwari, H. K., Armstrong, N. D., Benoit, B., Chen, C.-Y., Choi, K. W., *et al.* (2022). Development and validation of a trans-ancestry polygenic risk score for type 2 diabetes in diverse populations. *Genome Medicine*, *14*(1), 70.

Gibbs, R. A., Belmont, J. W., Hardenbol, P., Willis, T. D., Yu, F. L., Yang, H., Ch'ang, L.-Y., Huang, W., Liu, B., Shen, Y., *et al.* (2003). The international hapmap project.

Gong, J., Kim, D. M., Freeman, M. R., Kim, H., Ellis, L., Smith, B., Theodorescu, D., Posadas, E., Figlin, R., Bhowmick, N., *et al.* (2024). Genetic and biological drivers of prostate cancer disparities in black men. *Nature Reviews Urology*, *21*(5), 274–289.

Grinde, K. E., Qi, Q., Thornton, T. A., Liu, S., Shadyab, A. H., Chan, K. H. K., Reiner, A. P., & Sofer, T. (2019). Generalizing polygenic risk scores from europeans to hispanics/latinos. *Genetic epidemiology*, *43*(1), 50–62.

Guffanti, G., Kumar, P., Admon, R., Treadway, M. T., Hall, M. H., Mehta, M., Douglas, S., Arulpragasam, A. R., & Pizzagalli, D. A. (2019). Depression genetic risk score is associated with anhedonia-related markers across units of analysis. *Translational psychiatry*, *9*(1), 236.

Gupta, K. D., Gregory, G., Meiser, B., Kaur, R., Scheepers-Joynt, M., McInerny, S., Taylor, S., Barlow-Stewart, K., Antill, Y., Salmon, L., *et al.* (2021). Communicating polygenic risk scores in the familial breast cancer clinic. *Patient education and counseling*, *104*(10), 2512–2521.

Haines, J. L., Hauser, M. A., Schmidt, S., Scott, W. K., Olson, L. M., Gallins, P., Spencer, K. L., Kwan, S. Y., Noureddine, M., Gilbert, J. R., *et al.* (2005). Complement factor h variant increases the risk of age-related macular degeneration. *Science*, *308*(5720), 419–421.

Hardouin, G., Magrin, E., Corsia, A., Cavazzana, M., Miccio, A., & Semeraro, M. (2023). Sickle cell disease: From genetics to curative approaches. *Annual Review of Genomics and Human Genetics*, *24*(1), 255–275.

Hardy, J., & Singleton, A. (2009). Genomewide association studies and human disease. *New England Journal of Medicine*, *360*(17), 1759–1768.

Henne, S. K., Aldisi, R., Sivalingam, S., Hochfeld, L. M., Borisov, O., Krawitz, P. M., Maj, C., Nöthen, M. M., & Heilmann-Heimbach, S. (2023). Analysis of 72,469 uk biobank exomes links rare variants to male-pattern hair loss. *Nature Communications*, *14*(1), 5492.

Hiatt, R. A., & Brody, J. G. (2018). Environmental determinants of breast cancer. *Annual review of public health*, *39*(1), 113–133.

Ho, W.-K., Tan, M.-M., Mavaddat, N., Tai, M.-C., Mariapun, S., Li, J., Ho, P.-J., Dennis, J., Tyrer, J. P., Bolla, M. K., *et al.* (2020). European polygenic risk score for prediction of breast cancer shows similar performance in asian women. *Nature communications*, *11*(1), 3833.

Ho-Le, T. P., Tran, T. S., Nguyen, H. G., Center, J. R., Eisman, J. A., & Nguyen, T. V. (2023). Genetic prediction of lifetime risk of fracture. *The Journal of Clinical Endocrinology & Metabolism*, *108*(11), e1403–e1412.

Hrytsenko, Y., Shea, B., Elgart, M., Kurniansyah, N., Lyons, G., Morrison, A. C., Carson, A. P., Haring, B., Mitchell, B. D., Psaty, B. M., *et al.* (2024). Machine learning models for predicting blood pressure phenotypes by combining multiple polygenic risk scores. *Scientific Reports*, *14*(1), 12436.

Hu, C., Hart, S. N., Gnanaolivu, R., Huang, H., Lee, K. Y., Na, J., Gao, C., Lilyquist, J., Yadav, S., Boddicker, N. J., *et al.* (2021). A population-based study of genes previously implicated in breast cancer. *New England Journal of Medicine*, *384*(5), 440–451.

Huang, C., Chen, W., & Wang, X. (2023). Studies on the fat mass and obesity-associated (fto) gene and its impact on obesity-associated diseases. *Genes & Diseases*, *10*(6), 2351–2365.

Ibanez, L., Farias, F. H., Dube, U., Mihindukulasuriya, K. A., & Harari, O. (2019). Polygenic risk scores in neurodegenerative diseases: A review. *Current Genetic Medicine Reports*, *7*, 22–29.

Janssens, A. C. J. (2019). Validity of polygenic risk scores: Are we measuring what we think we are? *Human molecular genetics*, *28*(R2), R143–R150.

Jee, Y. H., Ho, W.-K., Park, S., Easton, D. F., Teo, S.-H., Jung, K. J., & Kraft, P. (2023). Polygenic risk scores for prediction of breast cancer in korean women. *International Journal of Epidemiology*, *52*(3), 796–805.

Jermy, B., Läll, K., Wolford, B. N., Wang, Y., Zguro, K., Cheng, Y., Kanai, M., Kanoni, S., Yang, Z., Hartonen, T., *et al.* (2024). A unified framework for estimating country-specific cumulative incidence for 18 diseases stratified by polygenic risk. *Nature Communications*, *15*(1), 5007.

Jiang, W., Chen, L., Girgenti, M. J., & Zhao, H. (2024). Tuning parameters for polygenic risk score methods using gwas summary statistics from training data. *Nature Communications*, *15*(1), 24.

Jin, J., Zhan, J., Zhang, J., Zhao, R., O'Connell, J., Jiang, Y., Aslibekyan, S., Auton, A., Babalola, E., Bell, R. K., *et al.* (2024). Mussel: Enhanced bayesian polygenic risk prediction leveraging information across multiple ancestry groups. *Cell Genomics*, *4*(4).

Jung, H., Jung, H.-U., Baek, E. J., Kwon, S. Y., Kang, J.-O., Lim, J. E., & Oh, B. (2024). Integration of risk factor polygenic risk score with disease polygenic risk score for disease prediction. *Communications Biology*, *7*(1), 180.

Kachuri, L., Chatterjee, N., Hirbo, J., Schaid, D. J., Martin, I., Kullo, I. J., Kenny, E. E., Pasaniuc, B., in Diverse Populations (PRIMED) Consortium Methods Working Group Auer Paul L. 20 Conomos Matthew P. 21 Conti David V. 22 23 Ding Yi 24 Wang Ying 19 25 26 Zhang Haoyu 27 28 Zhang Yuji 29, P. R. M., Witte, J. S., *et al.* (2024). Principles and methods for transferring polygenic risk scores across global populations. *Nature Reviews Genetics*, *25*(1), 8–25.

Kachuri, L., Graff, R. E., Smith-Byrne, K., Meyers, T. J., Rashkin, S. R., Ziv, E., Witte, J. S., & Johansson, M. (2020). Pan-cancer analysis demonstrates that integrating polygenic risk scores with modifiable risk factors improves risk prediction. *Nature communications*, *11*(1), 6084.

Kafkas, Ş., Althubaiti, S., Gkoutos, G. V., Hoehndorf, R., & Schofield, P. N. (2021). Linking common human diseases to their phenotypes; development of a resource for human phenomics. *Journal of biomedical semantics*, *12*, 1–15.

Kapoor, P. M., Mavaddat, N., Choudhury, P. P., Wilcox, A. N., Lindström, S., Behrens, S., Michailidou, K., Dennis, J., Bolla, M. K., Wang, Q., *et al.* (2021). Combined associations of a polygenic risk score and classical risk factors with breast cancer risk. *JNCI: Journal of the National Cancer Institute*, *113*(3), 329–337.

Kavousi, M., & Schunkert, H. (2022). Polygenic risk score: A tool ready for clinical use?

Khera, A. V., Chaffin, M., Aragam, K. G., Haas, M. E., Roselli, C., Choi, S. H., Natarajan, P., Lander, E. S., Lubitz, S. A., Ellinor, P. T., *et al.* (2018). Genome-wide polygenic scores for common diseases identify individuals with risk equivalent to monogenic mutations. *Nature genetics*, *50*(9), 1219–1224.

Khunsriraksakul, C., Markus, H., Olsen, N. J., Carrel, L., Jiang, B., & Liu, D. J. (2022). Construction and application of polygenic risk scores in autoimmune diseases. *Frontiers in immunology*, *13*, 889296.

Kim, E. S., Scharpf, R. B., Garcia-Closas, M., Visvanathan, K., Velculescu, V. E., & Chatterjee, N. (2023). Potential utility of risk stratification for multicancer screening with liquid biopsy tests. *NPJ Precision Oncology*, *7*(1), 39.

Kim, N. Y., Lee, H., Kim, S., Kim, Y.-J., Lee, H., Lee, J., Kwak, S. H., & Lee, S. (2024). The clinical relevance of a polygenic risk score for type 2 diabetes mellitus in the korean population. *Scientific Reports*, *14*(1), 5749.

Klimentidis, Y. C., Raichlen, D. A., Bea, J., Garcia, D. O., Wineinger, N. E., Mandarino, L. J., Alexander, G. E., Chen, Z., & Going, S. B. (2018). Genome-wide association study of habitual physical activity in over 377,000 uk biobank participants identifies multiple variants including cadm2 and apoe. *International journal of obesity*, *42*(6), 1161–1176.

Kloosterman, M., Santema, B. T., Roselli, C., Nelson, C. P., Koekemoer, A., Romaine, S. P., Van Gelder, I. C., Lam, C. S., Artola, V. A., Lang, C. C., *et al.* (2020). Genetic risk and atrial fibrillation in patients with heart failure. *European Journal of Heart Failure*, *22*(3), 519–527.

Knoppers, B. M. (2014). Framework for responsible sharing of genomic and health-related data. *The HUGO journal*, *8*(1), 3.

Koch, S., Schmidtke, J., Krawczak, M., & Caliebe, A. (2023). Clinical utility of polygenic risk scores: A critical 2023 appraisal. *Journal of Community Genetics*, *14*(5), 471–487.

Kontou, P. I., & Bagos, P. G. (2024). The goldmine of gwas summary statistics: A systematic review of methods and tools. *BioData Mining*, *17*(1), 31.

Kramer, I., Hooning, M. J., Mavaddat, N., Hauptmann, M., Keeman, R., Steyerberg, E. W., Giardiello, D., Antoniou, A. C., Pharoah, P. D., Canisius, S., *et al.* (2020). Breast cancer polygenic risk score and contralateral breast cancer risk. *The American Journal of Human Genetics*, *107*(5), 837–848.

Kujala, I., Vangipurapu, J., Maaniitty, T., Saraste, A., Kere, J., & Knuuti, J. (2024). Polygenic risk scores in predicting coronary artery disease in symptomatic patients. a validation study. *Journal of Atherosclerosis and Thrombosis*, 64623.

Kunkle, B. W., Grenier-Boley, B., Sims, R., Bis, J. C., Damotte, V., Naj, A. C., Boland, A., Vronskaya, M., Van Der Lee, S. J., Amlie-Wolf, A., *et al.* (2019). Genetic meta-analysis of diagnosed alzheimer s disease identifies new risk loci and implicates aβ, tau, immunity and lipid processing. *Nature genetics*, *51*(3), 414–430.

Kurniansyah, N., Goodman, M. O., Kelly, T. N., Elfassy, T., Wiggins, K. L., Bis, J. C., Guo, X., Palmas, W., Taylor, K. D., Lin, H. J., *et al.* (2022). A multi-ethnic polygenic risk score is associated with hypertension prevalence and progression throughout adulthood. *Nature communications*, *13*(1), 3549.

Lambert, S. A., Abraham, G., & Inouye, M. (2019). Towards clinical utility of polygenic risk scores. *Human molecular genetics*, *28*(R2), R133–R142.

Lambert, S. A., Gil, L., Jupp, S., Ritchie, S. C., Xu, Y., Buniello, A., McMahon, A., Abraham, G., Chapman, M., Parkinson, H., *et al.* (2021). The polygenic score catalog as an open database for reproducibility and systematic evaluation. *Nature Genetics*, *53*(4), 420–425.

Lee, C.-L., Yamada, T., Liu, W.-J., Hara, K., Yanagimoto, S., & Hiraike, Y. (2024). Interaction between type 2 diabetes polygenic risk and physical activity on cardiovascular outcomes. *European Journal of Preventive Cardiology*, zwae075.

Lee, S. H., Goddard, M. E., Wray, N. R., & Visscher, P. M. (2012). A better coefficient of determination for genetic profile analysis. *Genetic epidemiology*, *36*(3), 214–224.

Lewis, A. C., Green, R. C., & Vassy, J. L. (2021). Polygenic risk scores in the clinic: Translating risk into action. *Human Genetics and Genomics Advances*, *2*(4).

Lewis, C. M., & Vassos, E. (2020). Polygenic risk scores: From research tools to clinical instruments. *Genome medicine*, *12*(1), 44.

Linz, D., Gawalko, M., Betz, K., Hendriks, J. M., Lip, G. Y., Vinter, N., Guo, Y., & Johnsen, S. (2024). Atrial fibrillation: Epidemiology, screening and digital health. *The Lancet Regional Health–Europe*, *37*.

Liu, X., Littlejohns, T. J., Bešević, J., Bragg, F., Clifton, L., Collister, J. A., Trichia, E., Gray, L. J., Khunti, K., & Hunter, D. J. (2024). Incorporating polygenic risk into the leicester risk assessment score for 10-year risk prediction of type 2 diabetes. *Diabetes & Metabolic Syndrome: Clinical Research & Reviews*, *18*(4), 102996.

Locke, A. E., Kahali, B., Berndt, S. I., Justice, A. E., Pers, T. H., Day, F. R., Powell, C., Vedantam, S., Buchkovich, M. L., Yang, J., *et al.* (2015). Genetic studies of body mass index yield new insights for obesity biology. *Nature*, *518*(7538), 197–206.

Lu, X., Liu, Z., Cui, Q., Liu, F., Li, J., Niu, X., Shen, C., Hu, D., Huang, K., Chen, J., *et al.* (2022). A polygenic risk score improves risk stratification of coronary artery disease: A large-scale prospective chinese cohort study. *European heart journal*, *43*(18), 1702–1711.

Lu, X., Niu, X., Shen, C., Liu, F., Liu, Z., Huang, K., Wang, L., Li, J., Hu, D., Zhao, Y., *et al.* (2021). Development and validation of a polygenic risk score for stroke in the chinese population. *Neurology*, *97*(6), e619–e628.

Łukasiewicz, S., Czeczelewski, M., Forma, A., Baj, J., Sitarz, R., & Stanisławek, A. (2021). Breast cancer—epidemiology, risk factors, classification, prognostic markers, and current treatment strategies—an updated review. *Cancers*, *13*(17), 4287.

Ma, Y., & Zhou, X. (2021). Genetic prediction of complex traits with polygenic scores: A statistical review. *Trends in Genetics*, *37*(11), 995–1011.

Ma, Y., Su, B., Li, D., Cui, F., Tang, L., Wang, J., Tian, Y., & Zheng, X. (2023). Air pollution, genetic susceptibility, and the risk of atrial fibrillation: A large prospective cohort study. *Proceedings of the National Academy of Sciences*, *120*(32), e2302708120.

Mabey, B., Hughes, E., Kucera, M., Simmons, T., Hullinger, B., Pederson, H. J., Yehia, L., Eng, C., Garber, J., Gary, M., *et al.* (2024). Validation of a clinical breast cancer risk assessment tool combining a polygenic score for all ancestries with traditional risk factors. *Genetics in Medicine*, 101128.

Machairoudia, G., Kazantzis, D., Chatziralli, I., Kampoli, K., Ntavatzikos, A., Theodossiadis, P., & Koumarianou, A. (2024). Prevalence and risk factors of dry eye disease in patients treated with aromatase inhibitors for breast cancer. *Clinical and Experimental Optometry*, *107*(4), 415–419.

Maj, C., Salvi, E., Citterio, L., Borisov, O., Simonini, M., Glorioso, V., Barlassina, C., Glorioso, N., Thijs, L., Kuznetsova, T., *et al.* (2022). Dissecting the polygenic basis of primary hypertension: Identification of key pathway-specific components. *Frontiers in cardiovascular medicine*, *9*, 814502.

Mandla, R., Schroeder, P., Porneala, B., Florez, J. C., Meigs, J. B., Mercader, J. M., & Leong, A. (2024). Polygenic scores for longitudinal prediction of incident type 2 diabetes in an ancestrally and medically diverse primary care physician network: A patient cohort study. *Genome Medicine*, *16*(1), 63.

Mars, N., Kerminen, S., Feng, Y.-C. A., Kanai, M., Läll, K., Thomas, L. F., Skogholt, A. H., della Briotta Parolo, P., Neale, B. M., Smoller, J. W., *et al.* (2022). Genome-wide risk prediction of common diseases across ancestries in one million people. *Cell Genomics*, *2*(4).

Mars, N., Koskela, J. T., Ripatti, P., Kiiskinen, T. T., Havulinna, A. S., Lindbohm, J. V., Ahola-Olli, A., Kurki, M., Karjalainen, J., Palta, P., *et al.* (2020). Polygenic and clinical risk scores and their impact on age at onset and prediction of cardiometabolic diseases and common cancers. *Nature medicine*, *26*(4), 549–557.

Mars, N., Lindbohm, J. V., della Briotta Parolo, P., Widén, E., Kaprio, J., Palotie, A., & Ripatti, S. (2022). Systematic comparison of family history and polygenic risk across 24 common diseases. *The American Journal of Human Genetics*, *109*(12), 2152–2162.

Mars, N., Widén, E., Kerminen, S., Meretoja, T., Pirinen, M., della Briotta Parolo, P., Palta, P., Palotie, A., Kaprio, J., *et al.* (2020). The role of polygenic risk and susceptibility genes in breast cancer over the course of life. *Nature communications*, *11*(1), 6383.

Marston, N. A., Pirruccello, J. P., Melloni, G. E., Koyama, S., Kamanu, F. K., Weng, L.-C., Roselli, C., Kamatani, Y., Komuro, I., Aragam, K. G., *et al.* (2023). Predictive utility of a coronary artery disease polygenic risk score in primary prevention. *JAMA cardiology*, *8*(2), 130–137.

Martin, A. R., Kanai, M., Kamatani, Y., Okada, Y., Neale, B. M., & Daly, M. J. (2019). Clinical use of current polygenic risk scores may exacerbate health disparities. *Nature genetics*, *51*(4), 584–591.

Mavaddat, N., Michailidou, K., Dennis, J., Lush, M., Fachal, L., Lee, A., Tyrer, J. P., Chen, T.-H., Wang, Q., Bolla, M. K., *et al.* (2019). Polygenic risk scores for prediction of breast cancer and breast cancer subtypes. *The American Journal of Human Genetics*, *104*(1), 21–34.

Mavaddat, N., Pharoah, P. D., Michailidou, K., Tyrer, J., Brook, M. N., Bolla, M. K., Wang, Q., Dennis, J., Dunning, A. M., Shah, M., *et al.* (2015). Prediction of breast cancer risk based on profiling with common genetic variants. *Journal of the National Cancer Institute, 107*(5), djv036.

McElligott, B., Shi, Z., Rifkin, A. S., Wei, J., Zheng, S. L., Helfand, B. T., Woo, J. S., & Xu, J. (2023). Assessing the performance of genetic risk score for stratifying risk of post-sepsis cardiovascular complications. *Frontiers in Cardiovascular Medicine, 10*, 1076745.

Mega, J. L., Stitziel, N. O., Smith, J. G., Chasman, D. I., Caulfield, M. J., Devlin, J. J., Nordio, F., Hyde, C. L., Cannon, C. P., Sacks, F. M., *et al.* (2015). Genetic risk, coronary heart disease events, and the clinical benefit of statin therapy: An analysis of primary and secondary prevention trials. *The Lancet, 385*(9984), 2264–2271.

Merino, J., Guasch-Ferré, M., Li, J., Chung, W., Hu, Y., Ma, B., Li, Y., Kang, J. H., Kraft, P., Liang, L., *et al.* (2022). Polygenic scores, diet quality, and type 2 diabetes risk: An observational study among 35,759 adults from 3 us cohorts. *PLoS medicine, 19*(4), e1003972.

Milne, R., & Antoniou, A. (2011). Genetic modifiers of cancer risk for brca1 and brca2 mutation carriers. *Annals of oncology, 22*, i11–i17.

Mishra, A., Malik, R., Hachiya, T., Jürgenson, T., Namba, S., Posner, D. C., Kamanu, F. K., Koido, M., Le Grand, Q., Shi, M., *et al.* (2022). Stroke genetics informs drug discovery and risk prediction across ancestries. *Nature, 611*(7934), 115–123.

Miyazawa, K., Ito, K., Ito, M., Zou, Z., Kubota, M., Nomura, S., Matsunaga, H., Koyama, S., Ieki, H., Akiyama, M., *et al.* (2023). Cross-ancestry genome-wide analysis of atrial fibrillation unveils disease biology and enables cardioembolic risk prediction. *Nature genetics, 55*(2), 187–197.

Mojarrad, N. G., Plotnikov, D., Williams, C., Guggenheim, J. A., *et al.* (2020). Association between polygenic risk score and risk of myopia. *JAMA ophthalmology, 138*(1), 7–13.

Monti, R., Eick, L., Hudjashov, G., Läll, K., Kanoni, S., Wolford, B. N., Wingfield, B., Pain, O., Wharrie, S., Jermy, B., *et al.* (2024). Evaluation of polygenic scoring methods in five biobanks shows larger variation between biobanks than methods and finds benefits of ensemble learning. *The American Journal of Human Genetics, 111*(7), 1431–1447.

Morgan, M. D., Pairo-Castineira, E., Rawlik, K., Canela-Xandri, O., Rees, J., Sims, D., Tenesa, A., & Jackson, I. J. (2018). Genome-wide study of hair colour in uk biobank explains most of the snp heritability. *Nature communications, 9*(1), 5271.

Morieri, M. L., Gao, H., Pigeyre, M., Shah, H. S., Sjaarda, J., Mendonca, C., Hastings, T., Buranasupkajorn, P., Motsinger-Reif, A. A., Rotroff, D. M., *et al.* (2018). Genetic tools for coronary risk assessment in type 2 diabetes: A cohort study from the accord clinical trial. *Diabetes Care, 41*(11), 2404–2413.

Mujwara, D., Henno, G., Vernon, S. T., Peng, S., Di Domenico, P., Schroeder, B., Busby, G. B., Figtree, G. A., & Bottà, G. (2022). Integrating a polygenic risk score for coronary artery disease as a risk-enhancing factor in the pooled cohort equation: A cost-effectiveness analysis study. *Journal of the American Heart Association, 11*(12), e025236.

Mulder, N., Abimiku, A., Adebamowo, S. N., de Vries, J., Matimba, A., Olowoyo, P., Ramsay, M., Skelton, M., & Stein, D. J. (2018). H3africa: Current perspectives. *Pharmacogenomics and personalized medicine*, 59–66.

Murea, M., Ma, L., & Freedman, B. I. (2012). Genetic and environmental factors associated with type 2 diabetes and diabetic vascular complications. *The review of diabetic studies: RDS, 9*(1), 6.

Namba, S., Saito, Y., Kogure, Y., Masuda, T., Bondy, M. L., Gharahkhani, P., Gockel, I., Heider, D., Hillmer, A., Jankowski, J., *et al.* (2023). Common germline risk variants impact somatic alterations and clinical features across cancers. *Cancer Research, 83*(1), 20–27.

Neagu, A.-N., Jayaweera, T., Corrice, L., Johnson, K., & Darie, C. C. (2024). Breast cancer exposomics. *Life*, *14*(3), 402.

Nguyen, T. V., & Eisman, J. A. (2020). Post gwas polygenic risk score: Utility and challenges. *Journal of Bone and Mineral Research Plus*, *4*(11), e10411.

Nyberg, T., Brook, M. N., Ficorella, L., Lee, A., Dennis, J., Yang, X., Wilcox, N., Dadaev, T., Govindasami, K., Lush, M., *et al.* (2023). Canrisk-prostate: A comprehensive, externally validated risk model for the prediction of future prostate cancer. *Journal of Clinical Oncology*, *41*(5), 1092–1104.

O Sullivan, J. W., Raghavan, S., Marquez-Luna, C., Luzum, J. A., Damrauer, S. M., Ashley, E. A., O'Donnell, C. J., Willer, C. J., & Natarajan, P. (2022). Polygenic risk scores for cardiovascular disease: A scientific statement from the american heart association. *Circulation*, *146*(8), e93–e118.

Oh, J. J., Kim, E., Woo, E., Song, S. H., Kim, J. K., Lee, H., Lee, S., Hong, S. K., & Byun, S.-S. (2020). Evaluation of polygenic risk scores for prediction of prostate cancer in korean men. *Frontiers in Oncology*, *10*, 583625.

Panagiotou, N., Bersimis, F., Fotis, A., Ntoumou, E., Salata, E., Kallistratos, M., Psychogios, A., & Sagonas, A. (2023). P273: Adjusted polygenic score: Translation of a new concept for cardiovascular disease prevention and management. *Genetics in Medicine Open*, *1*(1).

Patel, A. P., Wang, M., Ruan, Y., Koyama, S., Clarke, S. L., Yang, X., Tcheandjieu, C., Agrawal, S., Fahed, A. C., Ellinor, P. T., *et al.* (2023). A multi-ancestry polygenic risk score improves risk prediction for coronary artery disease. *Nature Medicine*, *29*(7), 1793–1803.

Peng, H., Wang, S., Wang, M., Wang, X., Guo, H., Huang, J., & Wu, T. (2023). Lifestyle factors, genetic risk, and cardiovascular disease risk among breast cancer survivors: A prospective cohort study in uk biobank. *Nutrients*, *15*(4), 864.

Peng, J.-Y., Lee, Y.-K., Pham, R.-Q., Shen, X.-H., Chen, I.-H., Chen, Y.-C., & Fan, H.-S. (2024). Trends and age-period-cohort effect on incidence of male breast cancer from 1980 to 2019 in taiwan and the usa. *Cancers*, *16*(2), 444.

Perkins, D. O., Olde Loohuis, L., Barbee, J., Ford, J., Jeffries, C. D., Addington, J., Bearden, C. E., Cadenhead, K. S., Cannon, T. D., Cornblatt, B. A., *et al.* (2020). Polygenic risk score contribution to psychosis prediction in a target population of persons at clinical high risk. *American Journal of Psychiatry*, *177*(2), 155–163.

Polderman, T. J., Benyamin, B., De Leeuw, C. A., Sullivan, P. F., Van Bochoven, A., Visscher, P. M., & Posthuma, D. (2015). Meta-analysis of the heritability of human traits based on fifty years of twin studies. *Nature genetics*, *47*(7), 702–709.

Privé, F., Aschard, H., Carmi, S., Folkersen, L., Hoggart, C., O'Reilly, P. F., & Vilhjálmsson, B. J. (2022). Portability of 245 polygenic scores when derived from the uk biobank and applied to 9 ancestry groups from the same cohort. *The American Journal of Human Genetics*, *109*(1), 12–23.

Qanbari, S. (2020). On the extent of linkage disequilibrium in the genome of farm animals. *Frontiers in genetics*, *10*, 1304.

Qassim, A., Souzeau, E., Hollitt, G., Hassall, M. M., Siggs, O. M., & Craig, J. E. (2021). Risk stratification and clinical utility of polygenic risk scores in ophthalmology. *Translational Vision Science & Technology*, *10*(6), 14–14.

Qin, M., Wu, Y., Fang, X., Pan, C., & Zhong, S. (2024). Polygenic risk score predicts all-cause death in east asian patients with prior coronary artery disease. *Frontiers in Cardiovascular Medicine*, *11*, 1296415.

Raben, T. G., Lello, L., Widen, E., & Hsu, S. D. (2023). Biobank-scale methods and projections for sparse polygenic prediction from machine learning. *Scientific Reports*, *13*(1), 11662.

Ralapanawa, U., & Sivakanesan, R. (2021). Epidemiology and the magnitude of coronary artery disease and acute coronary syndrome: A narrative review. *Journal of epidemiology and global health, 11*(2), 169–177.

Riveros-Mckay, F., Weale, M. E., Moore, R., Selzam, S., Krapohl, E., Sivley, R. M., Tarran, W. A., Sørensen, P., Lachapelle, A. S., Griffiths, J. A., *et al.* (2021). Integrated polygenic tool substantially enhances coronary artery disease prediction. *Circulation: Genomic and Precision Medicine, 14*(2), e003304.

Roberts, E., Howell, S., & Evans, D. G. (2023). Polygenic risk scores and breast cancer risk prediction. *The Breast, 67*, 71–77.

Roberts, R., Chavira, J., & Venner, E. (2022). Genetic risk and its role in primary prevention of cad. *Journal of Translational Genetics and Genomics, 6*(4).

Robinson, J. F., & Katsanis, N. (2010). *Oligogenic disease*. In *Vogel and motulsky's human genetics* (243–262). Springer.

Ruotsalainen, S. E., Partanen, J. J., Cichonska, A., Lin, J., Benner, C., Surakka, I., 33, E.-V. M. T. O. 3. H. M. 3. P. R., *et al.* (2021). An expanded analysis framework for multivariate gwas connects inflammatory biomarkers to functional variants and disease. *European Journal of Human Genetics, 29*(2), 309–324.

Sally, I., Kuo, C., & Aliev, F. (2024). *Polygenic risk scores and comparative genomics: Best practices and statistical considerations*. In *Rigor and reproducibility in genetics and genomics* (91–113). Elsevier.

Shi, M., O'Brien, K., & Weinberg, C. (2020). Interactions between a polygenic risk score and non-genetic risk factors in young-onset breast cancer. *Scientific reports, 10*(1), 3242.

Shieh, Y., Hu, D., Ma, L., Huntsman, S., Gard, C. C., Leung, J. W., Tice, J. A., Vachon, C. M., Cummings, S. R., Kerlikowske, K., *et al.* (2016). Breast cancer risk prediction using a clinical risk model and polygenic risk score. *Breast cancer research and treatment, 159*, 513–525.

Simmons, T., Hughes, E., Pruss, D., Kucera, M., Roa, B., Judkins, T., Younus, A., Cummings, S. A., Jammulapati, S., Timms, K., *et al.* (2024). Improving a polygenic risk score (prs) for breast cancer (bc) risk assessment in diverse ancestries.

Sipeky, C., Talala, K. M., Tammela, T. L., Taari, K., Auvinen, A., & Schleutker, J. (2020). Prostate cancer risk prediction using a polygenic risk score. *Scientific Reports, 10*(1), 17075.

Slatkin, M. (2008). Linkage disequilibrium understanding the evolutionary past and mapping the medical future. *Nature Reviews Genetics, 9*(6), 477–485.

Slunecka, J. L., van der Zee, M. D., Beck, J. J., Johnson, B. N., Finnicum, C. T., Pool, R., Hottenga, J.-J., de Geus, E. J., & Ehli, E. A. (2021). Implementation and implications for polygenic risk scores in healthcare. *Human genomics, 15*(1), 46.

Smith, J. L., Tcheandjieu, C., Dikilitas, O., Iyer, K., Miyazawa, K., Hilliard, A., Lynch, J., Rotter, J. I., Chen, Y.-D. I., Sheu, W. H.-H., *et al.* (2024). Multi-ancestry polygenic risk score for coronary heart disease based on an ancestrally diverse genome-wide association study and population-specific optimization. *Circulation: Genomic and Precision Medicine, 17*(3), e004272.

Song, M., Kwak, S. H., & Kim, J. (2024). Risk prediction and interaction analysis using polygenic risk score of type 2 diabetes in a korean population. *Scientific Reports, 14*(1), 6790.

Spears, C., Xu, M., Shoben, A., Dason, S., Toland, A. E., & Byrne, L. (2024). Clinical features of prostate cancer by polygenic risk score. *Familial Cancer*, 1–7.

Stark, B., Johnson, C., & Roth, G. A. (2024). Global prevalence of coronary artery disease: An update from the global burden of disease study. *Journal of the American College of Cardiology, 83*(13_Supplement), 2320–2320.

Sugrue, L. P., & Desikan, R. S. (2019). What are polygenic scores and why are they important? *Jama, 321*(18), 1820–1821.

Sun, Q., Crowley, C. A., Huang, L., Wen, J., Chen, J., Bao, E. L., Auer, P. L., Lettre, G., Reiner, A. P., Sankaran, V. G., *et al.* (2022). From gwas variant to function: A study of 148,000 variants for blood cell traits. *Human Genetics and Genomics Advances*, *3*(1).

Sun, X., Verma, S. P., Jia, G., Wang, X., Ping, J., Guo, X., Shu, X.-O., Chen, J., Derkach, A., Cai, Q., *et al.* (2024). Case-case genome-wide analyses identify subtype-informative variants that confer risk for breast cancer. *Cancer Research*, *84*(15), 2533–2548.

Tada, H., & Takamura, M. (2024). Assessment timings of polygenic risk score for atherosclerotic cardiovascular disease. *Journal of Atherosclerosis and Thrombosis*, *31*(7), 1029–1030.

Tanigawa, Y., Qian, J., Venkataraman, G., Justesen, J. M., Li, R., Tibshirani, R., Hastie, T., & Rivas, M. A. (2022). Significant sparse polygenic risk scores across 813 traits in uk biobank. *PLoS Genetics*, *18*(3), e1010105.

Taniguchi, S., Ninomiya, K., Kushima, I., Saito, T., Shimasaki, A., Sakusabe, T., Momozawa, Y., Kubo, M., Kamatani, Y., Ozaki, N., *et al.* (2020). Polygenic risk scores in schizophrenia with clinically significant copy number variants. *Psychiatry and clinical neurosciences*, *74*(1), 35–39.

The International Schizophrenia Consortium. (2009). Common polygenic variation contributes to risk of schizophrenia and bipolar disorder. *Nature*, *460*(7256), 748–752.

Tian, J., Zhang, M., Zhang, F., Gao, K., Lu, Z., Cai, Y., Chen, C., Ning, C., Li, Y., Qian, S., *et al.* (2024). Developing an optimal stratification model for colorectal cancer screening and reducing racial disparities in multi-center population-based studies. *Genome Medicine*, *16*(1), 81.

Timmins, I. R., Consortium, P., & Dudbridge, F. (2024). Bayesian approach to assessing population differences in genetic risk of disease with application to prostate cancer. *Plos Genetics*, *20*(4), e1011212.

Torkamani, A., Wineinger, N. E., & Topol, E. J. (2018). The personal and clinical utility of polygenic risk scores. *Nature Reviews Genetics*, *19*(9), 581–590.

Toulopoulou, T., Zhang, X., Cherny, S., Dickinson, D., Berman, K. F., Straub, R. E., Sham, P., & Weinberger, D. R. (2019). Polygenic risk score increases schizophrenia liability through cognition-relevant pathways. *Brain*, *142*(2), 471–485.

Truong, B., Hull, L. E., Ruan, Y., Huang, Q. Q., Hornsby, W., Martin, H., Van Heel, D. A., Wang, Y., Martin, A. R., Lee, S. H., *et al.* (2024). Integrative polygenic risk score improves the prediction accuracy of complex traits and diseases. *Cell Genomics*, *4*(4).

Uffelmann, E., Huang, Q. Q., Munung, N. S., De Vries, J., Okada, Y., Martin, A. R., Martin, H. C., Lappalainen, T., & Posthuma, D. (2021). Genome-wide association studies. *Nature Reviews Methods Primers*, *1*(1), 59.

Vinuesa, C. G., Shen, N., & Ware, T. (2023). Genetics of sle: Mechanistic insights from monogenic disease and disease-associated variants. *Nature Reviews Nephrology*, *19*(9), 558–572.

Visscher, P. M., & Goddard, M. E. (2019). From ra fisher s 1918 paper to gwas a century later. *Genetics*, *211*(4), 1125–1130.

Visscher, P. M., Wray, N. R., Zhang, Q., Sklar, P., McCarthy, M. I., Brown, M. A., & Yang, J. (2017). 10 years of gwas discovery: Biology, function, and translation. *The American Journal of Human Genetics*, *101*(1), 5–22.

Wang, G., Zhao, D., Spring, D. J., & DePinho, R. A. (2018). Genetics and biology of prostate cancer. *Genes & development*, *32*(17-18), 1105–1140.

Wang, J. (2024). Biostatistical challenges in high-dimensional data analysis: Strategies and innovations. *Computational Molecular Biology*, *14*.

Wang, M., Menon, R., Mishra, S., Patel, A. P., Chaffin, M., Tanneeru, D., Deshmukh, M., Mathew, O., Apte, S., Devanboo, C. S., *et al.* (2020). Validation of a genome-wide polygenic score for coronary artery disease in south asians. *Journal of the American College of Cardiology*, *76*(6), 703–714.

Wang, S.-H., Huang, Y.-C., Cheng, C.-W., Chang, Y.-W., & Liao, W.-L. (2024). Impact of the trans-ancestry polygenic risk score on type 2 diabetes risk, onset age and progression among population in taiwan. *American Journal of Physiology-Endocrinology and Metabolism*, *326*(5), E547–E554.

Wang, X., Khurshid, S., Choi, S. H., Friedman, S., Weng, L.-C., Reeder, C., Pirruccello, J. P., Singh, P., Lau, E. S., Venn, R., *et al.* (2023). Genetic susceptibility to atrial fibrillation identified via deep learning of 12-lead electrocardiograms. *Circulation: Genomic and Precision Medicine*, *16*(4), 340–349.

Wang, Y., Namba, S., Lopera, E., Kerminen, S., Tsuo, K., Läll, K., Kanai, M., Zhou, W., Wu, K.-H., Favé, M.-J., *et al.* (2023). Global biobank analyses provide lessons for developing polygenic risk scores across diverse cohorts. *Cell Genomics*, *3*(1).

Wang, Y., Tsuo, K., Kanai, M., Neale, B. M., & Martin, A. R. (2022). Challenges and opportunities for developing more generalizable polygenic risk scores. *Annual review of biomedical data science*, *5*(1), 293–320.

Wolc, A., & Dekkers, J. C. (2022). Application of bayesian genomic prediction methods to genome-wide association analyses. *Genetics Selection Evolution*, *54*(1), 31.

Wong, C. K., Makalic, E., Dite, G. S., Whiting, L., Murphy, N. M., Hopper, J. L., & Allman, R. (2022). Polygenic risk scores for cardiovascular diseases and type 2 diabetes. *Plos one*, *17*(12), e0278764.

Wray, N. R., Lin, T., Austin, J., McGrath, J. J., Hickie, I. B., Murray, G. K., & Visscher, P. M. (2021). From basic science to clinical application of polygenic risk scores: A primer. *JAMA psychiatry*, *78*(1), 101–109.

Wright, A. F. (2001). Genetic variation: Polymorphisms and mutations. *e LS*.

Xia, C., Xu, Y., Li, H., He, S., & Chen, W. (2024). Benefits and harms of polygenic risk scores in organised cancer screening programmes: A cost-effectiveness analysis. *The Lancet Regional Health–Western Pacific*, *44*.

Xiang, R., Kelemen, M., Xu, Y., Harris, L. W., Parkinson, H., Inouye, M., & Lambert, S. A. (2024). Recent advances in polygenic scores: Translation, equitability, methods and fair tools. *Genome Medicine*, *16*(1), 33.

Xu, J., Lange, E. M., Lu, L., Zheng, S. L., Wang, Z., Thibodeau, S. N., Cannon-Albright, L. A., Teerlink, C. C., Camp, N. J., Johnson, A. M., *et al.* (2013). Hoxb13 is a susceptibility gene for prostate cancer: Results from the international consortium for prostate cancer genetics (icpcg). *Human genetics*, *132*, 5–14.

Yamada, Y., Yasukochi, Y., Kato, K., Oguri, M., Horibe, H., Fujimaki, T., Takeuchi, I., & Sakuma, J. (2018). Identification of 26 novel loci that confer susceptibility to early-onset coronary artery disease in a japanese population. *Biomedical reports*, *9*(5), 383–404.

Yanes, T., Tiller, J., Haining, C. M., Wallingford, C., Otlowski, M., Keogh, L., McInerney-Leo, A., & Lacaze, P. (2024). Future implications of polygenic risk scores for life insurance underwriting. *NPJ Genomic Medicine*, *9*(1), 25.

Yang, J., Benyamin, B., McEvoy, B. P., Gordon, S., Henders, A. K., Nyholt, D. R., Madden, P. A., Heath, A. C., Martin, N. G., Montgomery, G. W., *et al.* (2010). Common snps explain a large proportion of the heritability for human height. *Nature genetics*, *42*(7), 565–569.

Yang, Y., Tao, R., Shu, X., Cai, Q., Wen, W., Gu, K., Gao, Y.-T., Zheng, Y., Kweon, S.-S., Shin, M.-H., *et al.* (2022). Incorporating polygenic risk scores and nongenetic risk factors for breast cancer risk prediction among asian women. *JAMA network open*, *5*(3), e2149030–e2149030.

Ye, Y., Chen, X., Han, J., Jiang, W., Natarajan, P., & Zhao, H. (2021). Interactions between enhanced polygenic risk scores and lifestyle for cardiovascular disease, diabetes, and lipid levels. *Circulation: Genomic and Precision Medicine*, *14*(1), e003128.

Yiangou, K., Kyriacou, K., Kakouri, E., Marcou, Y., Panayiotidis, M. I., Loizidou, M. A., Hadjisavvas, A., & Michailidou, K. (2021). Combination of a 15-snp polygenic risk score and classical risk factors for the prediction of breast cancer risk in cypriot women. *Cancers, 13*(18), 4568.

Zeinali-Nezhad, N., Najafipour, H., Shadkam, M., & Pourhamidi, R. (2024). Prevalence and trend of multiple coronary artery disease risk factors and their 5-year incidence rate among adult population of kerman: Results from kercadr study. *BMC public health, 24*(1), 25.

Zhang, J., Chen, G., Wang, C., Wang, X., Qian, Z., Cai, M., Vaughn, M. G., Bingheim, E., Li, H., Gao, Y., *et al.* (2023). Associations of risk factor burden and genetic predisposition with the 10-year risk of atrial fibrillation: Observations from a large prospective study of 348,904 participants. *BMC medicine, 21*(1), 88.

Zhang, X., Rice, M., Tworoger, S. S., Rosner, B. A., Eliassen, A. H., Tamimi, R. M., Joshi, A. D., Lindstrom, S., Qian, J., Colditz, G. A., *et al.* (2018). Addition of a polygenic risk score, mammographic density, and endogenous hormones to existing breast cancer risk prediction models: A nested case–control study. *PLoS medicine, 15*(9), e1002644.

Zhao, J., Salter-Townshend, M., & O'Hagan, A. (2023). A simulation study for multifactorial genetic disorders to quantify the impact of polygenic risk scores on critical illness insurance. *European Actuarial Journal, 13*(2), 775–813.

Zirpoli, G. R., Pfeiffer, R. M., Bertrand, K. A., Huo, D., Lunetta, K. L., & Palmer, J. R. (2024). Addition of polygenic risk score to a risk calculator for prediction of breast cancer in us black women. *Breast Cancer Research, 26*(1), 2.

Zou, Y., Zhu, J., Song, C., Li, T., Wang, K., Shi, J., Ye, H., & Wang, P. (2024). A polygenetic risk score combined with environmental factors better predict susceptibility to hepatocellular carcinoma in chinese population. *Cancer Medicine, 13*(9), e7230.

7 Improving Healthcare Services using Machine Learning Techniques

Antonios Karaminas
Department of Business Administration, University of Piraeus, Greece

Polychronis Economou
Department of Civil Engineering, University of Patras, Greece
E-Mail: peconom@upatras.gr. ORCID: 0000-0001-6452-5920.

Sotiris Bersimis
Department of Business Administration, University of Piraeus, Greece
E-Mail: sbersim@unipi.gr. ORCID: 0000-0002-7448-1172.

7.1 INTRODUCTION

The science of statistics has been applied for decades in all sectors of society and is constantly evolving following the needs of each period. One of the fields where statistics has a special application with significant results is the field of health.

The first systematic attempt to collect data for analysis seems to be in the far 1662, with the Bills of Mortality, the weekly mortality statistics in London, designed to monitor burials, which kept track of London's birth and death rates (Brend, 1907). By analyzing this data in 1662, John Graunt, an English statistician who is considered the father of demography, made pioneering contributions to the fields of epidemiology and demography (Bellhouse, 1998). He examined mortality data from London, laying the foundation for the methodical gathering and examination of data pertaining to health.

Although health statistics have been collected and analyzed for centuries, the field was first officially formalized as a scientific discipline in the 19th century. An important development in the global cooperation on statistical issues, particularly health statistics, was the 1853 International Statistical Congress, which was held in Brussels. International guidelines for the gathering and dissemination of essential statistics were established as a result of this congress(Randeraad, 2011). William Farr, a statistician and epidemiologist of the 19th century from Britain, was instrumental in the creation of health statistics. He assisted in the creation of vital statistics systems and worked on the registration of births, deaths, and illnesses (Langmuir, 1976).

In the 20th century, and more specifically in 1948, the World Health Organization was founded, being essential to the statistics on global health (McCarthy, 2002). In parallel, health statistics were collected, stored, and analyzed with increasing sophistication with the introduction of computers and developments in data processing technologies. Tools for data analytics, data warehouses, and electronic health records (EHRs) have all improved the field.

During this period, the first models in the health sector also began to appear. The first and one of the most famous models to date is the Kaplan-Meier estimator. Edward L. Kaplan and Paul

DOI: 10.1201/9781032664330-7

Meier developed the Kaplan-Meier estimator in 1958, a technique for calculating the survival function from partial data, especially in cases where time-to-event outcomes are involved, which is widely used for survival analysis in medical and health-related research (Dudley *et al.*, 2016; Rich *et al.*, 2010). Later, another well-known model was created. David R. Cox introduced the Cox proportional hazards model, which was first published in 1972. Since its introduction, Cox's model for analyzing survival data has grown to be one of the most popular statistical techniques in the fields of epidemiology, clinical research, and many others (Kalbfleisch & Schaubel, 2023).

In the latter half of the 20th century, data analysis was dominated by conventional statistical techniques like analysis of variance and linear regression. However, the phrase machine learning (ML) began to take off as a more general term that covered a range of methods and algorithms outside of traditional statistics. ML gained momentum in the 21st century due to the explosion of big data and improvements in computational power (Cleophas, Zwinderman, *et al.*, 2015). In addition, when compared to traditional statistical methods, ML offers the advantages of flexibility and scalability (Rajula *et al.*, 2020).

ML and statistical learning are two closely related fields that are frequently used interchangeably. The main goal of ML is to create algorithms that let computers analyze, learn from, and forecast data. To do this, massive datasets and powerful computing power are frequently used to reveal intricate patterns and structures (Alpaydin, 2020). Statistical learning, on the other hand, focuses on applying statistical concepts to infer correlations between variables, frequently with the goal of developing understandable models that clarify the underlying procedures of data collection (James, 2013). Combining statistical learning's focus on model interpretability and theoretical foundations with ML's ability to handle large and complicated datasets can result in strong, reliable solutions that not only function well but also offer insightful information about the data.

Nowadays, the use of big data has played a crucial role in the development of economies, being at the centre of all modern industries. As the authors of the study Chen *et al.*, 2012 mentioned, "the Big Data era has quietly descended on many communities, from governments and e-commerce to health organizations".

Big data are characterized by eight components, or the 8 Vs: volume, velocity, variety, veracity, validity, volatility, value, and visualization. The enormous amount of data that is extracted every second comprises the Volume of Big Data. The fast rate at which new data is being generated defines Velocity. Variety is about all the available distinct types of data, such as pictures, text, medical records, etc. Veracity signifies the importance of data. Validity refers to the data's accuracy for the intended use. The duration for which the data must be retained is referred to as Volatility. Value is concentrated on knowledge extraction, decision-making, analytics, and statistical techniques. Lastly, the depiction of data from almost any category in a graphical format that is simple to read and comprehend is known as big data Visualization (Ahsaan & Mourya, 2019). At the same time, beyond the term of big data, there is also the term of high-dimensional data. The term high-dimensional data pertains to datasets that possess a significant number of features, or dimensions, in comparison to the total number of observations (Donoho *et al.*, 2000). This can result in problems such as the curse of dimensionality. High-dimensional data addresses the complexity brought on by several variables, whereas Big Data concentrates on the sheer amount of the dataset. Both necessitate distinct analytical techniques, however a dataset may simultaneously display both attributes (L'heureux *et al.*, 2017).

The proper use of big data could offer numerous advantages and extra value to industries. Decision support systems, needs discovery, and innovative products and services are some of the big data's benefits (Manyika *et al.*, 2011). The Healthcare industry is among all the industries where big data has been a valuable commodity. As it is noted by Dash *et al.*, 2019 the healthcare industry is producing data at such a rate that paves the way for significant results in its field. According to another study (Qayyum *et al.*, 2021), the four main areas of healthcare where ML techniques can be applied are clinical workflow, diagnosis, treatment, and prognosis. Indeed, several ML models based on health care have been developed in recent years, providing significant advantages. These models improved the management of healthcare services, managed to increase the quality provided while decreasing costs and expenditures at the same time, and, last but not least, led to more innovative and efficient research and development processes.

This study is a literature review of ML techniques in the healthcare industry. As there are plenty of different areas in which ML techniques could be applied, this study is focused on the prediction of no-shows in healthcare organizations and on the prediction of hospital readmissions. This study differs from other similar studies since it focuses on the collection and listing of models that have been developed from 2019 to 2023 and therefore consist of state-of-the-art techniques and algorithms. Another significant difference is that emphasis is placed on the disease. It is evident that the disease largely determines both the final ML model and its critical parameters (variables). The models of hospital readmissions are divided into categories regarding the disease. For example, there are models concentrated on hospital readmission for patients who suffer from heart failure (HF) or diabetes disease. Furthermore, this study documents the most suitable techniques for each reviewed model and records the most impactful features for the prediction of the outcomes.

As this study summarizes a long list of different algorithms that were used to create the following described models, could be beneficial to give a brief overview of the existing ML algorithms. Algorithms mainly are divided into two main categories, supervised and unsupervised. A definition of ML was given by Singh *et al.*, 2016 as "the construction of algorithms that are able to produce general patterns and hypotheses by using externally supplied instances to predict the fate of future instances". According to Ozgur, 2004, the main difference between supervised and unsupervised learning algorithms is that for supervised algorithms, a labeled training dataset is required, while for unsupervised, labeled documents are not necessary. Figure 7.1 shows indicative existing data analytics algorithms categorized as supervised and unsupervised and based on the type of the target variables, continuous or categorical. Most of these algorithms were used in the explored literature review of this study. Here we have to note that there are also other categories of models such as semi-supervised models and reinforcement learning models.

Furthermore, every ML algorithm hides simple or complex mathematical equations that are useful for the user to understand for the best use of these algorithms. For example, the study Biau, 2012 presented an analytical overview of the random forest (RF) algorithms explaining the mathematical and statistical properties of the methodology in detail. Also, Li, Fu & Li, 2020 implemented the extreme gradient boosting (XGBoost) algorithm for the creation of a model interpreting at the same time how it works and presenting its mathematical background. As well, Agatonovic-Kustrin & Beresford, 2000 demonstrated and explicated the Artificial Neural Networks (ANN) algorithms and their mathematical equations through their model.

The rest part of this paper is divided into five sections. Section 7.2 is a brief overview of the no-shows and hospital readmission issues. Section 7.3, includes the collecting methods of the articles that constituted in this paper and highlights in a graphic way the total number of found articles between 2019 and 2023, by country and category, and, finally, depicts the total number of published articles over this period. Section 7.4 is an extended literature review, and it is an overview of the explored articles. Section 7.5, summarizes the key findings, results, and future recommendations. Lastly, in Section 7.6 the main conclusions are presented.

7.2 OVERVIEW OF NO-SHOWS AND HOSPITAL READMISSIONS ISSUES

This study is an extensive literature review of the ML techniques used for addressing two crucial issues in the healthcare industry. More precisely, ML models aim to improve the appointment system by predicting the no-shows and to decrease the readmission rate in hospitals by detecting the key features. The reasons that this study focuses on these two issues are the significant impacts that these have, firstly, on the provided quality and, secondly, on the total healthcare expenditures that burden the healthcare system and, consequently, the citizens.

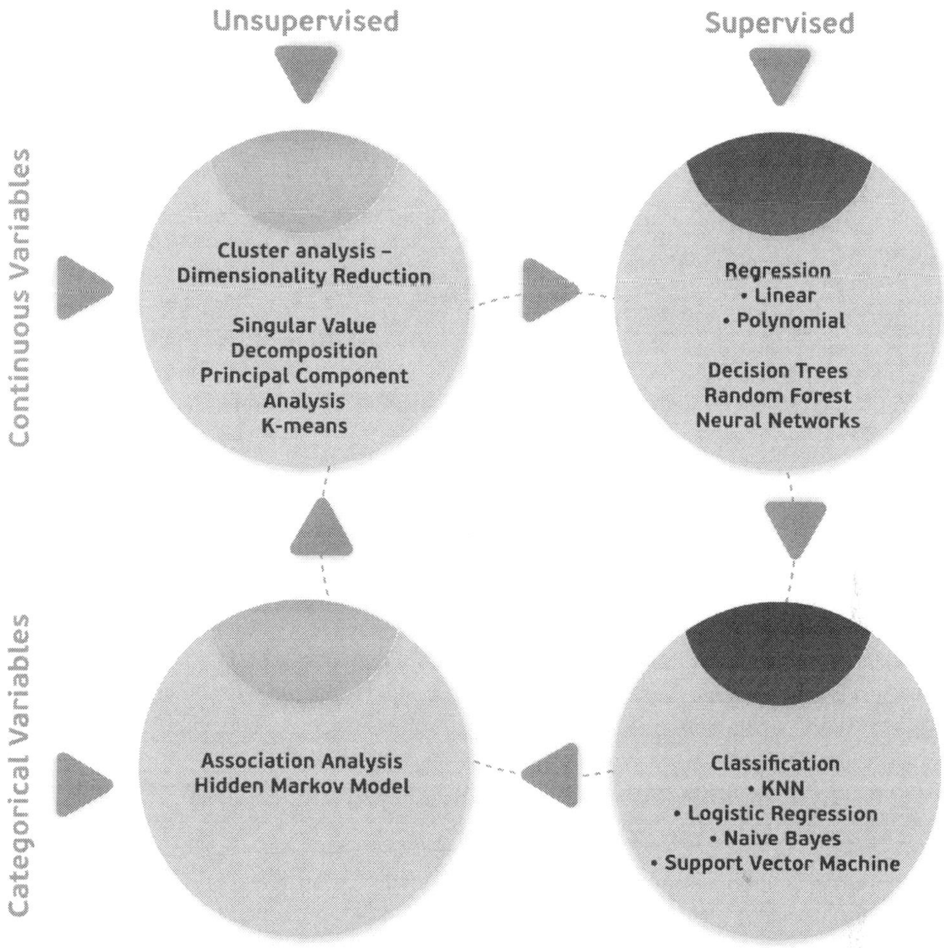

Figure 7.1 An overview of the algorithms used in the explored literature review

One of the essential issues medical centres, clinics, and hospitals have to deal with is no-shows and last-minute cancellations. According to AlMuhaideb *et al.*, 2019, the no-show term refers to the inability of a patient to attend a scheduled appointment or to cancel it. Also, they commented that the high rate of no-shows should be taken into consideration as it causes wasted capacity, under-utilization of expensive resources, and barriers to patient access. No-shows and late appointment cancellations were characterized by Ahmadi *et al.*, 2019 as common issues that negatively impact a healthcare organization's financial performance and quality of service.

Moreover, no-shows and last-minute cancellations completely deregulate the appointment planning system. As Srinivas & Ravindran, 2018 state, an insufficient appointment system decreases productivity and timely care access annually. An efficient appointment system should provide firstly an appointment, if not the desired date of the patients, in a short interval of time. Secondly, it could manage the appointments in a way that, on a daily basis, neither will there be large gaps between appointments nor will the system have to be overloaded, leading to massive delays (Liu *et al.*, 2010). In an attempt to account for no-shows, clinics frequently overbook, scheduling more patients than they can handle, which causes unforeseen delays and schedule disruptions. Rearranging a patient's appointment when they miss one is a challenging task for healthcare organizations (Jeffin *et al.*, 2022).

Therefore, it is crucial to comprehend and anticipate no-shows to adopt a proactive strategy (Nasir *et al.*, 2020). By predicting a patient's likelihood of not showing up, healthcare organizations can effectively deploy mitigation techniques (Alabdulkarim *et al.*, 2022). Determining the patient's profile and forecasting probable absences is the primary issue of no-show problem (Salazar *et al.*, 2021).

Frequent hospital readmissions are another aspect that provokes numerous negative effects on the healthcare industry, especially in hospitals. Has been observed that an essential percentage of patients suffering from any kind of disease have to be rehospitalized in the next 30 days after their discharge. Readmissions to hospitals are frequent and can provoke a variety of negative effects, including a higher chance of mortality and higher expenses (Hung, Li, *et al.*, 2020; Mišić *et al.*, 2020).

That phenomenon probably indicates low-quality hospital services, increasing the total expenditures of the healthcare industry at the same time. Davis *et al.*, 2022 mention that readmissions approximately cost the Canadian healthcare industry 2$ billion and the United States healthcare industry 26$ billion every year. Additionally, patients with HF are prone to repeated hospital admissions despite advancements in medical and percutaneous therapies. The average cost of each hospitalization is estimated to be $23,000, and overall yearly costs are expected to reach $70 billion by 2030 (Beecy *et al.*, 2020).

As a gauge of hospital quality and cost containment, hospitals throughout the United States are still being pressured to lower their readmission rates (Mahmoudi *et al.*, 2020). However, a significant amount of the massive annual costs incurred due to readmissions may have been avoided with the application of ML techniques.

7.3 LITERATURE REVIEW METHODOLOGY AND DESCRIPTIVE STATISTICS

This section describes precisely the methodology for the collection of the reviewed articles. It includes all the search criteria that led to the final selection of the articles, such as keywords, time range, and language. Furthermore, the present section encompasses some flowcharts that describe the whole process and depicts some insights into the origin of the articles and the total number of published articles per year.

The search queries for the no-show articles were constructed using the below-mentioned keywords in combination with the "AND" Boolean Operator. More precisely, the search criteria were "Machine Learning" AND "no-shows" AND "Healthcare". Similarly, the search queries for the hospital readmission articles were constructed, with the search criteria being "Machine Learning" AND "Prediction" AND "Hospital Readmissions". The articles were retrieved from Scopus. Figure 7.2 illustrates step by step the whole process of the selected articles for every issue, no-show, and hospital readmission. Also, the search results were limited to English-language articles published between 2019 and 2023 for the no-shows aspect and from 2020 to 2023 for the hospital readmissions aspect. The reason behind the choice of the selected time period for each aspect lies in the fact that there is a sudden increase in published articles and, consequently, in existing interest compared to previous years. Moreover, the study tried to capture state-of-the-approaches that handle the above-described cases. Hence, the selection of articles published in the most recent years was essential. The elimination of some articles was carried out either because the content of the article was not relevant to the topic of the specific study or because there was no free access.

In addition, Figure 7.3 and Figure 7.4 illustrate the nations with the highest interest in every issue in terms of the total number of publications and the total published articles per year in the time period that has been selected, respectively. More precisely, Figure 7.3 indicates that the United States has published the most number of articles, with Saudi Arabia and China coming in second

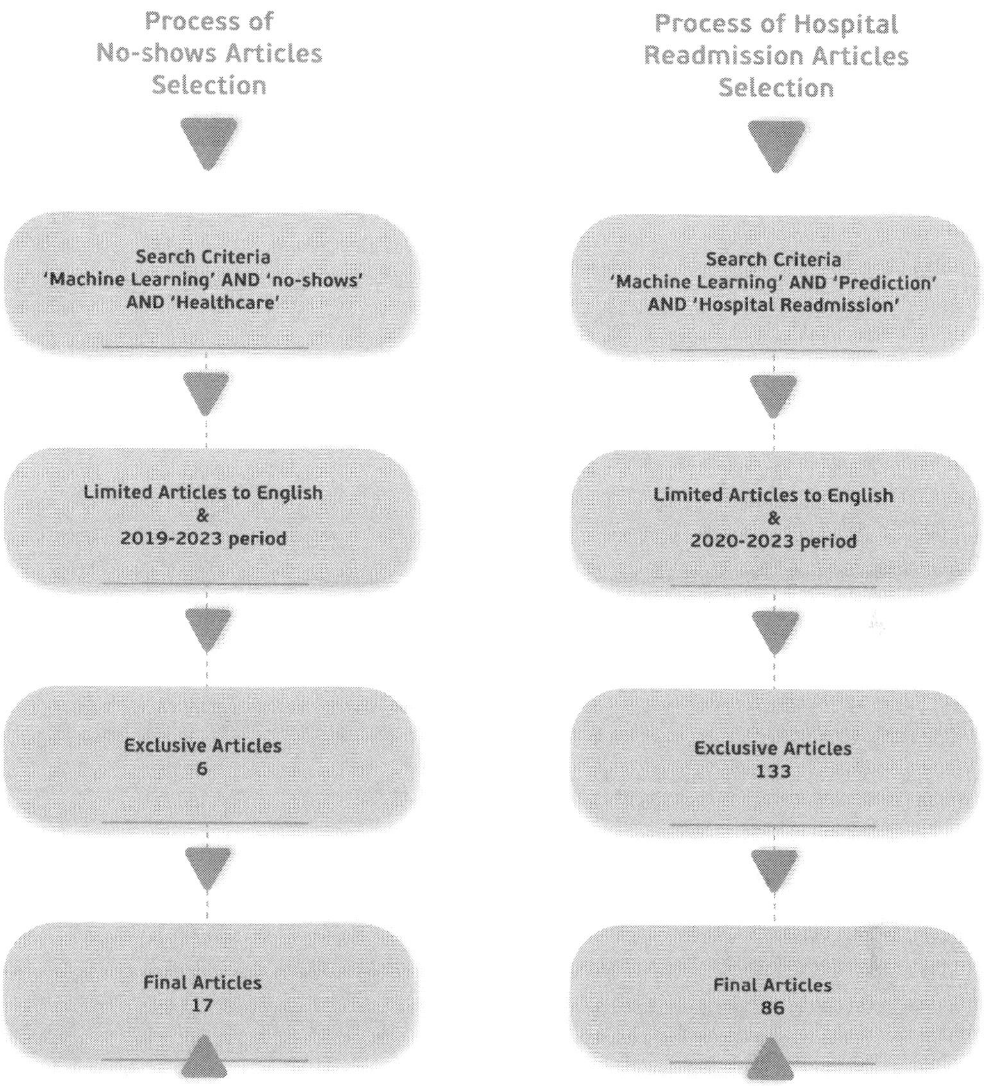

Figure 7.2 The process of the articles collection

for No-Shows and Hospital Readmissions correspondingly. Figure 7.4, depicts that the number of the annual publications for No-Shows is not steady, having its peak in 2022 with 8 publications. In contrast, the number of publications of Hospital Readmissions seems to have a continuous upward trend until 2022 with more than 70 publications in that year.

7.4 LITERATURE REVIEW

This section includes an overview of various models that have been developed in recent years in the healthcare industry that aim to solve two specific issues that plague the healthcare industry. In particular, various models are mentioned and described regarding the prediction of no-shows and the prediction of possible hospital readmissions. Consequently, this section is divided into two main

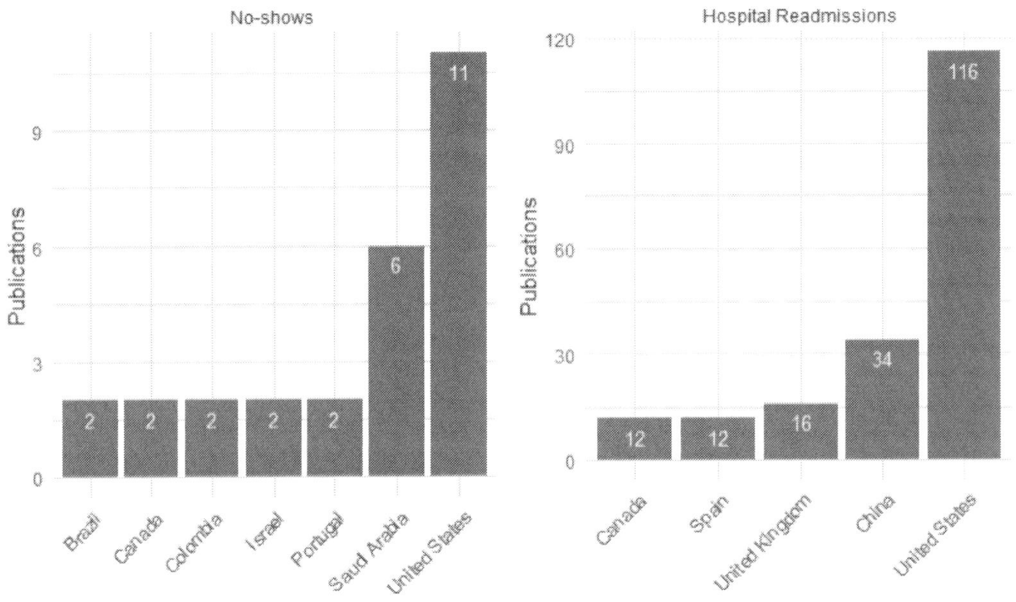

Figure 7.3 Top nations with the highest interest in hospital readmissions and no-shows.

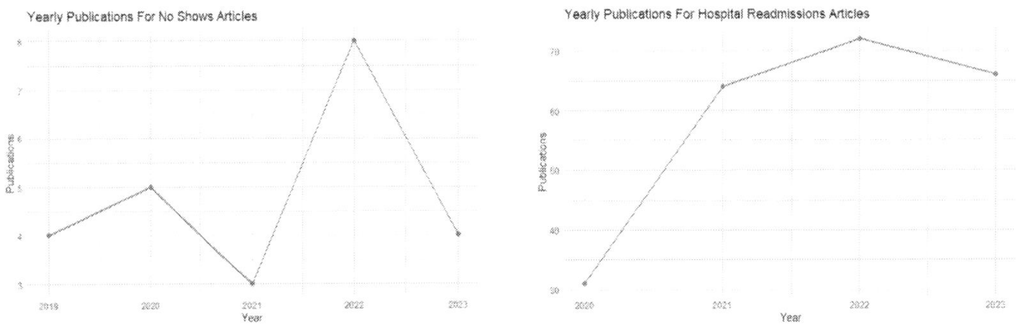

Figure 7.4 The number of articles published per field of interest per year, from 2019-2023

subsections. The first one describes the predictive models for no-shows issue, and the second one the predictive models for hospital readmissions.

7.4.1 LITERATURE REVIEW ON PREDICTION OF NO-SHOWS IN HEALTHCARE INDUSTRY CATEGORIZED ACCORDING TO ML MODEL

It has been observed that there is a significant percentage of people who, while they have made an appointment at a health facility, for various reasons, do not show up at their appointment without any notification. This often leads to the inability of health units to cover these time gaps that are created in their program on a daily basis. Consequently, a vicious circle of various problems is created.

Some people who are looking for an available appointment are hard to find or find one much later than their desired date, even if it is urgent. This happens because, in some exams, the balance between demand and supply leans significantly in the demand direction. However, at the same time, new appointment slots are opened every day, which are practically impossible to fill at the last minute since there is no relevant and timely notice. In parallel, the health units have resources that remain idle either for long or for short periods of time.

The above-described situation leads to the increase in the costs of an organization due to the employment of resources for periods of time when they do not produce any work and to the parallel decrease in their income due to the non-appearance of clients and the inability to cover the gaps with a new appointment. For these reasons, several prediction models of possible no-shows have been created by ML techniques in recent years. The following paragraphs describe the existing models in the literature.

The majority of the presented models are based on RF algorithms. Furthermore, plenty of the models have been implemented by ANN, and Gradient Boosting (GBM) algorithms. Light Gradient Boosting Machine (Light GBM), LR, bagging, balanced random, and Opposition-based Self-Adaptive Cohort Intelligence (OSACI) algorithms have also been used for the creation of models for no-show prediction. The following paragraphs present the reviewed articles, categorized according to the ML technique used to develop each model.

7.4.1.1 Random Forest

The authors of the study Ahmadi *et al.*, 2019 proposed a predictive model for no-shows and late appointment cancellations to assist in overbooking decisions. RF seems to perform better among the classifier algorithms used (Decision Trees (DT), RF, and Naïve Bayes). Simultaneously, a stacking or meta-ensemble model that they constructed based on different subsets of the features of a particular prediction method performed better than a single base classifier. Also, among all the available features, lead time, appointment type, and history of no-shows over the six months before the appointment date were shown to be the most crucial for the predictive outcome. Moreover Nasir *et al.*, 2020 demonstrated another model for the prediction of no-shows. Their approach took into account four classifiers and four different ways of data balancing, and in parallel, they used an error-cost-based threshold among the candidate model to identify the best harmony between specificity and sensitivity. The tested algorithms were ANN, support vector machine (SVM), RF, and LR, while they with imbalanced data using RUS (proportional depreciation/component), SMOTE (Synthetic Minority Over-sampling TEchnique), ADASYN (Adaptive Synthetic), and MWMOTE (Majority Weighted Minority Oversampling TEchnique) balancing methods. The most accurate and effective model in terms of offering a fair trade-off between identifying positive and negative cases was the RF model without balancing. The key factor for all used algorithms was the prior no-show, followed by age, waiting time, and trust score-based variables. Lastly, they remarked that none of the comorbidities significantly affect the likelihood of no-shows.

In addition, Salazar *et al.*, 2021 sought to determine the potential causes of patients' absenteeism and, most importantly, to forecast the probability that a particular patient would miss their scheduled appointments. They examined three ML algorithms: LR, DT, and RF, with the last one outperforming the others. They are limited to the execution of only these three approaches as these three techniques are the most appropriate for the used dataset, as the authors mentioned. Additionally, the results illustrate that the variables, appointment month, and whether or not the patient was contacted have the highest impact on the prediction of the outcomes.

Another study (Alabdulkarim *et al.*, 2022) focused on the prediction of no-shows for dental appointments. They carried out three models based on different ML techniques, but RF outperformed evaluation metrics compared to the other two algorithms that were utilized, LR and GBM. Lead time emerged as the most important feature after the implementation of an ablation test.

7.4.1.2 Gradient Boosting

A GBM algorithm used by Alshammari *et al.*, 2022 as their model's classifier. DT, RF, LR, and neural networks (NN) were implemented, too, but the evaluation metrics were not superior to GBM.

Their analysis revealed the distance between the patient's residence and the medical centre, the increased number of days between booking an appointment and the scheduled day, clinical medical services, and appointment time and day as the most crucial features of the model.

Plenty of different ML techniques were tested for the creation of another prediction no-shows model (Hamdan & Bakar, 2023). LR, DT, RF, GBM, Naive Bayes, K-nearest neighbor (KNN), and multilayer perceptron (MLP) techniques were applied. Additionally, multiple split data ratios were tested. In all cases, GBM had the highest performance according to the results of the evaluation metrics. The authors mentioned that the dataset's imbalance of class distribution has little effect on the boosting technique. Furthermore, the appointment month and the age variables were shown to have a significant impact on the prediction feature.

Similarly, the GBM algorithm yielded better results in their sturdy (Daghistani *et al.*, 2020). The rest of the candidate algorithms, in that case, were RF, LR, SVM, and MLP. Hold-out and tenfold cross-validation methods were used for the evaluation of the models. The study's findings suggest that addressing the lead-time factor might diminish outpatient no-show rates.

In addition, Jeffin *et al.*, 2022 is another model based on GBM which outperformed the results of LR, DT, RF, and linear SVM. The study used a variety of variables, including demographics, appointment features, and clinical traits like alcoholism and handicap, without being able to determine which of these had the biggest influence on the prediction outcome.

Models for the prediction of no-shows were developed, taking into consideration the COVID-19 impact on appointments too. The COVID-19 pandemic has made it more difficult for healthcare systems to handle late or missed appointments, which has resulted in inefficient use of medical resources and higher healthcare costs. An extensive examination of no-shows for appointments made in 2020 for breast imaging in an Israel medical network was carried out (Ozery-Flato *et al.*, 2021). When the COVID-19 and pre-COVID-19 periods were compared, it was found that the COVID-19 no-show rate had significantly increased. This research is unique in that it is the first to integrate pandemic-associated predictors with pre-pandemic known predictors of no-show. The GBM algorithm was utilized to train the no-show prediction model. Additionally, the study used average linkage and the Euclidean distance between the pairwise Pearson correlations as the distance metric to obtain an agglomerative hierarchical clustering of time-series data. The cancer diagnostics phase, which comprises five categories ranging from basic screening to biopsy-related procedures, was found to be the strongest predictor. Furthermore, a patient's past no-shows history and longer wait times were linked to no-shows.

7.4.1.3 Light GBM

On the other hand, the research Babayoff *et al.*, 2022 noted that GBDTs attain cutting-edge results in a range of ML tasks. However, the authors suggested some alternative methods, XGBoost and Light GBM, as GBDTs need significant time to detect the ideal split point. The research aimed to predict patient no-shows and appointment duration by implementing generic supervised ML models with the goal of optimizing appointment scheduling in hospital outpatient clinics. The MLP technique was used in addition to the first two methods described. Nevertheless, Light GBM was superior, according to the results. The findings also pointed out that temperature, rain, clinic distance, and patient no-show rate had the highest effects on the prediction.

7.4.1.4 Artificial Neural Network

Further, the study Alshammari *et al.*, 2020 compared deep feed-forward neural networks (DNN), AdaBoost, and Naive Bayes algorithm for the creation of a no-show prediction model. The findings of the experiments revealed that the model based on the DNN classifier was superior according to its evaluation metrics. They pointed out that Deep Learning techniques are applied to problems that are challenging to resolve through traditional methods. Hence, deep learning techniques offer the best method for handling large volumes and high dimensional data, presenting a complete picture embedded in extensive data and revealing unidentified structures. Finally, they added that deep learning techniques have some cons, too, as they won't be able to offer detailed advice on how to control the risk factors that affect no-show rates, owing to the hidden layers in the NN, which the DNN is unable to prioritize the features.

Additionally, Ferro *et al.*, 2020, using ML models, created a DSS to enhance healthcare accessibility by handling no-show issues. Among the three used algorithms, ANN, RF, and LR, the ANN had better performance on average. They cited that NNs are very accurate for classification tasks. The main disadvantage, though, is that they took into consideration black box models. The authors mentioned that it is challenging to decipher what information in the input data leads them to make their decisions because of their nested non-linear structure. Therefore, Layer-wise Relevance Propagation (LRP), was employed, resulting in a heat-map visualization that makes the model's conclusions instantly clear. In parallel, the algorithm's parameters were properly tuned. Furthermore, according to their analysis, socioeconomic data showed that could potentially increase the performance of the models, as variables such as low-income zones and crime incidents seem to affect the no-shows issue.

A predict-then-schedule framework was suggested by Salah & Srinivas, 2022 that estimates no-show risk and service duration using ML. Three ML algorithms were used: DNN, RF, and XGBoost. All three performed comparably, with DNN outperforming the others. The patient's age, marital status, work status, insurance group, primary language, traveling distance, appointment details, care delivery model, service type, medical history, history of chronic conditions, patient behaviour, service time history, and no-show history are among the feature categories. Finally, they indicated that there are two overbooking strategies that fall under their category of the no-show adjustment decisions approach. The first strategy involves booking one patient per slot and reducing the consultation time in proportion to the likelihood that the patient won't show up. The second involves booking two or more patients for the same slot without changing the estimated time between appointments. According to the study, implementing this strategy in a sequential framework can increase productivity by 60%.

7.4.1.5 Other Machine Learning Models

The study Dunstan *et al.*, 2023 concentrated on the prediction of no-shows in a pediatric clinic and developed models for each specialty, examining plenty of ML techniques and taking into account cost-effectiveness metrics. Hence, multiple models for the prediction of no-shows for 25 different specialties have been implemented. RF, balanced random, bagging, and LR classifiers performed better for 7, 7, 9, and 2 specialties, respectively. The study also presented a visualized heat map with the impact of features for each specialty. The findings demonstrated a well-known variation in specializations' ability to predict features.

Also, a different algorithm was suggested by Aladeemy *et al.*, 2020 to select the optimal feature subset for the prediction of no-shows, known as OSACI with OBL initialization. Contrasting this algorithm with other known algorithms, Genetic Algorithm (GA), Particle Swarm Optimization (PSO), Differential Evolution (DE), and SACI, the OSACI algorithm outperformed them in terms of dimensionality reduction and convergence speed, achieving comparable evaluation metrics. Moreover, all the above algorithms perform better evaluation metrics than the standard statistical analysis for feature selection. Lastly, the authors declared that their proposed algorithm carries out a good AUC (Area Under Curve) score only with two features: lead status, and confirmation status.

Moreover, the study AlMuhaideb *et al.*, 2019 presented a predictive model for no-show appointments at a hospital. For the creation of the model, two different ML techniques were tried. These are JRip, and Hoeffding Trees. According to the authors, the first one was selected due to its compact size models, which facilitate human comprehension. The second algorithm was chosen because of its ability to effectively manage large datasets with reasonable time and costs, as well

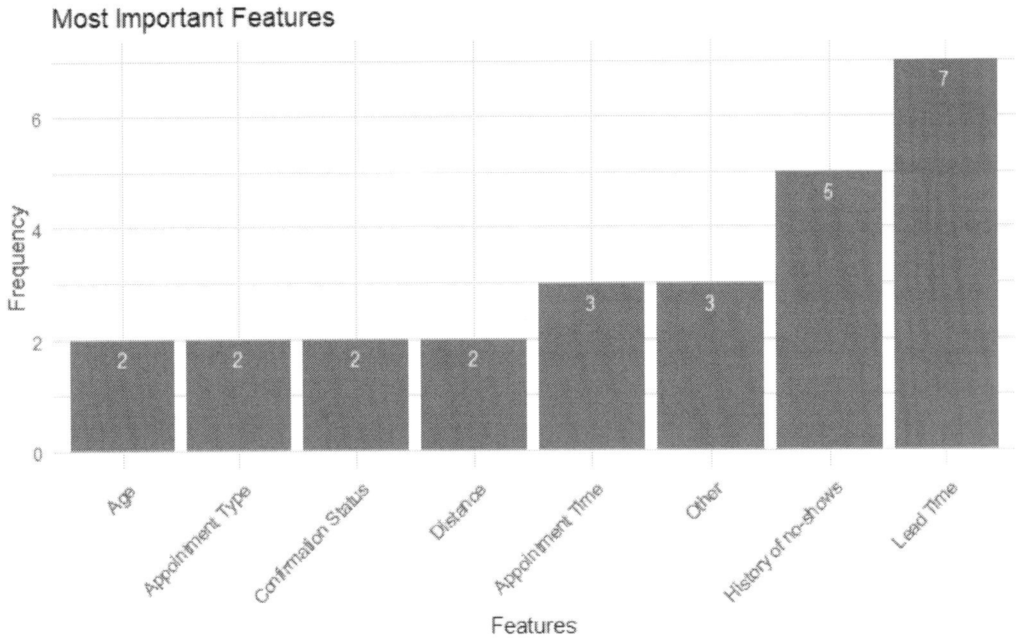

Figure 7.5 The key features of the described predictive no-show models.

as its ability to adapt to changes in concepts that may result in no-shows. However, the Hoeffding Trees Algorithm showed better evaluation metrics than the JRip algorithm. They pointed out that the most essential features of the model were the history of no-shows, appointment location, and specialty, while age, day of the week, slot description, time of appointment, gender, and nationality seem to be less relevant with the no-shows phenomenon.

7.4.1.6 Key Features of No-shows models

Figure 7.5 illustrates the number of appearances of specific features among the key predictors in the above-described models.

A previous literature review of the applied ML techniques for no-show appointments has been implemented by Salazar *et al.*, 2022. Their findings suggested that DT algorithms were the most suitable algorithms for the construction of these specific models. In parallel, the age, whether or not the patient missed a prior appointment, and the time between the appointment and the patient's scheduling were the three most important factors that determined the no-show rate.

Table 7.1 gathers all the reviewed articles for no-show prediction, summarizing their techniques and key features.

7.4.2 LITERATURE REVIEW ON PREDICTION OF POSSIBLE READMISSIONS CATEGORIZED ACCORDING TO THE DISEASE

As already mentioned, another issue that has been noticed and pointed out in recent times in the healthcare industry is the readmission of patients to hospitals, even within the next 30 days from the day they were discharged. This phenomenon indicates, to a large extent, poor management and assessment of the health status of the patients, which can be fatal several times, and at the same time, it leads the health industry to bigger expenditures. Again, the use of ML techniques and the creation of models based on them led to the prediction of such cases and, consequently, to the improvement

Table 7.1
Important findings for the no-shows

Article	Year	Important Keys for the prediction model Important Predictors
Random Forest		
Ahmadi *et al.*, 2019	2019	Lead Time Appointment Type History of no-shows
Nasir *et al.*, 2020	2020	History of no-show Age Waiting Time
Salazar *et al.*, 2021	2021	Appointment Month Contacted Patient
Alabdulkarim *et al.*, 2022	2022	Lead Time
Gradient Boosting		
Daghistani *et al.*, 2020	2020	Lead Time
Ozery-Flato *et al.*, 2021	2021	Cancer Diagnostics Phase History of no-show Lead Time
Alshammari *et al.*, 2022	2022	Distance Lead Time Appointment Time and Day
Jeffin *et al.*, 2022	2022	Not Mentioned
Hamdan & Bakar, 2023	2023	Appointment Month Age
Artificial Neural Network		
Alshammari *et al.*, 2020	2020	Not Mentioned
Ferro *et al.*, 2020	2020	Not Mentioned
Salah & Srinivas, 2022	2022	Not Mentioned
Hoeffding Trees Algorithm		
AlMuhaideb *et al.*, 2019	2019	History of no-shows Appointment Location Specialty
Balanced Random		
Dunstan *et al.*, 2023	2023	Varies among specialties
Bagging		
Dunstan *et al.*, 2023	2023	Varies among specialties
Logistic Regression		
Dunstan *et al.*, 2023	2023	Varies among specialties
Light GBM		
Babayoff *et al.*, 2022	2022	Temperature Rain Distance History of no-show
OSACI		
Aladeemy *et al.*, 2020	2020	Lead Time Confirmation Status

of the above problems, simultaneously increasing the quality of the offered services and reducing the total costs of the organizations.

The majority of the reviewed articles in this study were focused on three specific diseases. These are heart diseases, diabetes, and orthopedic issues. Nevertheless, there are also plenty of articles focused on all-cause disorders, intensive care units (ICU), Chronic Obstructive Pulmonary Disease (COPD), preterm infants, dental, COVID-19, stroke, severe sepsis, pediatric pulmonary hypertension, peripheral arterial disease, and sickle cell disease (SCD).

The authors of the study Chen, Madanian, *et al.*, 2022 who conducted a literature review on hospital readmission issues focused on the period 2017-2020. The findings emerged the following target diseases in this particular period: HF, Diabetes, ICU, Stroke, COPD, Psychiatric disease, SCD, Sepsis, Acute Myocardial Infarction (AMI), Ankle fracture surgery, LSLPSF, MDD, M/SUDs, SLE, Rheumatology, PCI, Surgeries of CAE, AFB/AAR, and FDAB, Lumbar laminectomy and last but not least diseases in general. Simultaneously, they cited the crucial factors of each one of the targeted diseases. Moreover, they summarized the performed steps from the collection and preprocessing of the dataset to the execution of the models and the evaluation of the metrics. Similarly, Mahmoudi *et al.*, 2020 demonstrated another review article focused on hospital readmissions and mentioned that the majority of existing models are related to heart disease and diabetes. Also, they pointed out that a range of clinical data indicating the severity of the patient's condition during the index admission, as well as healthcare encounter history, were the most important predictors across all models. Additionally, Schiltz *et al.*, 2020 examined in their study the importance of limitations in instrumental activities of daily living (IADL) in predicting hospital readmissions. The results of their experiments showed that IADL limitations are essential predictors of 30-day readmission through a variety of ML. More precisely, RF analysis determined ADLs and IADL limitations to be the most crucial predictors.

In addition, Huang *et al.*, 2021 summarized the research on ML techniques and how well they work to forecast hospital readmissions in the United States. The findings demonstrated that regularized LR, SVM, KNN, and tree-based techniques are the most frequently used algorithms to predict hospital readmissions. Also, Saranya & Pravin, 2020 presented a comprehensive article review focused on ML techniques for the prediction of heart disease and breast cancer. They claimed that utilizing these methods on medical data has improved risk prediction models to identify individuals who would benefit from disease management programs, which lowers hospital readmission rates and healthcare costs.

The following paragraphs, which describe all the reviewed models, are categorized by the disease or the disorder that patients suffered that led to the rehospitalization. Furthermore, each subsection includes a table that summarizes the year of publication of the article, the algorithm used for the creation of the final model, and the key features for the prediction.

7.4.2.1 Heart Diseases

Hospital readmission models for people suffering from heart disease are among the most common in the whole literature, as mentioned in the above paragraphs. However, as heart diseases can be divided into numerous different ailments, the rest part of this subsection is split according to the specific heart disease. More precisely, the described models are related to HF, AMI, atrial fibrillation, carotid artery stenting (CAS), coronary heart disease, and cardiac arrest.

For the creation of the next predictive models, multiple different ML algorithms were tested. According to the analysis, the techniques with the best performances in the majority were XGBoost, SVM, ANN, and RF. However, when stacking and ensemble methods were used, these outperformed every other compared technique. Lastly, most of the models demonstrated that length of stay in the hospital, age, and previous hospitalization history are among the most impactful variables for the prediction of the outcome. Table 7.2 highlights the techniques used and the important predictors of all the described no-show models.

Table 7.2
Important findings for the readmission aspect for diseases associated with the heart

			Important Keys for the prediction model	
Article	**Year**	**Readmission Days**	**Technique**	**Important Predictors**
Heart Failure				
Stehlik *et al.*, 2020	2020	90	SBM	Not Mentioned
Beecy *et al.*, 2020	2020	30 & 180	XGBoost	Discharge to home Serum chemistry values Quantitative ECG variables Socioeconomic variables
Wang *et al.*, 2021b	2021	30 & 90	Bi-LSTM	Not Mentioned
Lv *et al.*, 2021	2021	365	SVM	Diabetes Chronic renal disease Gender
Polo Friz *et al.*, 2022	2022	30	XGBoost	Beats Per Minute Previous Hospitalizations Aspartate Aminotransferase
Bat-Erdene *et al.*, 2022	2022	6,12,24 months	DNN	Not Mentioned
Sharma *et al.*, 2022	2022	30	XGBoost	Previous Hospitalization History Hemoglobin Levels Number of physicians Visits
Ru *et al.*, 2023	2023	30, 90 & 365	XGBoost & RF	Not Mentioned
Sarijaloo *et al.*, 2021	2021	90	LR	Not Mentioned
Miswan *et al.*, 2021	2021	30	MLP	Not Mentioned
Rahman *et al.*, 2023	2023	30	Stacking Model	Admission method LACE score Decreased hemoglobin Type of heart failure
Rizinde *et al.*, 2023	2023	20	RF	District of Residence Shortness of breath Max diastolic and systolic blood pressure
Ben-Assuli *et al.*, 2023	2023	30	XGBoost	Selected the top 50 features by four different ML methods
Sabouri *et al.*, 2023	2023	30 & 90	SVM & KNN	Discharge CR (30d-Boruta) WRF (30d-Boruta) UricAcid (30d-Boruta) Base Cr (90d-MRMR) Wide QRS (90d-MRMR)
Rajkumar *et al.*, 2023	2023	30	Ensemble Machine Learning	Not Mentioned
Acute myocardial infarction				
Zhang, Qiu, *et al.*, 2020	2020	30	Stacking-based Model	Age Length of stay
Brown *et al.*, 2022	2022	30	GBM	Not Mentioned
Matheny *et al.*, 2021	2021	30	LASSO	Discharge location
Atrial Fibrillation				
Hung, Lauren, *et al.*, 2020	2020	30	KNN	Age Total Hospital discharge Number of Diagnosis Number of Chronic Conditions
Carotid Artery Stenting				
Amritphale *et al.*, 2021	2021	30	DNN	Hospital urban-rural designation Lipid disorders Age Length of stay
Coronary Heart Disease				
Song *et al.*, 2023	2023	30	XGBoost	Fracture Hypertension Length of stay Aspirin D-dimer
Cardiac Arrest				
Chi *et al.*, 2021	2021	30	LSTM	Not Mentioned
Cardiovascular Diseases				
Alzeer *et al.*, 2021	2021	30	DT	Age Hospitalization Days Hypertension COPD Renal Failure Liver Disease Obesity Fluid Electrolyte Disorder Depression

7.4.2.1.1 Heart Failure

Another study (Beecy *et al.*, 2020) intended to create a predictive model based on an integrated big data approach that incorporates clinical data, psychosocial factors, demographics, and both invasive and noninvasive hemodynamic and physiological parameters to accurately describe the risk of 30-day and 180-day unplanned readmissions or all-cause mortality in patients with HF. XGBoost was used to build three prediction models: the index admission model, the index discharge model, and the feature-aggregated model. The difference between them is the used variables. The model results were compared with the HOSPITAL score (a verified predictive tool to pinpoint patients most likely to experience avoidable hospital readmissions) for avoidable readmissions, with the evaluation metrics of the presented method outperforming those of the HOSPITAL approach. For the 30-day readmission discharge to home, serum chemistry values and quantitative electrocardiogram (ECG) variables were the three models' main features. However, for 180-day outcomes, socioeconomic variables became more significant.

Furthermore, Polo Friz *et al.*, 2022 created another model for the prediction of hospital readmission in HF patients. They executed experiments using only ensemble learning algorithms and compared them with the LACE. To be more precise, the algorithms used were AdaBoost, GBM, XGBoost, and Random. All of the ML algorithms that were used outperformed the LACE, while XGBoost performed the best evaluation metrics of all. Furthermore, they applied the MICE (Multivariate Imputation via Chained Equations) method to acquire a balanced, fully valued dataset and the LASSO (Least Absolute Shrinkage and Selection Operator) algorithm to get the most critical variables. The results showed that the variables with the most impact in the model were, Beats Per Minute, previous Hospitalizations, and aspartate aminotransferase.

The study's Lv *et al.*, 2021 object was to predict different outcomes of HF patients using ML techniques. Among the study's goals was also the prediction of hospital readmission up to 1 year after the discharge. For the construction of the model, five ML techniques had been taken into account. SVM had the highest performance with respect to evaluation metrics. The other candidate algorithms were LR, ANN, RF, and XGBoost. Finally, according to the results, the features with the highest impact on the model were diabetes, chronic renal disease, and gender.

Another study (Sarijaloo *et al.*, 2021) for HF readmission cases focused on creating a comprehensive, ML-based risk prediction model for 90-day acute HF readmission by integrating various classes of risk factors identified in the electronic health record. Hence, they carried out Univariate analysis to detect those factors with the highest association with 90-day acute HF readmission. After that, through ML algorithms, the final selection of the variables was implemented. Finally, a LR algorithm was used for the prediction of the outcome.

Also, Wang *et al.*, 2021b aimed to develop multiple deep-learning models to predict four types of outcomes for patients with HF and reduced ejection fraction (HFrEF), including 30-day and 90-day readmissions after HF. The results showed superior evaluation metrics when compared to non-DL models (LR, RF, XGBoost). However, the performance metrics are not so promising, with the deep learning model's performance in forecasting a 90-day readmission significantly improved when the patient's age was limited to less than 65 years.

In addition, a DNN model was proposed by Bat-Erdene *et al.*, 2022 for the prediction of HF rehospitalization 6,12,24 months after discharge for patients with AMI. They remarked that the proposed method outperforms all the classical, traditional ML methods, such as LR, SVM, AdaBoost, GBM, and RF. In their research, they did not report any important features, as the process of feature selection is eliminated by DL algorithms, which incrementally extract high-level features from the data.

Another predicting model for 30-day readmissions in patients with HF was developed by Sharma *et al.*, 2022 and was contrasted with the known LACE score. They implemented different approaches using a variety of ML techniques. In more detail, they tested XGBoost, GBM, AdaBoost, CatBoost, LGBM, linear SVM, Gaussian naïve Bayes, RF, DT, L1 LR, and NN. Despite the fact that the ML models they examined outperformed the LACE score, not even the best-performing ML model, the XGBoost classifier, performed well enough as a classifier to predict readmission risk. However, the most crucial predictors for the model were previous hospitalization history, hemoglobin levels, and number of physician visits. Finally, they highlighted that combining ML with natural language processing (NLP) to leverage additional clinical data may enhance prediction accuracy.

Moreover, the study Ru *et al.*, 2023 examined ML prediction models for worsening HF events (WHFEs) following HF hospitalizations, as well as 30-day, 90-day, and 365-day hospital readmissions. The study offered a thorough analysis and comparison of various feature engineering techniques, including BERT, clinical classification software (CCS), and the use of raw codes, for predicting these HF outcomes. Compared to other medical code processing techniques, tree-based ensemble, boosting, and LR algorithms all obtained a higher AUC with CCS frequency features within a given prediction model. The findings demonstrated that, overall, XGBoost prediction models outperformed other models, with the exception of the 90-day readmission prediction, where RF with CCS frequency outperforms the other models. MLP NN and LR were the other possible algorithms. They did, however, note that it might be too close to make deterministic judgments based on the AUC and AUPR (area under the precision-recall curve) differences amongst the different algorithms and feature engineering combinations. Future research is necessary to validate and enhance the model using longitudinal electronic health records, which have longer follow-up periods and are more thorough and comprehensive. The feature engineering process referred to in this paragraph includes a series of different procedures, such as feature transformation, feature generation, feature extraction, and feature analysis (Dong & Liu, 2018). Feature engineering is the practice of modelling a problem domain in a way that facilitates learning techniques. In addition, a lot of raw data types need extensive feature engineering before they can be utilized with ML (Duboue, 2020).

In addition, Miswan *et al.*, 2021 introduced a novel, reliable framework for predicting hospital readmissions that combines ML classifiers with feature selection algorithms. Grey relational analysis (GRA) was used in the Grey-LASSO feature selection method to determine each patient's relational grade towards readmission, and to assess the degree of correlation for various data sequences. It is a component of the grey system theory, which was utilized to analyze the relationship between systems under uncertain conditions. The dataset was reduced by 54%-65% using the suggested selection feature method. The models were compared both with and without the use of the new proposed Grey-LASSO technique. The best performance was shown by the Multi-Layer Perceptron with Grey-LASSO. The other ML algorithms that were employed were GBM, RF, and Naive Bayes.

The objectives of the research of Rajkumar *et al.*, 2023 were to compare metrics, implement an ensemble model with submodels for structured data, assess the effect of principal component analysis (PCA) based optimized data manipulation on shorter readmissions, and quantitatively confirm the causal relationship between resource intensity weight (RIW) value and expected length of stay (ELOS) for a holistic economic viewpoint. The ensemble algorithm was made up of multiple submodels, such as Light GBM, XGBoost, and RF classifiers. In order to ascertain the relationship between RIW and ELOS, linear regression was also carried out. The study's findings show that these models have the capacity to precisely and reliably predict readmissions with a balanced degree of recall, which may enable medical professionals to identify patients who are at risk of readmission and take preventative action. Lastly, the tuned ensemble model is the same as the LGBM model when compared to the submodels because, even though recall is preferred, class 1's precision-recall trade-off helps avoid an excessive number of false positives.

Also, plenty of different machine-learning algorithms were evaluated by Rizinde *et al.*, 2023 for the prediction of 20-day hospital readmissions for people who suffer from HF. They take into consideration the following ML algorithms: multi-layer perceptron, KNN, LR, DT, SVM. The evaluation of the models showed that the RF algorithm outperformed all the others achieving a high score of 94% in terms of AUC. At the same time, the analysis of the dataset illustrated the most impactful features for the model. District of residence, shortness of breath and maximum systolic and diastolic blood pressure were the top three significant variables.

Moreover, Stehlik *et al.*, 2020 presented a non-invasive remote monitoring system based on a general ML method of similarity-based modelling (SBM), which evaluates the effectiveness of a customized analytical platform that uses continuous data streams to forecast hospital readmission following HF. The goal of this study, which was achieved, was to prove that non-invasive remote monitoring can predict rehospitalization as accurately as implanted devices.

Another study, Sabouri *et al.*, 2023, noted also the importance of outcome prediction of patients with HF. In their study, they proposed some models for the prediction of different outcomes. Among them, was the prediction of hospital readmissions 30 and 90 days after the discharge. For the selection of the most significant features they used three techniques, namely, Boruta, RFE (Recursive Feature Elimination), and MRMR (Minimum Redundancy Maximum Relevant). Simultaneously, numerous ML techniques were tested for the prediction of hospital readmissions. The results illustrated that among all the tested algorithms, the SVM technique in combination with Boruta feature selection performed the highest evaluation metrics for the 30-day hospital readmission. In parallel, the KNN technique with the MRMR-LR feature selection outperformed all the other for the 90-day hospital readmission.

In addition, Ben-Assuli *et al.*, 2023 referred in their study that although there are numerous models for the prediction of hospital readmissions for HF patients, only a few of them have compared the features extracting using different techniques, and even fewer have compared the predictive ability of the feature sets chosen by ML algorithms and human experts. Consequently, this study aimed to extract a feature set using four different ML algorithms and contrast its predictive ability with a feature set chosen by experts. For the selection of the feature set, LR (Ridge), LR (LASSO), RF (Gini index), and RF (Information gain) algorithms were used. The analysis showed that the two feature sets differed significantly even in the chosen variables. For example, in the ML feature set, neither medications nor comorbidities nor personal characteristics were listed, in contrast to the expert's list. The XGBoost algorithm was applied in three different feature sets for the prediction model. One was derived from the ML techniques, the other one by human experts, and the last one was a combination of the first two. The evaluation of the models showed that the model based on human-machine collaboration outperforms the other two in terms of the AUC criterion. Second, came the predictive model, whose features have been selected through ML algorithms, while the human experts' predictive model came last.

Furthermore, this project's Rahman *et al.*, 2023 goal was to use traditional ML models to predict emergency readmissions among HF patients who had been discharged. Thirteen traditional ML models and three feature selection methods – XGBoost, RF, and ExtraTree – were examined. The top three models were those based on the following three algorithms: GBM, AdaBoost, and CatBoost. These three top-performing models' predictions were used to train a stacking ML model for final classification. All other models were outperformed by the final suggested model. Ultimately, the analysis revealed that the admission method, LACE score, decreased hemoglobin, and type of HF were the top three predictors of the model.

Finally, Shin *et al.*, 2021 carried out a literature review to compare the performance between ML methods and conventional statistical models for predicting HF readmission and mortality. Through their analysis, they pointed out that ML techniques performed better for readmission and mortality prediction among HF patients than CSMs. Tree-type ML algorithms were the most commonly used supervised ML technique, while LR modelling was the most often used traditional statistical method.

7.4.2.1.2 Acute myocardial infarction

For the forecast of unplanned hospital readmission for all causes of patients who suffer from AMI, Zhang, Qiu, *et al.*, 2020 suggested a stacking-based model Further, the most vital feature sets were selected using the SelectFromModel (SFM) feature selection method. The proposed stacking-based model outperformed the other eight ML classifiers in terms of evaluation metrics, suggesting that the stacking model can combine the benefits of various individual models to produce predictions more accurately. The age and the length of stay were depicted as the most significant features of the prediction model by the analysis.

Additionally, Matheny *et al.*, 2021 tested different ML risk prediction models using a standard data set from EHRs.These models were created to forecast readmission in individuals suffering from an AMI. Elastic net (EN), LASSO, ridge regression, RF, and GBM were the algorithms that were employed. Due to a number of factors, including the fact that the AUROCs did not differ statistically and that the main focus was on the proximity of the calibration plots to the diagonal fit line, they ultimately selected LASSO as the best-fit model. They noted that although choosing LASSO as the best model among the options may seem odd given that it did not have the highest AUROC, it seems to represent a sensible trade-off between calibration performance and discrimination. Lastly, according to the analysis, the discharge location feature seems to have the highest impact on the model.

At the same time, Brown *et al.*, 2022 examined whether the integration of social risk factors could improve the performance of prediction of 30-day readmission after an AMI. The social risk factors were derived from clinical notes and extracted using NLP. For the implementation of the model, five machine-learning techniques were tested. These five techniques were the EN, LASSO, ridge regression, RF, and GBM. The model based on GBM, including NSRF, performed better evaluation metrics. However, the results showed that the addition of the chosen NLP-derived social risk factors did not statistically improve the derivation models' performance.

7.4.2.1.3 Atrial Fibrillation

Predicting the 30-day readmission status for patients with atrial fibrillation undergoing catheter ablation was the goal of the Hung, Lauren, *et al.*, 2020 study. The classification of the data is accomplished by supervised ML approaches. These approaches included KNN, SVM, and DT classifier, with KNN achieving the best performance. RF was utilized to select the most significant predictor variables according to their relative importance.

7.4.2.1.4 Carotid Artery Stenting

Another research focused on hospital readmissions after CAS. More precisely, Amritphale *et al.*, 2021 sought to characterize the frequency and reasons for unscheduled readmissions that occurred within 30 days after CAS. The model with the best performance was based on the DNN ML algorithm. LR, SVM, RF, and DT models were evaluated too. The findings pointed out the most significant predictors of the following variables: hospital urban-rural designation, lipid disorders, age, and length of stay.

7.4.2.1.5 Coronary Heart Disease

The purpose of Song *et al.*, 2023 was to use ML algorithms to predict unplanned readmissions within 7 days in older patients with coronary heart disease. Predictive models were established using five ML algorithms: LR, RF, MLP, XGBoost, and categorical boosting. The most significant predictors of the risk of 7-day unplanned readmissions were found to be fracture, hypertension, length of stay, aspirin, and D-dimer, according to the SHAP method. Utilizing the top ten variables, a compact XGB model was constructed and demonstrated strong predictive capabilities. According to the study's findings, the XGB model had the best predictive capability and the most promise for clinical application.

7.4.2.1.6 Cardiac Arrest

An HVec model using long short-term memory (LSTM) networks for the prediction of mortality and readmission up to 30 days after the discharge of in-hospital cardiac arrest (IHCA) was developed by Chi *et al.*, 2021. They also compared the multitasking learning strategy with the single-task learning, inferring that the first strategy illustrates better results. According to the authors, this was the first study to predict mortality and readmission after IHCA events.

7.4.2.1.7 Cardiovascular Diseases

Developing a machine-learning classification model to identify cardiovascular patients with a high risk of readmission was the primary objective of this study Alzeer *et al.*, 2021. The results showed that ischemic heart disease was identified as the most prevalent cardiovascular disease (HF, Cardiac Arrhythmias, Ischemic Heart Disease, and Valvular Diseases). For the implementation of the model, out of all tested ML techniques (LR, DT, XGBoost, RF, CatBoost), DT was used as illustrated the highest performance. Moreover, plenty of key features have been mentioned in the study. These are age, hospitalization days, hypertension, COPD, renal failure, liver disease, obesity, fluid electrolyte disorder, and depression.

7.4.2.2 Diabetes

The next most common predictive models for hospital readmissions in literature are associated with diabetes. Plenty of different models have been implemented, with RF and ANN being the most ordinary selected machine-learning techniques. However, a model based on the stacking method showed better performance when this was compared with other algorithms. Moreover, the analysis of the models illustrates that the number of medications and the number of inpatient visits are the two most crucial factors for the prediction. More detailed information for each model described in the following paragraphs and relating to diabetes is in Table 7.3

The authors of the study Farajollahi *et al.*, 2021 built a predictive model for hospital readmissions for diabetes patients, too. A deep learning technique and five conventional ML techniques were employed in the creation of the model. These are DT, RF, KNN, AdaBoost, XgBoost, and DNN. Applying the PCA for data reduction, three different databases were created, including 10, 15, and 20 features. The model with the highest accuracy was based on DNN and applied to the dataset with the 20 features.

Another study's objectives were to create an early diabetes readmission prediction model and pinpoint the critical variables that influence diabetic patients' readmissions. In parallel with ML classifiers, a firefly bio-inspired technique was used to keep the most essential features for the model. RF classifier outperformed not only all the other used classifiers of this study but also plenty of other approaches in the existing literature review. The variables with the highest impact on the prediction were the number of lab procedures, the number of medications, and the time in hospital (Aslam *et al.*, 2021).

Also, Alajmani & Jambi, 2020 applied, linear discriminant analysis (LDA), KNN, AdaBoost, GBM, and RF algorithms for the creation of the predictive model. All the techniques used performed similar evaluation metrics except for LDA, which had very low accuracy. The two used ensemble-based learning algorithms, RF and AdaBoost, were suggested, with RF conducted with the best accuracy. At the same time, the number of lab procedures, medications, and inpatients were the most significant variables. Similarly, Shang *et al.*, 2021 performed another approach for the prediction of readmission for people who suffer from diabetes. RF was considered the most suitable classifier for the model. The number of hospitalizations, age, length of hospital stay, and sex were

Table 7.3
Important findings for the readmission aspect for diseases associated with diabetes

			Important Keys for the prediction model	
Article	**Year**	**Readmission Days**	**Technique**	**Important Predictors**
Diabetes				
Alajmani & Jambi, 2020	2020	30	RFAdaBoost	Number of lab procedures Number of medications Number of inpatients
Farajollahi *et al.*, 2021	2020	30	DNN	Not Mentioned
Aslam *et al.*, 2021	2021	30	RF	Number of lab procedures Number of medications Time in Hospital
Wang *et al.*, 2021a	2021	30	XGBoost	Not Mentioned
Shang *et al.*, 2021	2021	30	RF	Number of Hospitalizations Age Length of hospital stay Sex
Xiang *et al.*, 2021	2021	30	EasyEnsemble	Discharge disposition Repaglinide Number of inpatient visits
Zaky *et al.*, 2021	2021	30	CNN	Not Mentioned
Lu & Uddin, 2022	2022	30	Stacking-based Model	Number of inpatients Primary diagnosis Discharge to home with home service Number of emergencies

the key features of the model. Further, Wang *et al.*, 2021a, trying to predict hospital readmissions for diabetes patients, implemented five ML classifiers. XGBoost's evaluation metrics outperformed those of LR, ANN, Naïve Bayes, and SVM.

Moreover, the results of two readmission prediction models were compared by Zaky *et al.*, 2021. One is based on normalization techniques, and the other on non-normalization techniques. Two types of data normalization were performed (z-score & min-max normalization). The results indicated, firstly, that z-score normalization outperformed the min-max normalization and, secondly, that the non-normalization technique presented better evaluation metrics overall. They deployed KNN, LR, SVM, and CNN algorithms for the classification of the model. The CNN algorithm performed the most promising evaluation metrics regarding accuracy, precision, and recall.

Additionally, Xiang *et al.*, 2021, implemented various ML techniques taking into account plenty of different approaches to handle the imbalance issue. Random oversampling, random undersampling, synthetic minority over-sampling technique, synthetic minority over-sampling technique with Tomek link, and undersampling NearMiss methods are all the used approaches. The tested ML algorithms were LR, multiple-layer deep ANN, and EasyEnsemble (EE). Because of their consistent and evenly distributed performance across the training and testing sets, the LR and EE models outperformed the ANN model in comparison. Additionally, in two of the three metrics, the LR and EE models outperformed the ANN model in terms of prediction accuracy. The testing dataset's best performance came from the EE.

Finally, Lu & Uddin, 2022 created a stacking classifier by combining five models for the prediction of unplanned readmission of diabetes patients. The RF, Adaboost, XGBoost, and CatBoost were used as base learners, while LR was used as the meta-learner. This proposed model is also explicable and comprehensible, in contrast to other existing black box models, as they mentioned and pointed out that with the advent of explainable artificial intelligence (XAI) techniques, we can now transform the models into white boxes. Also, the proposed model

Table 7.4
Important findings for the readmission aspect for people hospitalized with orthopedics issues

			Important Keys for the prediction model	
Article	Year	Readmission Days	Technique	Important Predictors
Total Joint Replacement				
Slezak *et al.*, 2021	2021	30	Light GBM	ASA mFI-5 Age
Lumbar Fusion				
Shah *et al.*, 2021	2021	30	XGBoost	Angina Pectoris Metastatic Cancer Number of comorbidities Workers' compensation insurance
Thoracolumbar Fractures				
Nunes *et al.*, 2022	2022	30	LDA	Hospital Stay Charlson comordity score Age
Total Knee Arthroplasty				
Klemt *et al.*, 2023	2023	90	ANN/SVM/KNN/EN	Drug abuse Surgical operative time ASA score
Posterior Cervical Fusion				
Karabacak & Margetis, 2023b	2023	30	Light GBM	Preoperative serum sodium Heigth Preoperative serum creatinine
Cervical Disc Arthroplasty				
Karabacak & Margetis, 2023c	2023	30	Light GBM	Race Height Age
Cervical Laminoplasty				
Karabacak & Margetis, 2023a	2023	30	Light GBM	Height Surgical Specialty Sex

outperformed all the other models based on KNN, SVM, DT, RF, AdaBoost, XGBoost, and CatBoost classifiers. Lastly, the most vital indicators were the number of inpatients, the primary diagnosis, discharge to home with home service, and the number of emergencies.

7.4.2.3 Orthopedics issues

Among the most common models of patient hospital readmission are those involving orthopedic issues. More specifically, there are predictive models in the literature for patient hospital readmissions after total joint replacement, total knee arthroplasty, posterior cervical fusion, cervical disc arthroplasty, cervical laminoplasty, lumbar fusion, and thoracolumbar fractures.

The analysis of the following described models depicted the Light GBM algorithm with the highest performance in most cases. Also, the age of the patient seems to be of the highest importance for the prediction of the outcome in the majority of the models. Table 7.4 illustrates the dominant technique for each model described in the following paragraphs, as well as the key variables.

The main goal of the Slezak *et al.*, 2021 research was to evaluate the potential predictive value of mFI-5 versus ASA classification for 30-day readmission after total joint replacement. ML techniques such as Light GBM, RF, LR, and XGBoost were used. Because the Light GBM model computed more quickly than the XGB model, which showed comparable performance, they decided that it was the best final model overall. Age and the mFI-5 frailty index are the next most predictive factors for potential readmission following surgery, according to feature importance analysis, with the assigned ASA class having the highest predictive power.

The aim of the Klemt *et al.*, 2023 study was to create and validate ML models to predict unplanned 90-day readmissions after total knee replacements. Four ML algorithms, ANN, SVM, KNN, and EN, were investigated. However, regarding AUC and Brier Score, there was no discernible difference in their performance. Each of the four models performed exceptionally well. Furthermore, the analysis revealed that the ASA score, drug abuse, and surgical operative time were the model's most influential features. admission method, LACE score, decreased hemoglobin, and type of HF were the top three predictors of the model.

Furthermore, Karabacan and Margetis focused their research on predictive outcomes after cervical operations. More precisely, they proposed some models for the outcome prediction of posterior cervical fusion, cervical laminoplasty, and cervical arthroplasty (Karabacak & Margetis, 2023a, 2023b, 2023c). One of the outcomes that their models tried to predict was the readmission of patients 30 days after their discharge. For posterior cervical fusion and cervical disc arthroplasty, the algorithms used were the same: XGBoost, Light GBM, CatBoost, and RF. For the cervical laminoplasty TabPFN algorithm was also used. However, the models based on the Light GBM algorithm were superior. Finally, the three most significant variables for each one of the above-mentioned models were preoperative serum creatinine, height, and preoperative serum sodium for cervical fusion, height, surgical specialty, and sex for cervical laminoplasty and race, height, and age for cervical disc arthroplasty.

Moreover, a model for the prediction of hospital readmission for lumbar spinal fusion up to 30 days after discharge was built by Shah *et al.*, 2021. They examined multiple ML techniques, with the highest performance gained by the XGBoost model. The other candidate techniques were LR, Adaboost, GBM, and RF. The research findings showed that the following variables had the most significant influence on the model: angina pectoris, metastatic cancer, number of comorbidities, and workers'compensation insurance.

Lastly, Nunes *et al.*, 2022 focused their study on the prediction of 30-day rehospitalization after surgery for thoracolumbar fractures. Using a Cox proportional hazards model, they estimated the relationship between 30-day readmission and different predictors. Furthermore, they checked the performance of plenty of different ML techniques, with LDA carrying out the best performance. The other candidate techniques were XGBoost, recursive partitioning, regression trees, a generalized linear model with regularization, conditional inference tree, C5.0 DT, regularized discriminant analysis, naive Bayes, RF, and NN. Finally, the most influential predictors were hospital stay, Charlson comorbidity score, and age.

7.4.2.4 All-Cause

Some studies did not focus just on patient hospital readmissions due to one specific disease but tried to predict, through a unified model, all-cause hospital readmissions.

The next paragraphs present all these specific models searched in the literature. The analysis showed that GBM algorithms look to fit better for the prediction of the outcome, as the majority of the models perform better when these are applied. Additionally, The number of inpatient visits is a common predictor in plenty of models. Table 7.5 presents briefly the described models.

The study Li, Yao & Échevin, 2020 presented a predictive model for the probability of all-cause 30-day readmission at hospital admission and discharge. The results of the implemented models vary not only in terms of evaluation metrics but also in terms of the importance of explanatory variables. Five techniques were used and evaluated, namely, LR, naive Bayes, RF, deep learning, and XGBoost. The performance of all of the above algorithms is better in the prediction model

Table 7.5
Important findings for the readmission aspect for diseases associated with all-cause hospitalizations

Article	Year	Readmission Days	Important Keys for the prediction model	
			Technique	Important Predictors
All-Cause				
Li, Yao & Échevin, 2020	2020	30	XGBoost	Destination DRG Diagnostic of Major Disease / Condition
Zhang, Yin, *et al.*, 2020	2020	30	F-CNNF-LSTM	Not Mentioned
Hilton *et al.*, 2020	2020	30	GBM	Primary Diagnosis Days between current admission and previous discharge Number of past admissions
Zhao *et al.*, 2021	2021	30	GBM	Census Region Number of inpatient visits Insurance
Zhou *et al.*, 2021	2021	30	XGBoost	Not Mentioned
Mohanty *et al.*, 2022	2022	30	CatBoost	Discharge to a Rehabilitation Facility Number of Diagnoses Prior Readmits
Ryu & Kim, 2021	2021	30	GBM	Not Mentioned
Lo *et al.*, 2021	2021	14	CatBoost	Number of inpatient diagnoses in the past year Number of unique inpatient diagnoses in the past year DRG
Jaotombo *et al.*, 2020	2020	30	RF	Category of Disease Charlson Comorbidity Index Length of Hospital Stay
Davis *et al.*, 2022	2022	30	GBM	Not Mentioned

at discharge than in the prediction model at admission. The algorithm with the highest prediction was XGBoost, and the key features were the destination, DRG, and diagnostic of major disease or condition.

Also, Zhang, Yin, *et al.*, 2020 mentioned that although temporal structured data is frequently used in research studies for predictive modelling, unstructured clinical notes often contain potentially useful information that is overlooked. Therefore, they suggested two general-purpose multi-modal NN architectures that combine structured data and sequential unstructured notes to improve patient representation learning, assessing the suggested models' effectiveness on three risk prediction tasks, including hospital readmission. The findings demonstrate that the suggested models perform better when unstructured clinical notes are combined with structured data than when unstructured notes or structured data are used alone. Both of these NN models performed better than baseline methods.

Implementing GBM algorithms the authors of the study Hilton *et al.*, 2020 created another model for the prediction of all-cause hospital readmissions. SHAP values were used to interpret cohort and personalized model prediction in order to extract the essential features needed to create the dataset. Similarly, Jaotombo *et al.*, 2020 used five distinct types of predictive models based on LR as a baseline, CART, RF, GBM, and one hidden-layer NN for the prediction of all-cause hospital readmissions. For the evaluation of the models, the H-measure and the AUC were used, with the model based on RF outperforming all the others. Moreover, the analysis revealed that the features, category of disease, Charlson Comorbidity Index, and length of hospital stay were the most significant features.

In addition, Zhao *et al.*, 2021 claimed that existing approaches for readmission prediction are focused on inpatient care with enhanced care transition and post-discharge interventions. Thus, their goal was to create a model that could predict unplanned 30-day hospital readmissions early, allowing healthcare providers to pay extra attention to patients who have been identified as high-risk. So, the model was based on index admission data and previous historical data up to one year prior to the index admission. Six ML classifiers were tested. Among them, XGBoost performed better evaluation metrics. The other candidate classifiers were LR, Naïve Bayes, alternative DT, RF, and NN. Furthermore, the above candidates' performance was superior to the three baseline algorithms' (Hospital Score, LACE index, LACE-rt index). Lastly, their analysis led to 13 significant variables, useful for prediction purposes, that have never been reported in the literature.

Moreover, the study Zhou *et al.*, 2021 examined whether the inclusion of written discharge documentation features could bolster the performance of pediatric 30-day unplanned readmission. For that reason, three models were created. Only administrative data was included in the first, clinical and administrative data were included in the second, and written discharge documentation and clinical and administrative data were included in the final one. The techniques of stepwise LR, RF, GBM, and EN were applied. In contrast, the administrative-only model's prediction was greatly outperformed by the clinical variables' inclusion, and the administrative and clinical variables' combined model was further outperformed by the written discharge documentation variables. Also, the GBM model was the highest-performing prediction model as determined by the mean receiver operating curve (ROC) statistic (C-statistic).

Another model based on ML algorithms for 30-day all-cause unplanned hospital readmission for elderly people was developed by Mohanty *et al.*, 2022. Out of five candidate classifiers (RF, XGBoost, LR, CatBoost, and a stacking classifier based on an XGBoost-based meta-learner), CatBoost presented better evaluation metrics. In parallel, due to imbalance issues, three resampling techniques were tested, with the most suitable being the under-sampling technique. The other two tested techniques were over-sampling and no-sampling. The key factors, according to their study, were the discharge to a rehabilitation facility, the number of diagnoses, and the prior readmits.

With EHR-based clinical data converted to CDM, the Ryu & Kim, 2021 study aimed to develop and validate a patient-level prediction model for all-cause 30-day hospital readmission using ML techniques. The study's research hypothesis is that the predictive model created with CDM can easily be applied to other institutions and examine clinical factors that may have an impact on hospital readmissions within a 30-day period. Additionally, in an effort to enhance the model's performance, this study sought to determine the best time to retrieve the data. Models were created and various algorithms were tested in order to forecast the occurrence of an outcome; GBM demonstrated the best overall performance. The RF, AdaBoost, and LASSO algorithms were the others. Ultimately, the findings demonstrated that using data collected 30 days prior to the discharge date produced the best results.

The study Lo *et al.*, 2021 focused on the prediction of all-cause readmissions 14 days after the discharge, remarking that this is probably the first attempt at the prediction of 14-day readmission. LR, RF, XGBoost, and Catboost were the four ML algorithms that were employed. The model with the best average performance was Catboost. The dataset consisted of demographics, diagnosis-related variables, healthcare services during admission, laboratory tests upon discharge, and variables related to the discharge process for analysis. Among them, the most influential variables for the model were the total count of inpatient diagnoses in the past year, the number of unique inpatient diagnoses in the past year, and DRG. Furthermore, the results showed that ML algorithms are capable of accurately and discriminatingly predicting a patient's likelihood of an unplanned readmission after 14 days.

Furthermore, Davis *et al.*, 2022 tried to predict possible readmission of all-cause using LR and GBM machine models. For the creation of their dataset, they extracted extra features from each patient's longitudinal health records using Word2Vec, an unsupervised technique from NLP. One interesting point of this study is that they ran the models three times for every different algorithm used, and they compared them with their AUC results. One time, it used only manual features, such as sex and age; one time, it used only the extracted features from Word2Vec; and one last time, it used the combination of both manual and extracted features. Overall, in both models, the combination of features gives better prediction results. Two ML algorithms, LR and GBM, were tested. The GBM illustrated better performance.

7.4.2.5 Intensive Care Unit & Emergency Department

In the literature, there are also numerous hospital readmission models in ICU and Emergency Departments. However, there is not any particular algorithm that stands out. Plenty of different algorithms have been selected as the most appropriate for various models. On the other hand, there are some features that showed up as the most crucial predictor in different models. These are the age, the number of hospital admissions, and the length of stay. Table 7.6 is a summary of the used technique and the important predictors of each described model for rehospitalization at the ICU.

A study Loreto *et al.*, 2020 focused on the prediction of possible readmission in the ICU, making the hypothesis that the collected data at the time of admission was sufficient for an accurate prediction. Their final model performed better than any other compared to similar approaches used in literature reviews in the ICU domain. Taking into consideration eight ML techniques, Naive Bayes, J48, RF, Sequential Minimal Optimization (SMO), JRip, AdaBoost, Logit Boost, and Iterative classifier, AdaBoost was selected, although RF achieved the best scores for most metrics. The reason was its best performance in terms of recall rate, as higher sensitivity is desired. The length of hospital stay before unit admission, the admission source, and the admission type were the most discriminative attributes.

An ensemble model, that was carried out by Artetxe *et al.*, 2020, was created by two algorithms, an extreme learning machine (ELM) and a DT, with the purpose of modelling heavily imbalanced datasets. The results of the study's experiments demonstrated that their method outperformed other well-known classification algorithms and that a more reliable classifier for the hospital readmissions problem was created when the ensemble model and random undersampling were combined.

A model based on Multi-layer Perceptron, an ANN algorithm, was produced by Battula & Balaganesh, 2020. The main aim of the authors was the analysis of patients' historical data and the prediction of potential readmissions. Their analysis is based on the MIMIC-III database, which contains health-related data for patients who are hospitalized in the ICU. In order to deal with the time this algorithm needs to preprocess medical data, they combined it with the firefly optimization, a technique capable of diminishing the classification time. Further, the proposed model, according to the analysis, demonstrated better accuracy in comparison with other existing mechanisms.

Also, a study Wu *et al.*, 2021 presented real-time 30-day readmission risk score generation in Singapore's electronic medical records (EMR) system. Patients with a high risk of readmission are flagged by the proposed system to hospital care providers. EMR storyboards and flow sheets were set up with the real-time high-risk flag to accommodate the requirements of the post-discharge, emergency department, inpatient medical, nursing, and case management teams. They developed the risk score model using the parameters that were available at the time of admission. This makes it possible to forecast patient outcomes risks in real-time and to identify high-risk patients for flagging to frontline clinical teams so that decisions can be made in real-time, such as including preventive interventions by the second day of the patient's index admission. For the classification of the

Table 7.6

Important findings for the readmission aspect for people hospitalized at Intensive Care & Emergency Departments

			Important Keys for the prediction model	
Article	**Year**	**Readmission Days**	**Technique**	**Important Predictors**
Intensive Care Units				
Loreto *et al.*, 2020	2020	Not Mentioned	AdaBoost	Length of hospital stay before unit admission Admission Source Admission type
Artetxe *et al.*, 2020	2020	30	Ensemble Algorithms	Not Mentioned
Battula & Balaganesh, 2020	2020	30	MLP	Not Mentioned
Wu *et al.*, 2021	2021	30	LR	Functional Dependency Medical Social Referral (Previous Year) Number of hospital admissions (Previous Year) Staying Status
Xie *et al.*, 2022	2022	7,14,21,30,60, and 90	SERAP	Number of emergency admission last year Age Malignancy history Renal disease history Serum creatinine level Serum albumin
Hegselmann *et al.*, 2022	2022	3-Day	EBM	Tubus Exist Age
Pishgar *et al.*, 2022	2022	30-Day	NN	Charlson Severity Score Elixhauser Severity Score Admission Events
Deng *et al.*, 2022	2022	30	LSTM	Blood urea nitrogen GCS score Ethnicity
González-Nóvoa *et al.*, 2023	2023	30	XGBoost	Length of Stay PAO2max WBmax KLmax
Kessler *et al.*, 2023	2023	30	LSTM	Not Mentioned

patients to high or low risk, five different ML techniques were used. These are LR, conditional DT, generalized linear model trees, RF, and GBM. Even though RF showed better evaluation metrics, LR was preferred for the interpretation of the model taking into account its benefits of relatively simple implementation, computational efficiency, transparency, and ease of interpretation. Stepwise selection over penalized regression was used in order to retain the significant predictors. Patient's functional dependency, medical social referral in the previous year, number of hospital admissions in the previous year, and staying status (1-room flat, institution, or homeless) are the most important features of the developed model.

Through their study Xie *et al.*, 2022, the authors proposed the SERAP (Score for Emergency ReAdmission Prediction) tool, an interpretable ML-based system. SERAP tool takes into account the total number of emergency admissions last year, age, malignancy history, renal disease history, serum creatinine level, and serum albumin. Also, comparing the SERAP with other existing scores (LACE $ HOSPITAL), the results indicate that SERAP is a better performance tool. Further, they mentioned that SERAP includes just six variables, which makes it simpler to apply in clinical settings. Additionally, at any point during the 90-day period, SERAP could forecast the readmission rate, such as 7-day, 14-day, 21-day, 30-day, 60-day, and 90-day.

Moreover, Hegselmann *et al.*, 2022 presented an intrinsically comprehensible model for forecasting a 3-day ICU readmission. This model was based on explainable boosting machines, known as EBMs. Experiments showed that the EBM model, including only 67 risk functions, outperformed recurrent neural networks (RNNs) trained on time series data and was able to match the performance of the most advanced GBMs trained on 1,423 features, which can be regarded as black box models due to their complexity. Also, according to the findings, key features of the model were the tubes' existence and age. The tube's existence variable includes the duration in days since the endotracheal tube was last used.

Also, Pishgar *et al.*, 2022 developed an NN predictive model for the hospital readmissions of ICU patients who suffer from HF. They combined time and demographic features along with severity scores at the time of admission. After variable ablation performance, the features, Charlson severity score, Elixhauser severity score, and admission events were depicted as most significant. Finally, the proposed NN model was compared with other six ML techniques (RF, XGBoost, CatBoost, SVM, DT and KNN), and presented better evaluation metrics.

Additionally, Deng *et al.*, 2022 targeted to use of time-series deep learning models to forecast readmission incidences in ICUs. The presented model combined three popular scoring systems, known as APACHE II (Knaus *et al.*, 1985), SOFA (Vincent *et al.*, 1996), and SAPS II (Le Gall *et al.*, 1993), while the used features consisted of time-series vital signs, laboratory tests, medication, and procedures. Three candidate deep learning models were examined in the particular study. These are the RNN, gated recurrent unit (GRU), and LSTM. Also, the result of the analysis demonstrates the following three key features for the model: blood urea nitrogen, GCS score, and ethnicity. Furthermore, the LSTM model presented better evaluation metrics in contrast to the other two approaches. However, overall, the results of the models could be improved by taking into consideration more variables than the ones used in this study.

The goal of another study González-Nóvoa *et al.*, 2023 was to use explainability techniques and state-of-the-art artificial intelligence algorithms to improve the prediction of early ICU patient readmission. The research employed a novel approach that combined Bayesian techniques with an XGBoost predictor model for optimization. Techniques based on Shapley's Additive Explanation were used to explain how the model functions internally. The most influential variable of the model showed to be the length of stay, PAO2max, WBmax, and KLmax. According to the study's findings, this method could greatly enhance the prediction of readmission for ICU patients.

A deep learning model, LSTM, based on LSTM, trained on time series data, has been developed to help doctors decide which patients to discharge from ICUs by creating a predictive 30-day readmission model (Kessler *et al.*, 2023). LSTM is a unique type of RNN, capable of learning long-term dependencies. They compared the performance of their LSTM with that of a feedforward neural network, a RF , additional trees, and a LR model in order to highlight its advantages. The model's performance was higher with the use of the LSTM approach.

7.4.2.6 Other Diseases and Disorders

Beyond the above five categories, there are several models aimed at predicting patient hospital readmissions due to other diseases and disorders. The following paragraphs summarize and present all those sub-cases found in the literature. More specifically, the models presented below concern the following diseases, disorders, and special cases: COPD & respiratory disease, preterm infants, dental issues, vascular diseases, COVID-19, stroke, severe sepsis, pediatric pulmonary hypertension, peripheral arterial disease, SCD, Postoperative, surgical elderly patient, esophagectomy, and cancer.

Although plenty of different algorithms have been tested, the majority of the models seem to present better evaluation metrics and predictive performance when XGBoost and RF algorithms are applied. Tables 7.7 and 7.8 demonstrate briefly the particular models, their techniques, and the key predictors.

A study Bonomo *et al.*, 2022 mentioned that COPD is a major factor in hospital readmissions, pointing out that approximately 16 million adults in the US suffer from COPD, which is the third leading cause of death in the country after heart disease and cancer. Hence, they created a model based on RF algorithm for the prediction of 90-day risk for readmission among inpatients with acute exacerbation of COPD (AECOPD). The variables with the greatest significance for the model were, length of stay, hemoglobin, White Blood Cell Count, Heart Rate, and age.

The study's Fakhraei *et al.*, 2023 objectives were to identify patients with AECOPD and predict 30-day readmissions in patients discharged following AECOPD by creating two clinical prediction models using data from EMR. Four algorithms, which were superior when compared to the HOSPITAL score, were used to develop alternative models: LR, recursive partitioning, LDA, and LASSO. The best-performing model, however, was a two-variable regression model that included the number of admissions with COPD and the number of admissions without the disease in the prior year. Individuals who were readmitted for an AECOPD within 30 days of admission had higher admission heart and respiratory rates, shorter lengths of stay, and were more likely to be labeled as discharged against medical advice.

Also, Reed *et al.*, 2021 objected to comparing two ML methods to expert-opinion-driven LR modelling for the prediction of 30-day rehospitalization of preterm infants. The proposed two models were LASSO and RF. They highlighted in their study that the existing model takes into consideration only 10 features as predictors, while LASSO and RF models consist of 75 features. The analysis illustrated that the RF model outperformed the existing model. However, the LASSO approach had a performance similar to that of LR. Moreover, they pinpointed that although RF outperformed the other two, the overall performance of all used models was relatively low.

Another study, focused on dental diseases, revealed that 11% of patients with all-cause dental disorders (ACDC) were likely to be readmitted to hospitals within 30 days of their discharge. The purpose of their study was the creation of an ML model for the prediction of possible readmissions. Out of five applied methods, LR, DT, SVM, KNN, and ANN, the last one (ANN) showed the most promising results in terms of the AUC metric. However, the precision was low among all methods used. Total charges, Age, and Length of stay were the most important features.

Moreover, about 18.9% of dental patients were readmitted within 90 days, according to the study Li *et al.*, 2021. The authors mentioned that although Dental readmissions could result in annual cost savings of more than $500 million, few studies have addressed the connection between hospital readmission and dental health. In order to create prediction models for all-cause 90-day hospital readmissions for dental patients, this study evaluated ML as a tool. To create predictive models, five machine ML classification algorithms were used: DT, LR, SVM, KNN, and ANNs. Each model performed similarly, with ANN outperforming the others by a small margin. The number of diagnoses, age, number of chronic conditions, number of procedures, length of hospital stays, primary expected payer, and severity of illness are all factors that affect the total costs.

For the prediction of hospital readmission, 90 days after the discharge for patients suffering from vascular diseases Amato *et al.*, 2020 implemented plenty of different algorithms. These are RF, LDA, Penalized Discriminant Analysis, and Stochastic GBM, with the Shrinkage Discriminant Analysis outperforming all the others. At the same time, they employed the Local Interpretable Model-Agnostic Explanations (LIME) algorithm, which offers information on risk factors that contributed to predictions for specific patients to gain a better understanding of how the model classified individual patients. In addition, the three most significant features for the prediction were the length of stay log, Van Walraven score, and Charlson score categoric.

Table 7.7

Important findings for the readmission aspect for people hospitalized due to other diseases & disorders (Part I).

		Important Keys for the prediction model		
Article	**Year**	**Readmission Days**	**Technique**	**Important Predictors**
COPD & Respiratory Disease				
Bonomo *et al.*, 2022	2022	90	RF	Length of stay Hemoglobin White Blood Cell Count Heart Rate Age
Fakhraei *et al.*, 2023	2023	30	LR	Number of admissions with COPD in prior year Number of admissions without COPD in prior year
Preterm Infants				
Reed *et al.*, 2021	2021	30	RF	Not Mentioned
Dental				
Hung, Li, *et al.*, 2020	2020	30	ANN	Total Charges Age Length of Stay
Li *et al.*, 2021	2021	90	ANN	Number of Diagnosis Age Number of chronic conditions Number of procedures Length of hospital stays Primary expected payer Severity of illness
Vascular Diseases				
Amato *et al.*, 2020	2020	90	SDA	Length of Stay Van Walraven Score Charlson Score Categoric
Covid-19				
Rodriguez *et al.*, 2021	2021	7	EN Logistic	Leukocytes Disorder of respiratory system Respiratory Rate Temperature
Afrash *et al.*, 2022	2022	30	XGBoost	Covid status ICU admission Oxygen Therapy CRP on admission
Shanbehzadeh *et al.*, 2022	2022	30	WWO	17 features are reported as crucial
Stroke				
Darabi *et al.*, 2021	2021	30	XGBoost	NIHHS score greater than 24 Insert of indwelling urinary catheter Hypercoagulable State Percutaneous Gastrostomy Insurance type
Chen, Chung, *et al.*, 2022	2022	30	ANN	Post-acute rate Nasogastric Tube Stroke Type
Severe Sepsis				
Burdick *et al.*, 2020	2020	30	GBM	Not Mentioned
Pediatric Pulmonary Hypertension				
Duan *et al.*, 2022	2022	30	CatBoost	Age Length of Stay Congenital Heart Surgery Non-medical Order

Table 7.8

Important findings for the readmission aspect for people hospitalized due to other diseases & disorders (Part II).

		Important Keys for the prediction model		
Article	Year	Readmission Days	Technique	Important Predictors
Peripheral Arterial Disease				
Cox *et al.*, 2022	2022	30	RF	Open wound/wound infection Major reintervention of treated arterial segment Elective surgery Claudication Diabetes
Sickle Cell Disease				
Patel *et al.*, 2021	2021	30	LR / RF	Reticulocytosis Past Year inpatient visits Sodium
Postoperative				
Mišić *et al.*, 2020	2020	30	GBM	Number of lab subtypes ordered Maximum delta for glucose Number of admissions in previous years
Surgical Elderly Patient				
Li *et al.*, 2022	2022	30	RF + XGBoost	Operation Duration White Blood Cell Count BMI Total Bilirubin Concentration Blood Glucose Concentration
Esophagectomy				
Bolourani *et al.*, 2021	2021	30	RF	Severity score Malnutrition Open Thoracotomy Pneumonia
Cancer				
Jin *et al.*, 2022	2022	90	LR	Post-surgery hospitalization length Tumor malignancy Discharge to Rehab
Masum *et al.*, 2022	2022	90	BI-LSTM	Age Laparoscopic procedure Stoma performed preoperative nodal stage Operation time & mode Previous surgery type Length of stay Specific procedure
Colorectal Surgery				
Chen, Joisa, *et al.*, 2022	2022	30	NN	Ileus Index admission length of stay Organ-space surgical site infection present at time of surgery Ostomy placement
Multimorbid				
Le Lay *et al.*, 2022	2022	30 & 365	RF	Not Mentioned

ML techniques to predict three outcomes for COVID-19 cases, including readmission applied by Rodriguez *et al.*, 2021. Three classifiers were tested: L1 penalized regression, elastic-net LR, and GBM. Their evaluation metrics showed that the logistic EN outperformed the other two for the readmission predicting model. The SHAP method was examined to detect important features, with the results indicating the leukocytes, disorder of respiratory system, respiratory rate, and temperature as the model's key features.

In addition, according to Afrash *et al.*, 2022, reducing COVID-19 readmissions may be able to maintain hospital capacity in a period during which hospital resource shortages caused by the COVID-19 pandemic have taken the health systems by surprise. The aim of their study was to detect the most impactful characteristics that lead to COVID-19 readmissions and examine the performance of ML algorithms in the prediction of COVID-19 readmissions based on the chosen features. The used classifiers were HistGradientBoosting, Bagging, XGBoost, SVM (linear & RBF), and MLP, with XGBoost's evaluation metrics outperforming all of them. The most significant and pertinent features are chosen using the LASSO feature selection method, and these are COVID status, ICU admission, Oxygen therapy, and CRP on admission. The HistGradientBoosting which is presented for the first time in this study, was presented and compared with other algorithms by the authors of the study Nhat-Duc & Van-Duc, 2023.

A different approach for the prediction of hospital readmissions for people with COVID-19 proposed by Shanbehzadeh *et al.*, 2022. They presented a hybrid metaheuristic algorithm known as water wave optimization (WWO) in combination with a NN. The performance of the proposed algorithms compared with all the other tested ML algorithms and the results demonstrated the WWO to be more accurate. The rest of the competitive algorithms were DT, SVM, and KNN.

Also, the authors of the study Darabi *et al.*, 2021 claimed that stroke cases are associated with a high probability of readmission. Consequently, they built a model based on an ML classifier to predict readmission after an ischemic stroke. This was the first effort to use ML techniques to forecast ischemic stroke patients' 30-day readmission. The results of their analysis showed that the combination of XGBoost and ROSE sampling produced the best results in terms of AUC, specificity, and positive predictive value (PPV). National Institutes of Health Stroke Scale (NIHHS) score greater than 24, insert of an indwelling urinary catheter, hypercoagulable state, percutaneous gastrostomy, and insurance type had the highest significance.

Additionally, Chen, Chung, *et al.*, 2022 noted that ML algorithms for 30-day stroke readmission prediction are seldom discussed, highlighting that 21% of stroke patients are readmitted within 30 days, which costs billions of dollars per year to the healthcare system. Additionally, they pointed out that the 30-day readmission mortality rate following a stroke is over 2.5 times higher than the index admissions rate. Therefore, they developed a model for the prediction of this kind of readmission 30 days after the discharge. Five classifiers were tested, namely ANN, KNN, RF, SVM, naive Bayes classifier, and Cox regression. The model based on the ANN classifier performed better results. The experiments showed that the variables post-acute rate, nasogastric tube, and stroke type have the most crucial impact on the prediction of the outcomes.

One more study Burdick *et al.*, 2020 aimed to assess the impact of an ML algorithm for severe sepsis prediction on 30-day readmission, length of hospital stay, and in-hospital mortality. The algorithm used was the GBM, which reduced by 22.7% the total hospital readmissions. The algorithm examined age and the following patient's vital indicators: systolic blood pressure, diastolic blood pressure, heart rate, temperature, respiratory rate, and oxygen saturation.

Furthermore, study Duan *et al.*, 2022 dealt with the readmission issue in diagnoses with pulmonary hypertension. They cited that the unplanned readmission cases of pediatric pulmonary hypertension come up to approximately 26.3%. Through the development of a ML model, they targeted to handle that issue, predicting the 30-day unplanned hospital readmission. The model was developed using the CatBoost classifier. Other classifiers were tested, fourteen in total, but CatBoost presented superior performances. To find important clinical variables, eliminate redundant and unnecessary data, and increase the discriminative ability of ML models, LASSO was used. Age, length of stay, congenital heart surgery, and non-medical order discharge were detected as the most valuable features of the model.

Another model for unplanned hospital readmissions focused on patients undergoing lower extremity infra-inguinal endovascular intervention for peripheral artery disease (Cox *et al.*, 2022). SVM, MLP, GBM, and RF classifiers were tested, with the last one presenting better evaluation metrics. Furthermore, in order to verify that their model is equally accurate for different demographic groups, the set was stratified by three categories: race, sex, and age, and DeLong's test was applied for the evaluation and detection of any statistically significant differences. Open wound/wound infection, major reintervention of treated arterial segments, elective surgery, claudication, and diabetes were identified as the features with the highest impact on the outcome prediction.

The goal of the Patel *et al.*, 2021 study is to predict the likelihood of readmission SCD patients 30 days after discharge. They stated that research revealed a significant portion of SCD patients (33.4%) were readmitted to the hospital within 30 days. Three common ML algorithms were used: RF, SVM, and LR. SVM was outperformed by RF and LR, which showed similar evaluation metrics. Furthermore, compared to LACE and HOSPITAL scores, all suggested algorithms performed better. The difference between the LR model and RF model was on the selected important variables. For LR, the most essential features in descending order were, reticulocytosis, number of inpatient visits in the past year, and sodium. Respectively, the number of inpatient visits in the past year, total days since the last inpatient visit, and total inpatient visits in the study period are depicted as the variables with the highest meaning in RF.

Another approach for the prediction of hospital readmissions for postoperative emergency occasions was performed by Mišić *et al.*, 2020. Their approach was based on three ML algorithms: L1 LR, RF, and GBM. Moreover, they compared the performance of these algorithms against two existing models, known as HOSPITAL and LACE. These algorithms were applied to two different datasets. The first included only the general data, while the second included the lab data too. The used algorithms presented better evaluation metrics when lab data was added to the model. The performance of the algorithms was based on AUC and Brier scores and comparing also the plots of receiver-operator characteristic curves and their precision-recall curves. GBM attained the best performance. Lastly, the most essential predictors for the GBM algorithm were the number of lab subtypes ordered, the maximum delta for glucose, and the number of admissions in previous years.

Some authors focused their study Li *et al.*, 2022 on surgical elderly patient populations. They examined whether the implementation of ML algorithms could used to forecast postoperative unplanned 30-day readmission in old surgical patients effectively. The outcomes were encouraging, with a hybrid model presenting overall better performance. The hybrid model was constructed by the combination of RF and XGBoost classifiers. The other tested classifiers were RF, LR, XGBoost, LGBM, and MLP. In addition, the analysis illustrated the following variables as the most significant for the model: operation duration, white blood cell count, BMI, total bilirubin concentration, and concentration.

The prediction of hospital readmissions for patients undergoing esophagectomy was the subject of another study (Bolourani *et al.*, 2021). They cited that 19% of patients who had esophagectomy procedures were readmitted within 30 days, indicating a higher risk of mortality and more severe illness. Additionally, their analysis showed that patients with malnutrition or COPD had a higher possibility of being readmitted early. Using the RF technique, two models were constructed: one by RF and one by RF, followed by the NearMiss algorithm. Also, some conventional statistical techniques, like univariate LR, pairwise correlation matrix, and LASSO regularization, were used to identify potential readmission predictors prior to the application of ML techniques. The features that were found to have the greatest predictive significance were pneumonia, open thoracotomy, malnourishment, and severity score. In addition, the first model – which is based solely on RF has the best accuracy, 98.7% specificity, and 23% sensitivity, whereas the second one has a higher sensitivity of 71.7% and a lower specificity of 51.4%. However, the authors denoted that in a clinical setting, it is crucial to identify all patients who are at risk of respiratory failure, so a model that places a high priority on sensitivity is appropriate.

According to Jin *et al.*, 2022, previous research has tried to assess predictors of outcome following intradural tumor resection, but the majority of these studies have been single institution series that have only examined a small number of patients and have limited cross-regional and healthcare system interpretability. Therefore, they displayed a model based on LR, regularized by LASSO penalty for prediction of 90-day unplanned readmission, taking into consideration over 5,000 patients receiving surgery for intradural spinal tumors. The post-surgery hospitalization length, tumor malignancy, and discharge to rehab were the most influential variables for the model.

Furthermore, Masum *et al.*, 2022 in their study tried to predict patients' outcomes after colorectal cancer surgery. Consequently, three different models were constructed. One is for the prediction of mortality, one is for the prediction of length of stay, and finally, one is for the prediction of hospital readmission. For the creation of the readmission model, five ML techniques were tested, and their evaluation metrics were compared. In particular, RF, KNN, SVM, MLP, and BI-LSTM were applied with the last one to present higher performance. Lastly, out of all the used variables, those with the highest impact in the prediction of the readmission outcome were age, laparoscopic procedure, stoma performed, preoperative nodal stage, operation time and mode, previous surgery type, length of stay, and the specific procedure.

Another study Chen, Joisa, *et al.*, 2022 focused their research on 30-day hospital readmissions after colorectal surgery. Their approach relied on three ML techniques, RF, GBM, and NN, while an LR model was constructed for contrast reasons. According to AUROC, NN outperformed the other algorithms. However, the difference between NN and RF, and GBM is minor in terms of AUROC. Also, for the comparison of the models, the AUPRC was used, with the NN-based models outperforming the rest. Additionally, four features were denoted as crucial for the predictive model. These are the ileus, index admission length of stay, organ-space surgical site infection present at the time of surgery, and ostomy placement.

Moreover, a study Le Lay *et al.*, 2022 occupied with the hospital readmission issues of the multimorbid population. Their study's goal was to pinpoint the characteristics of the multimorbid population and, in parallel, to predict 30-day and 365-day hospital readmissions and the total length of stay. They opted to incorporate multimorbidity by utilizing diagnoses information and two multimorbidity scores: the Calderon-Larrañaga score and the hospital frailty risk score. DT, RF, and KNN techniques were tested for the development of the model, with RF illustrating better performance.

7.5 DISCUSSION

The healthcare industry faces many significant issues due to unpredictable factors that have a crucial impact, not only on a financial level but also on the level of provided quality. Despite this, with the rise of big data in recent years, it's now more possible than ever to create models to predict future situations. The models that have been developed from time to time have shown excellent results, reducing the enormous costs that burden the healthcare industry and, at the same time, improving to a significant extent the quality of its provided services. Of course, the developed models vary in terms of the range of medical services that they are trying to cover. However, this paper focuses on four specific services.

Moreover, most of the researched studies are recent, while it is difficult to find models based on the data analytics algorithms in the past. This shows the potential and the dynamic future of ML algorithms in the healthcare industry. To assume something like this, the research results must offer significant insights and provide extra value to the industry. Indeed through the analysis of the explored literature review, it turns out the crucial contribution of the data analytics techniques.

In parallel, another important finding of the study is that for the same purpose, there are several models with different approaches. During the process of searching and evaluating academic articles, it was observed that plenty of different machine-learning algorithms have been used for similar models. So, this study is not focused on models built in a specific algorithm. On the contrary, research has been done on models with different machine-learning algorithms. The selected and described models in this model include both simple ML algorithms such as LR and DT, but also deep learning algorithms known as ANN. The results indicate that each one has specific strengths and weaknesses. However, depending on the database and its characteristics, a different algorithm gives better evaluation metrics each time. Table 7.9 gathers some of the most significant properties and limitations of the described algorithms. As can be seen, the choice of the appropriate algorithm depends to a great extent on the data we have at our disposal. It depends, on the type of data, their volume and their complexity.

The research of the existing models for the no-show and hospital readmissions predictions brought to light some challenges and limitations, as well as some opportunities that could be derived. More specifically, the completeness and quality of the data are among the main challenges in creating no-show prediction models. Inaccurate or missing data can greatly impact how well the models work. Furthermore, there may be differences in the accessibility of pertinent data, such as behavioural indicators or socioeconomic characteristics, throughout institutions, which could account for variations in model accuracy. At the same time, no-show prediction models are useful in some healthcare contexts but are frequently not very generalizable to other contexts or populations. No-show rates can be significantly impacted by factors that vary greatly across urban and rural regions, between demographic groups, or even between different kinds of clinics located in the same hospital. On the other hand, healthcare organizations can customize interventions to meet the requirements of each patient by identifying those who are at a greater risk of missing appointments. Reminders for appointments, for instance, might be helpful for certain patients, while more flexible scheduling would be necessary for others. Furthermore, by combining data from various populations, cooperative efforts between several healthcare organizations might increase the model's generalizability. This may result in more resilient models that function effectively in a variety of environments.

Regarding the hospital readmission prediction models, numerous interrelated elements, such as the patient's health status, post-discharge care, social support, and the healthcare system, interact to influence hospital readmissions. It is difficult to include all of these components in a prediction model, and models can overlook significant but unmeasured variables. Also, interpreting advanced ML models can be challenging. Healthcare organizations are faced with a dilemma since they must comprehend and have faith in the model's predictions in order to act appropriately. However, healthcare organizations could more efficiently identify high-risk patients and arrange follow-up care by incorporating predictive algorithms into care coordination platforms. This may entail making certain that patients receive the right kind of assistance at home, have timely access to their prescriptions, and schedule follow-up visits.

Overall, healthcare services are being revolutionized by ML approaches that make it possible to provide more individualized, precise, and efficient care. Large-scale medical data, such as imaging, genetic data, and patient records, are analyzed by these methods in order to spot trends and anticipate outcomes that human doctors would miss. In order to improve patient outcomes, ML algorithms, for instance, can forecast disease risks. This allows for early intervention and customized treatment approaches. Furthermore, by aiding in the interpretation of medical images and more accurately identifying anomalies than conventional techniques, ML improves diagnostic accuracy. Additionally, by anticipating patient no-shows or readmissions, ML-driven predictive models optimize hospital operations and improve resource allocation while cutting expenses. ML systems can adjust to changing medical knowledge by continuously learning from new data, guaranteeing that healthcare services stay state-of-the-art.

Table 7.9
Properties and Limitations of the most described algorithms

ML algorithm	Properties	Limitations
LR	High interpretability	May not capture complex patterns
	Computationally efficient	Assumes a linear relationship
		Limited to binary or multi-class classification
DT	High interpretability	Instability. Small changes could make decision trees unstable
	Fitting complex datasets	Poorly on imbalanced datasets
	Fast to train	Computationally expensive
RF	Handle missing values	Less Interpretable compared to simpler models
	Performs well with high-dimensional datasets	Requires more computational resources
	More robust to outliers than individual decision trees	Can be biased towards the majority class in imbalanced datasets
XGBoost	Often achieves high predictive accuracy	Can be sensitive to noisy data
	Highly efficient in terms of both training and prediction time	Require significant computational resources and memory
	Provides a range of hyper-parameters and customization options	Difficult interpretation
LightGBM	Highly efficient in terms of speed and memory usage	Require careful tuning
	Scales well with large datasets	Straggle with very small datasets or datasets with few features
	Supports categorical features	Not high interpretability
KNN	Easy to understand and implement	Computationally expensive
	Handle multi-class classification and regression problems	Poorly performance with high-dimensional data
		Sensitive to the choice of k and the distance metric
SVM	Can be used for both linear and non-linear classifications	High computational cost
	Prevent overfitting	Sensitive to the choice of kernel
	Performs well in high-dimensional spaces	Less interpretable than simpler models
ANN	Capable of modelling complex non-linear	Requires a large amount of data for training
	Can automatically learn features from raw data	Training is computationally intensive
	Ability to handle large scale datasets	Prone to overfitting
DNN	Capable of modelling complex relationships	Perform poorly with small datasets
	Achieves state of the art performance	Training can be computationally intensive
	Can automatically learn and extract features from raw data	Prone to overfitting

7.6 CONCLUSION

To sum up, this study is a literature review of existing models based on ML techniques in the healthcare industry and is focused on two crucial aspects: no-show cases and hospital readmission incidences. From the analysis of the existing articles so far, it seems that with the contribution of ML algorithms, not only can significant improvements occur, but also great opportunities could arise for the development of the healthcare industry. Although the specific study is mainly focused on analytical techniques that were used in these two specific domains, the healthcare industry is multidimensional and offers countless services. Consequently, the data analytic techniques cannot be limited to the services that have been mentioned in this study. Therefore, some future studies could deal with different models and applications based on ML algorithms that aim to improve other issues. Lastly, in this study, the results indicate that the use of ML algorithms led to the creation of models and applications that significantly improve the quality of services provided in clinics and hospitals, reducing at the same time the enormous healthcare services expenditures.

REFERENCES

Afrash, M. R., Kazemi-Arpanahi, H., Shanbehzadeh, M., Nopour, R., & Mirbagheri, E. (2022). Predicting hospital readmission risk in patients with covid-19: A machine learning approach. *Informatics in medicine unlocked*, *30*, 100908.

Agatonovic-Kustrin, S., & Beresford, R. (2000). Basic concepts of artificial neural network (ann) modeling and its application in pharmaceutical research. *Journal of pharmaceutical and biomedical analysis*, *22*(5), 717–727.

Ahmadi, E., Garcia-Arce, A., Masel, D. T., Reich, E., Puckey, J., & Maff, R. (2019). A metaheuristic-based stacking model for predicting the risk of patient no-show and late cancellation for neurology appointments. *IISE Transactions on Healthcare Systems Engineering*, *9*(3), 272–291.

Ahsaan, S. U. l., & Mourya, A. K. (2019). Big data analytics: Challenges and technologies. *Annals of the Faculty of Engineering Hunedoara*, *17*(4), 75–79.

Alabdulkarim, Y., Almukaynizi, M., Alameer, A., Makanati, B., Althumairy, R., & Almaslukh, A. (2022). Predicting no-shows for dental appointments. *PeerJ Computer Science*, *8*, e1147.

Aladeemy, M., Adwan, L., Booth, A., Khasawneh, M. T., & Poranki, S. (2020). New feature selection methods based on opposition-based learning and self-adaptive cohort intelligence for predicting patient no-shows. *Applied Soft Computing*, *86*, 105866.

Alajmani, S., & Jambi, K. (2020). Assessing advanced machine learning techniques for predicting hospital readmission. *International Journal of Advanced Computer Science and Applications*, *11*(2).

AlMuhaideb, S., Alswailem, O., Alsubaie, N., Ferwana, I., & Alnajem, A. (2019). Prediction of hospital no-show appointments through artificial intelligence algorithms. *Annals of Saudi medicine*, *39*(6), 373–381.

Alpaydin, E. (2020). *Introduction to machine learning*. MIT press.

Alshammari, A., Alaboodi, H., & Alshammari, R. (2022). The prediction of pediatric outpatient no-show visits by employing machine learning framework. *International Journal of Advanced Computer Science and Applications*, *13*(12).

Alshammari, R., Daghistani, T., & Alshammari, A. (2020). The prediction of outpatient no-show visits by using deep neural network from large data. *International Journal of Advanced Computer Science and Applications*, *11*(10).

Alzeer, A. H., Althemery, A., Alsaawi, F., Albalawi, M., Alharbi, A., Alzahrani, S., Alabdulaali, D., Alabdullatif, R., & Tash, A. (2021). Using machine learning to reduce unnecessary rehospitalization of cardiovascular patients in saudi arabia. *International Journal of Medical Informatics*, *154*, 104565.

Amato, A. C. M., Santos, R. V. d., Saucedo, D. Z., & Amato, S. J. d. T. A. (2020). Machine learning in prediction of individual patient readmissions for elective carotid endarterectomy, aortofemoral bypass/aortic aneurysm repair, and femoral-distal arterial bypass. *SAGE Open Medicine*, *8*, 2050312120909057.

Amritphale, A., Chatterjee, R., Chatterjee, S., Amritphale, N., Rahnavard, A., Awan, G. M., Omar, B., & Fonarow, G. C. (2021). Predictors of 30-day unplanned readmission after carotid artery stenting using artificial intelligence. *Advances in therapy*, *38*, 2954–2972.

Artetxe, A., Graña, M., Beristain, A., & Ríos, S. (2020). Balanced training of a hybrid ensemble method for imbalanced datasets: A case of emergency department readmission prediction. *Neural Computing and Applications*, *32*, 5735–5744.

Aslam, N., Khan, I. U., Alkhalifah, S., Al-Sadiq, S. A., Bughararah, S. W., Al-Otabi, M. A., & Al-Odinie, Z. M. (2021). Predicting diabetic patient hospital readmission using optimized random forest and firefly evolutionary algorithm. *International Journal on Advanced Science, Engineering and Information Technology*, *11*(5), 1876–1883.

Babayoff, O., Shehory, O., Geller, S., Shitrit-Niselbaum, C., Weiss-Meilik, A., & Sprecher, E. (2022). Improving hospital outpatient clinics appointment schedules by prediction models. *Journal of Medical Systems*, *47*(1), 5.

Bat-Erdene, B.-I., Zheng, H., Son, S. H., & Lee, J. Y. (2022). Deep learning-based prediction of heart failure rehospitalization during 6, 12, 24-month follow-ups in patients with acute myocardial infarction. *Health Informatics Journal*, *28*(2), 14604582221101529.

Battula, B. P., & Balaganesh, D. (2020). Prediction of hospital re-admission using firefly based multi-layer perceptron. *Ingénierie des Systèmes d'Information*, *25*(4).

Beecy, A. N., Gummalla, M., Sholle, E., Xu, Z., Zhang, Y., Michalak, K., Dolan, K., Hussain, Y., Lee, B. C., Zhang, Y., *et al.* (2020). Utilizing electronic health data and machine learning for the prediction of 30-day unplanned readmission or all-cause mortality in heart failure. *Cardiovascular Digital Health Journal*, *1*(2), 71–79.

Bellhouse, D. R. (1998). London plague statistics in 1665. *Journal of Official Statistics*, *14*(2), 207.

Ben-Assuli, O., Heart, T., Klempfner, R., & Padman, R. (2023). Human-machine collaboration for feature selection and integration to improve congestive heart failure risk prediction. *Decision Support Systems*, 113982.

Biau, G. (2012). Analysis of a random forests model. *The Journal of Machine Learning Research*, *13*(1), 1063–1095.

Bolourani, S., Tayebi, M. A., Diao, L., Wang, P., Patel, V., Manetta, F., & Lee, P. C. (2021). Using machine learning to predict early readmission following esophagectomy. *The Journal of Thoracic and Cardiovascular Surgery*, *161*(6), 1926–1939.

Bonomo, M., Hermsen, M. G., Kaskovich, S., Hemmrich, M. J., Rojas, J. C., Carey, K. A., Venable, L. R., Churpek, M. M., & Press, V. G. (2022). Using machine learning to predict likelihood and cause of readmission after hospitalization for chronic obstructive pulmonary disease exacerbation. *International Journal of Chronic Obstructive Pulmonary Disease*, 2701–2709.

Brend, W. A. (1907). Bills of mortality. *Transactions of the Medico-Legal Society for the year...* (1), 140–154.

Brown, J. R., Ricket, I. M., Reeves, R. M., Shah, R. U., Goodrich, C. A., Gobbel, G., Stabler, M. E., Perkins, A. M., Minter, F., Cox, K. C., *et al.* (2022). Information extraction from electronic health records to predict readmission following acute myocardial infarction: Does natural language processing using clinical notes improve prediction of readmission? *Journal of the American Heart Association*, *11*(7), e024198.

Burdick, H., Pino, E., Gabel-Comeau, D., McCoy, A., Gu, C., Roberts, J., Le, S., Slote, J., Pellegrini, E., Green-Saxena, A., *et al.* (2020). Effect of a sepsis prediction algorithm on patient mortality, length of stay and readmission: A prospective multicentre clinical outcomes evaluation of real-world patient data from us hospitals. *BMJ health & care informatics*, *27*(1).

Chen, H., Chiang, R. H. L., & Storey, V. C. (2012). Business intelligence and analytics: From big data to big impact. *MIS quarterly*, 1165–1188.

Chen, K. A., Joisa, C. U., Stitzenberg, K. B., Stem, J., Guillem, J. G., Gomez, S. M., & Kapadia, M. R. (2022). Development and validation of machine learning models to predict readmission after colorectal surgery. *Journal of Gastrointestinal Surgery*, *26*(11), 2342–2350.

Chen, T., Madanian, S., Airehrour, D., & Cherrington, M. (2022). Machine learning methods for hospital readmission prediction: Systematic analysis of literature. *Journal of Reliable Intelligent Environments*, *8*(1), 49–66.

Chen, Y.-C., Chung, J.-H., Yeh, Y.-J., Lou, S.-J., Lin, H.-F., Lin, C.-H., Hsien, H.-H., Hung, K.-W., Yeh, S.-C. J., & Shi, H.-Y. (2022). Predicting 30-day readmission for stroke using machine learning algorithms: A prospective cohort study. *Frontiers in Neurology*, *13*, 875491.

Chi, C.-Y., Ao, S., Winkler, A., Fu, K.-C., Xu, J., Ho, Y.-L., Huang, C.-H., & Soltani, R. (2021). Predicting the mortality and readmission of in-hospital cardiac arrest patients with electronic health records: A machine learning approach. *Journal of medical Internet research*, *23*(9), e27798.

Cleophas, T. J., Zwinderman, A. H., *et al.* (2015). Machine learning in medicine-a complete overview.

Cox, M., Panagides, J., Tabari, A., Kalva, S., Kalpathy-Cramer, J., & Daye, D. (2022). Risk stratification with explainable machine learning for 30-day procedure-related mortality and 30-day unplanned readmission in patients with peripheral arterial disease. *Plos one*, *17*(11), e0277507.

Daghistani, T., AlGhamdi, H., Alshammari, R., & AlHazme, R. H. (2020). Predictors of outpatients' no-show: Big data analytics using apache spark. *Journal of Big Data*, *7*, 1–15.

Darabi, N., Hosseinichimeh, N., Noto, A., Zand, R., & Abedi, V. (2021). Machine learning-enabled 30-day readmission model for stroke patients. *Frontiers in neurology*, *12*, 638267.

Dash, S., Shakyawar, S. K., Sharma, M., & Kaushik, S. (2019). Big data in healthcare: Management, analysis and future prospects. *Journal of big data*, *6*(1), 1–25.

Davis, S., Zhang, J., Lee, I., Rezaei, M., Greiner, R., McAlister, F. A., & Padwal, R. (2022). Effective hospital readmission prediction models using machine-learned features. *BMC Health Services Research*, *22*(1), 1415.

Deng, Y., Liu, S., Wang, Z., Wang, Y., Jiang, Y., & Liu, B. (2022). Explainable time-series deep learning models for the prediction of mortality, prolonged length of stay and 30-day readmission in intensive care patients. *Frontiers in Medicine*, *9*, 933037.

Dong, G., & Liu, H. (2018). *Feature engineering for machine learning and data analytics*. CRC press.

Donoho, D. L., *et al.* (2000). High-dimensional data analysis: The curses and blessings of dimensionality. *AMS math challenges lecture*, *1*(2000), 32.

Duan, M., Shu, T., Zhao, B., Xiang, T., Wang, J., Huang, H., Zhang, Y., Xiao, P., Zhou, B., Xie, Z., *et al.* (2022). Explainable machine learning models for predicting 30-day readmission in pediatric pulmonary hypertension: A multicenter, retrospective study. *Frontiers in Cardiovascular Medicine*, *9*, 919224.

Duboue, P. (2020). *The art of feature engineering: Essentials for machine learning*. Cambridge University Press.

Dudley, W. N., Wickham, R., & Coombs, N. (2016). An introduction to survival statistics: Kaplan-meier analysis. *Journal of the advanced practitioner in oncology*, *7*(1), 91.

Dunstan, J., Villena, F., Hoyos, J. P., Riquelme, V., Royer, M., Ramírez, H., & Peypouquet, J. (2023). Predicting no-show appointments in a pediatric hospital in chile using machine learning. *Health Care Management Science*, 1–17.

Fakhraei, R., Matelski, J., Gershon, A., Kendzerska, T., Lapointe-Shaw, L., Kaneswaran, L., & Wu, R. (2023). Development of multivariable prediction models for the identification of patients admitted to hospital with an exacerbation of copd and the prediction of risk of readmission: A retrospective cohort study using electronic medical record data. *COPD: Journal of Chronic Obstructive Pulmonary Disease*, *20*(1), 274–283.

Farajollahi, B., Mehmannavaz, M., Mehrjoo, H., Moghbeli, F., & Sayadi, M. J. (2021). Predicting hospital readmission of diabetic patients using machine learning. *Frontiers in Health Informatics*, *10*(1), 74.

Ferro, D. B., Brailsford, S., Bravo, C., & Smith, H. (2020). Improving healthcare access management by predicting patient no-show behaviour. *Decision Support Systems*, *138*, 113398.

González-Nóvoa, J. A., Campanioni, S., Busto, L., Fariña, J., Rodríguez-Andina, J. J., Vila, D., Íñiguez, A., & Veiga, C. (2023). Improving intensive care unit early readmission prediction using optimized and explainable machine learning. *International Journal of Environmental Research and Public Health*, *20*(4), 3455.

Hamdan, A. F. A., & Bakar, A. A. (2023). Machine learning predictions on outpatient no-show appointments in a malaysia major tertiary hospital. *The Malaysian Journal of Medical Sciences: MJMS*, *30*(5), 169.

Hegselmann, S., Ertmer, C., Volkert, T., Gottschalk, A., Dugas, M., & Varghese, J. (2022). Development and validation of an interpretable 3 day intensive care unit readmission prediction model using explainable boosting machines. *Frontiers in Medicine*, *9*, 960296.

Hilton, C. B., Milinovich, A., Felix, C., Vakharia, N., Crone, T., Donovan, C., Proctor, A., & Nazha, A. (2020). Personalized predictions of patient outcomes during and after hospitalization using artificial intelligence. *NPJ digital medicine*, *3*(1), 51.

Huang, Y., Talwar, A., Chatterjee, S., & Aparasu, R. R. (2021). Application of machine learning in predicting hospital readmissions: A scoping review of the literature. *BMC medical research methodology*, *21*(1), 1–14.

Hung, M., Lauren, E., Hon, E., Xu, J., Ruiz-Negrón, B., Rosales, M., Li, W., Barton, T., O'Brien, J., & Su, W. (2020). Using machine learning to predict 30-day hospital readmissions in patients with atrial fibrillation undergoing catheter ablation. *Journal of personalized medicine*, *10*(3), 82.

Hung, M., Li, W., Hon, E. S., Su, S., Su, W., He, Y., Sheng, X., Holubkov, R., & Lipsky, M. S. (2020). Prediction of 30-day hospital readmissions for all-cause dental conditions using machine learning. *Risk Management and Healthcare Policy*, 2047–2056.

James, G. (2013). An introduction to statistical learning.

Jaotombo, F., Pauly, V., Auquier, P., Orleans, V., Boucekine, M., Fond, G., Ghattas, B., & Boyer, L. (2020). Machine-learning prediction of unplanned 30-day rehospitalization using the french hospital medico-administrative database. *Medicine*, *99*(49).

Jeffin, J., Senith, S., Kirubaraj, A. A., & Ramson, S. J. (2022). Machine learning for prediction of clinical appointment no-shows. *International Journal of Mathematical, Engineering and Management Sciences*, *7*(4), 558.

Jin, M. C., Ho, A. L., Feng, A. Y., Medress, Z. A., Pendharkar, A. V., Rezaii, P., Ratliff, J. K., & Desai, A. M. (2022). Prediction of discharge status and readmissions after resection of intradural spinal tumors. *Neurospine*, *19*(1), 133.

Kalbfleisch, J. D., & Schaubel, D. E. (2023). Fifty years of the cox model. *Annual Review of Statistics and its Application*, *10*, 1–23.

Karabacak, M., & Margetis, K. (2023a). Development of personalized machine learning-based prediction models for short-term postoperative outcomes in patients undergoing cervical laminoplasty. *European Spine Journal*, 1–11.

Karabacak, M., & Margetis, K. (2023b). Interpretable machine learning models to predict short-term postoperative outcomes following posterior cervical fusion. *Plos one*, *18*(7), e0288939.

Karabacak, M., & Margetis, K. (2023c). Machine learning-based prediction of short-term adverse postoperative outcomes in cervical disc arthroplasty patients. *World Neurosurgery*.

Kessler, S., Schroeder, D., Korlakov, S., Hettlich, V., Kalkhoff, S., Moazemi, S., Lichtenberg, A., Schmid, F., & Aubin, H. (2023). Predicting readmission to the cardiovascular intensive care unit using recurrent neural networks. *Digital Health*, *9*, 20552076221149529.

Klemt, C., Tirumala, V., Habibi, Y., Buddhiraju, A., Chen, T. L.-W., & Kwon, Y.-M. (2023). The utilization of artificial neural networks for the prediction of 90-day unplanned readmissions following total knee arthroplasty. *Archives of Orthopaedic and Trauma Surgery*, *143*(6), 3279–3289.

Knaus, W. A., Draper, E. A., Wagner, D. P., & Zimmerman, J. E. (1985). Apache ii: A severity of disease classification system. *Critical care medicine*, *13*(10), 818–829.

Langmuir, A. D. (1976). William farr: Founder of modern concepts of surveillance. *International journal of epidemiology*, *5*(1), 13–18.

Le Gall, J.-R., Lemeshow, S., & Saulnier, F. (1993). A new simplified acute physiology score (saps ii) based on a european/north american multicenter study. *Jama*, *270*(24), 2957–2963.

Le Lay, J., Alfonso-Lizarazo, E., Augusto, V., Bongue, B., Masmoudi, M., Xie, X., Gramont, B., & Célarier, T. (2022). Prediction of hospital readmission of multimorbid patients using machine learning models. *Plos one*, *17*(12), e0279433.

L'heureux, A., Grolinger, K., Elyamany, H. F., & Capretz, M. A. M. (2017). Machine learning with big data: Challenges and approaches. *Ieee Access*, *5*, 7776–7797.

Li, L., Wang, L., Lu, L., & Zhu, T. (2022). Machine learning prediction of postoperative unplanned 30-day hospital readmission in older adult. *Frontiers in Molecular Biosciences*, *9*, 910688.

Li, M., Fu, X., & Li, D. (2020). Diabetes prediction based on xgboost algorithm. *IOP conference series: materials science and engineering*, *768*(7), 072093.

Li, Q., Yao, X., & Échevin, D. (2020). How good is machine learning in predicting all-cause 30-day hospital readmission? evidence from administrative data. *Value in Health*, *23*(10), 1307–1315.

Li, W., Lipsky, M. S., Hon, E. S., Su, W., Su, S., He, Y., Holubkov, R., Sheng, X., & Hung, M. (2021). Predicting all-cause 90-day hospital readmission for dental patients using machine learning methods. *BDJ open*, *7*(1), 1.

Liu, N., Ziya, S., & Kulkarni, V. G. (2010). Dynamic scheduling of outpatient appointments under patient no-shows and cancellations. *Manufacturing & Service Operations Management*, *12*(2), 347–364.

Lo, Y.-T., Liao, J. C.-h., Chen, M.-H., Chang, C.-M., & Li, C.-T. (2021). Predictive modeling for 14-day unplanned hospital readmission risk by using machine learning algorithms. *BMC medical informatics and decision making*, *21*, 1–11.

Loreto, M., Lisboa, T., & Moreira, V. P. (2020). Early prediction of icu readmissions using classification algorithms. *Computers in biology and medicine*, *118*, 103636.

Lu, H., & Uddin, S. (2022). Explainable stacking-based model for predicting hospital readmission for diabetic patients. *Information*, *13*(9), 436.

Lv, H., Yang, X., Wang, B., Wang, S., Du, X., Tan, Q., Hao, Z., Liu, Y., Yan, J., & Xia, Y. (2021). Machine learning–driven models to predict prognostic outcomes in patients hospitalized with heart failure using electronic health records: Retrospective study. *Journal of medical Internet research*, *23*(4), e24996.

Mahmoudi, E., Kamdar, N., Kim, N., Gonzales, G., Singh, K., & Waljee, A. K. (2020). Use of electronic medical records in development and validation of risk prediction models of hospital readmission: Systematic review. *bmj*, *369*.

Manyika, J., Chui, M., Brown, B., Bughin, J., Dobbs, R., Roxburgh, C., & Hung Byers, A. (2011). Big data: The next frontier for innovation, competition, and productivity.

Masum, S., Hopgood, A., Stefan, S., Flashman, K., & Khan, J. (2022). Data analytics and artificial intelligence in predicting length of stay, readmission, and mortality: A population-based study of surgical management of colorectal cancer. *Discover Oncology*, *13*(1), 11.

Matheny, M. E., Ricket, I., Goodrich, C. A., Shah, R. U., Stabler, M. E., Perkins, A. M., Dorn, C., Denton, J., Bray, B. E., Gouripeddi, R., *et al.* (2021). Development of electronic health record–based prediction models for 30-day readmission risk among patients hospitalized for acute myocardial infarction. *JAMA network open*, *4*(1), e2035782–e2035782.

McCarthy, M. (2002). A brief history of the world health organization. *The Lancet*, *360*(9340), 1111–1112.

Mišić, V. V., Gabel, E., Hofer, I., Rajaram, K., & Mahajan, A. (2020). Machine learning prediction of postoperative emergency department hospital readmission. *Anesthesiology*, *132*(5), 968–980.

Miswan, N. H., Chan, C. S., & Ng, C. G. (2021). Hospital readmission prediction based on improved feature selection using grey relational analysis and lasso. *Grey Systems: Theory and Application*, *11*(4), 796–812.

Mohanty, S. D., Lekan, D., McCoy, T. P., Jenkins, M., & Manda, P. (2022). Machine learning for predicting readmission risk among the frail: Explainable ai for healthcare. *Patterns*, *3*(1).

Nasir, M., Summerfield, N., Dag, A., & Oztekin, A. (2020). A service analytic approach to studying patient no-shows. *Service Business*, *14*, 287–313.

Nhat-Duc, H., & Van-Duc, T. (2023). Comparison of histogram-based gradient boosting classification machine, random forest, and deep convolutional neural network for pavement raveling severity classification. *Automation in Construction*, *148*, 104767.

Nunes, A. A., Pinheiro, R. P., Costa, H. R. T., & Defino, H. L. A. (2022). Predictors of hospital readmission within 30 days after surgery for thoracolumbar fractures: A mixed approach. *The International Journal of Health Planning and Management*, *37*(3), 1708–1721.

Ozery-Flato, M., Pinchasov, O., Dabush-Kasa, M., Hexter, E., Chodick, G., Guindy, M., & Rosen-Zvi, M. (2021). Predictive and causal analysis of no-shows for medical exams during covid-19: A case study of breast imaging in a nationwide israeli health organization. *AMIA Annual Symposium Proceedings*, *2021*, 930.

Ozgur, A. (2004). Supervised and unsupervised machine learning techniques for text document categorization. *Unpublished Master's Thesis, İstanbul: Boğaziçi University.*

Patel, A., Gan, K., Li, A. A., Weiss, J., Nouraie, M., Tayur, S., & Novelli, E. M. (2021). Machine-learning algorithms for predicting hospital re-admissions in sickle cell disease. *British Journal of Haematology*, *192*(1), 158–170.

Pishgar, M., Theis, J., Del Rios, M., Ardati, A., Anahideh, H., & Darabi, H. (2022). Prediction of unplanned 30-day readmission for icu patients with heart failure. *BMC Medical Informatics and Decision Making*, *22*(1), 1–12.

Polo Friz, H., Esposito, V., Marano, G., Primitz, L., Bovio, A., Delgrossi, G., Bombelli, M., Grignaffini, G., Monza, G., & Boracchi, P. (2022). Machine learning and lace index for predicting 30-day readmissions after heart failure hospitalization in elderly patients. *Internal and Emergency Medicine*, *17*(6), 1727–1737.

Qayyum, A., Qadir, J., Bilal, M., & Al-Fuqaha, A. (2021). Secure and robust machine learning for healthcare: A survey. *IEEE Reviews in Biomedical Engineering*, *14*, 156–180.

Rahman, M. S., Rahman, H. R., Prithula, J., Chowdhury, M. E., Ahmed, M. U., Kumar, J., Murugappan, M., & Khan, M. S. (2023). Heart failure emergency readmission prediction using stacking machine learning model. *Diagnostics*, *13*(11), 1948.

Rajkumar, E., Nguyen, K., Radic, S., Paa, J., Geng, Q., *et al.* (2023). Machine learning and causal approaches to predict readmissions and its economic consequences among canadian patients with heart disease: Retrospective study. *JMIR Formative Research*, *7*(1), e41725.

Rajula, H. S. R., Verlato, G., Manchia, M., Antonucci, N., & Fanos, V. (2020). Comparison of conventional statistical methods with machine learning in medicine: Diagnosis, drug development, and treatment. *Medicina*, *56*(9), 455.

Randeraad, N. (2011). The international statistical congress (1853—1876): Knowledge transfers and their limits. *European History Quarterly*, *41*(1), 50–65.

Reed, R. A., Morgan, A. S., Zeitlin, J., Jarreau, P.-H., Torchin, H., Pierrat, V., Ancel, P.-Y., & Khoshnood, B. (2021). Machine-learning vs. expert-opinion driven logistic regression modelling for predicting 30-day unplanned rehospitalisation in preterm babies: A prospective, population-based study (epipage 2). *Frontiers in Pediatrics*, *8*, 585868.

Rich, J. T., Neely, J. G., Paniello, R. C., Voelker, C. C. J., Nussenbaum, B., & Wang, E. W. (2010). A practical guide to understanding kaplan-meier curves. *Otolaryngology—Head and Neck Surgery*, *143*(3), 331–336.

Rizinde, T., Ngaruye, I., & Cahill, N. D. (2023). Comparing machine learning classifiers for predicting hospital readmission of heart failure patients in rwanda. *Journal of Personalized Medicine*, *13*(9), 1393.

Rodriguez, V. A., Bhave, S., Chen, R., Pang, C., Hripcsak, G., Sengupta, S., Elhadad, N., Green, R., Adelman, J., Metitiri, K. S., *et al.* (2021). Development and validation of prediction models for mechanical ventilation, renal replacement therapy, and readmission in covid-19 patients. *Journal of the American Medical Informatics Association*, *28*(7), 1480–1488.

Ru, B., Tan, X., Liu, Y., Kannapur, K., Ramanan, D., Kessler, G., Lautsch, D., & Fonarow, G. (2023). Comparison of machine learning algorithms for predicting hospital readmissions and worsening heart failure events in patients with heart failure with reduced ejection fraction: Modeling study. *JMIR Formative Research*, *7*, e41775.

Ryu, B., & Kim, S. (2021). Development of prediction models for unplanned hospital readmission within 30 days based on common data model: A feasibility study. *Methods of Information in Medicine*, *60*, e65–e75.

Sabouri, M., Rajabi, A. B., Hajianfar, G., Gharibi, O., Mohebi, M., Avval, A. H., Naderi, N., & Shiri, I. (2023). Machine learning based readmission and mortality prediction in heart failure patients. *Scientific Reports*, *13*(1), 18671.

Salah, H., & Srinivas, S. (2022). Predict, then schedule: Prescriptive analytics approach for machine learning-enabled sequential clinical scheduling. *Computers & Industrial Engineering*, *169*, 108270.

Salazar, L. H. A., Leithardt, V. R. Q., Parreira, W. D., da Rocha Fernandes, A. M., Barbosa, J. L. V., & Correia, S. D. (2021). Application of machine learning techniques to predict a patient's no-show in the healthcare sector. *Future Internet*, *14*(1), 3.

Salazar, L. H. A., Parreira, W. D., Fernandes, A. M. d. R., & Leithardt, V. R. Q. (2022). No-show in medical appointments with machine learning techniques: A systematic literature review. *Information*, *13*(11), 507.

Saranya, G., & Pravin, A. (2020). A comprehensive study on disease risk predictions in machine learning. *International Journal of Electrical and Computer Engineering*, *10*(4), 4217.

Sarijaloo, F., Park, J., Zhong, X., & Wokhlu, A. (2021). Predicting 90 day acute heart failure readmission and death using machine learning-supported decision analysis. *Clinical cardiology*, *44*(2), 230–237.

Schiltz, N. K., Dolansky, M. A., Warner, D. F., Stange, K. C., Gravenstein, S., & Koroukian, S. M. (2020). Impact of instrumental activities of daily living limitations on hospital readmission: An observational study using machine learning. *Journal of general internal medicine*, *35*, 2865–2872.

Shah, A. A., Devana, S. K., Lee, C., Bugarin, A., Lord, E. L., Shamie, A. N., Park, D. Y., van der Schaar, M., & SooHoo, N. F. (2021). Prediction of major complications and readmission after lumbar spinal fusion: A machine learning–driven approach. *World neurosurgery*, *152*, e227–e234.

Shanbehzadeh, M., Yazdani, A., Shafiee, M., & Kazemi-Arpanahi, H. (2022). Predictive modeling for covid-19 readmission risk using machine learning algorithms. *BMC Medical Informatics and Decision Making*, *22*(1), 139.

Shang, Y., Jiang, K., Wang, L., Zhang, Z., Zhou, S., Liu, Y., Dong, J., & Wu, H. (2021). The 30-days hospital readmission risk in diabetic patients: Predictive modeling with machine learning classifiers. *BMC medical informatics and decision making*, *21*(2), 1–11.

Sharma, V., Kulkarni, V., McAlister, F., Eurich, D., Keshwani, S., Simpson, S. H., Voaklander, D., & Samanani, S. (2022). Predicting 30-day readmissions in patients with heart failure using administrative data: A machine learning approach. *Journal of Cardiac Failure*, *28*(5), 710–722.

Shin, S., Austin, P. C., Ross, H. J., Abdel-Qadir, H., Freitas, C., Tomlinson, G., Chicco, D., Mahendiran, M., Lawler, P. R., Billia, F., *et al.* (2021). Machine learning vs. conventional statistical models for predicting heart failure readmission and mortality. *ESC heart failure*, *8*(1), 106–115.

Singh, A., Thakur, N., & Sharma, A. (2016). A review of supervised machine learning algorithms. *2016 3rd international conference on computing for sustainable global development (INDIACom)*, 1310–1315.

Slezak, J., Butler, L., & Akbilgic, O. (2021). The role of frailty index in predicting readmission risk following total joint replacement using light gradient boosting machines. *Informatics in Medicine Unlocked*, *25*, 100657.

Song, X., Tong, Y., Luo, Y., Chang, H., Gao, G., Dong, Z., Wu, X., & Tong, R. (2023). Predicting 7-day unplanned readmission in elderly patients with coronary heart disease using machine learning. *Frontiers in Cardiovascular Medicine*, *10*.

Srinivas, S., & Ravindran, A. R. (2018). Optimizing outpatient appointment system using machine learning algorithms and scheduling rules: A prescriptive analytics framework. *Expert Systems with Applications*, *102*, 245–261.

Stehlik, J., Schmalfuss, C., Bozkurt, B., Nativi-Nicolau, J., Wohlfahrt, P., Wegerich, S., Rose, K., Ray, R., Schofield, R., Deswal, A., *et al.* (2020). Continuous wearable monitoring analytics predict heart failure hospitalization: The link-hf multicenter study. *Circulation: Heart Failure*, *13*(3), e006513.

Vincent, J. .-., Moreno, R., Takala, J., Willatts, S., De Mendonça, A., Bruining, H., Reinhart, C. K., Suter, P. M., & Thijs, L. G. (1996). The sofa (sepsis-related organ failure assessment) score to describe organ dysfunction/failure: On behalf of the working group on sepsis-related problems of the european society of intensive care medicine (see contributors to the project in the appendix).

Wang, D., *et al.* (2021a). A comparison of machine learning methods to predict hospital readmission of diabetic patient. *Studies of Applied Economics*, *39*(4).

Wang, Z., Chen, X., Tan, X., Yang, L., Kannapur, K., Vincent, J. L., Kessler, G. N., Ru, B., & Yang, M. (2021b). Using deep learning to identify high-risk patients with heart failure with reduced ejection fraction. *Journal of Health Economics and Outcomes Research*, *8*(2), 6.

Wu, C. X., Suresh, E., Phng, F. W. L., Tai, K. P., Pakdeethai, J., D'souza, J. L. A., Tan, W. S., Phan, P., Lew, K. S. M., Tan, G. Y.-H., *et al.* (2021). Effect of a real-time risk score on 30-day readmission reduction in singapore. *Applied clinical informatics*, *12*(02), 372–382.

Xiang, X., Liu, C., Zhang, Y., Xiang, W., & Fang, B. (2021). Predictive modeling of 30-day readmission risk of diabetes patients by logistic regression, artificial neural network, and easyensemble. *Asian Pacific Journal of Tropical Medicine*, *14*(9), 417–428.

Xie, F., Liu, N., Yan, L., Ning, Y., Lim, K. K., Gong, C., Kwan, Y. H., Ho, A. F. W., Low, L. L., Chakraborty, B., *et al.* (2022). Development and validation of an interpretable machine learning scoring tool for estimating time to emergency readmissions. *EClinicalMedicine*, *45*.

Zaky, E. H., Soliman, M. M., Elkholy, A., & Ghali, N. I. (2021). Enhanced predictive modelling for 30-day readmission diabetes patients based on data normalization analysis. *International Journal of Intelligent Engineering & Systems*, *14*(4).

Zhang, D., Yin, C., Zeng, J., Yuan, X., & Zhang, P. (2020). Combining structured and unstructured data for predictive models: A deep learning approach. *BMC medical informatics and decision making*, *20*(1), 1–11.

Zhang, Z., Qiu, H., Li, W., & Chen, Y. (2020). A stacking-based model for predicting 30-day all-cause hospital readmissions of patients with acute myocardial infarction. *BMC medical informatics and decision making*, *20*, 1–13.

Zhao, P., Yoo, I., Naqvi, S. H., *et al.* (2021). Early prediction of unplanned 30-day hospital readmission: Model development and retrospective data analysis. *JMIR medical informatics*, *9*(3), e16306.

Zhou, H., Albrecht, M. A., Roberts, P. A., Porter, P., & Della, P. R. (2021). Using machine learning to predict paediatric 30-day unplanned hospital readmissions: A case-control retrospective analysis of medical records, including written discharge documentation. *Australian Health Review*, *45*(3), 328–337.

Section III

Statistical Methods in Sports and Industrial Systems Assurance

8 Health Budgets using Data Analytical Techniques

Elisavet Nika
University of Piraeus, Department of Business Administration, Piraeus, Greece
E-Mail: elisavet.nika@yahoo.gr.

Thomas Tsiampalis
Harokopio University of Athens, School of Health Sciences & Education, Department of Nutrition and Dietetics, Athens, Greece
University of Thessaly, School of Physical Education, Sports and Dietetics, Department of Nutrition and Dietetics, Trikala, Greece
E-Mail: ttsiampalis@gmail.com. ORCID: 0000-0002-4625-460X.

Dimitrios Georgakellos
University of Piraeus, Department of Business Administration, Piraeus, Greece
E-Mail: dgeorg@unipi.gr. ORCID: 0000-0002-7956-4727.

8.1 INTRODUCTION

8.1.1 WHY ACCURATE HEALTH BUDGETING IS CRUCIAL?

Accurate health budgeting is crucial for several reasons, as it plays a significant role in ensuring the effectiveness, efficiency, and sustainability of healthcare systems. The financial stability and the long-term strategic planning of any healthcare system depends on the accuracy of health budgets. This stability is essential for long-term planning, preventing budget shortfalls, and avoiding disruptions in healthcare services provided by the government to the citizens (Amiri *et al.*, 2021).

The financial stability engendered by precise health budgeting is indispensable for preventing budget shortfalls. Adequate funding ensures uninterrupted delivery of essential healthcare services to citizens, shielding them from potential disruptions in care provision due to financial constraints. This stability fosters public confidence in the healthcare system's ability to meet the populace's needs consistently and reliably. As highlighted by Amiri *et al.*, 2021, the significance of accurate extends beyond immediate fiscal considerations. It serves as a linchpin for long-term planning, fostering resilience in healthcare systems and enabling proactive measures to address emerging health challenges. By aligning budget allocations with healthcare priorities and population needs, healthcare systems can optimize resource utilization, enhance service delivery, and ultimately, improve health outcomes for individuals and communities alike.

Healthcare – unlike any other industry – is characterized by the lack of perfectly competitive markets (Arrow, 1978) and information asymmetry between stakeholders on the suitability and value of services being provided. In such a scenario, any market-driven allocation of resources is likely to be inefficient. Moreover, the complexity of public healthcare spending challenges the

DOI: 10.1201/9781032664330-8

definitive associations between spending and healthcare outcomes (Singh, 2014). Accurate enables efficient resource allocation within healthcare systems, facilitating the optimal deployment of staff, equipment, and supplies in accordance with the genuine needs of the population. By aligning budget allocations with actual requirements, healthcare providers can deliver superior services to patients, ensuring that resources are utilized effectively to meet individual healthcare needs.

Precise health budgeting promotes equity in resource distribution, thereby fostering equal access to healthcare services across diverse demographic groups. This equitable allocation of resources is instrumental in addressing disparities in healthcare outcomes, contributing significantly to the advancement of health equity objectives. Accurate health budgeting enhances the readiness of healthcare systems to respond swiftly and effectively to public health emergencies, ranging from disease outbreaks like COVID-19 to natural disasters exacerbated by climate change. Adequate funding enables healthcare systems to mobilize resources promptly, bolstering their capacity to mitigate the impact of unforeseen challenges and safeguard public health.

In summary, accurate health budgeting is indispensable for optimizing the performance of healthcare systems, facilitating the delivery of high-quality, accessible, and sustainable healthcare services to populations. By serving as the foundation for strategic planning, emergency preparedness, and the implementation of innovative healthcare solutions, effective health budgeting contributes profoundly to the overall well-being of communities and populations, promoting resilience and progress in the face of evolving healthcare dynamics.

8.1.2 THEORETICAL FRAMEWORK AND EXISTING MODELS OF HEALTH BUDGETING

This study is grounded in theoretical perspectives on health economics and healthcare resource allocation, with particular reference to the frameworks proposed by Martin *et al.*, 2008 and Mitton & Donaldson, 2001. Traditional theories in healthcare budgeting often emphasize incremental or historical budgeting, where resources are allocated based on past expenditures. However, this approach has been criticized for failing to account for changes in population needs, healthcare advancements, and the emergence of new diseases, which can lead to inefficiencies and misallocations.

Martin *et al.*, 2008 emphasizes the importance of priority-setting within healthcare budgets, advocating for the integration of evidence-based methodologies to maximize health outcomes within limited budgets. Mitton & Donaldson, 2001 contribute a framework focused on decision-making and resource allocation, highlighting the need for transparency and systematic approaches in allocating funds. These frameworks underscore the limitations of traditional budgeting methods and provide a foundation for our study's data-driven approach, which aims to address these gaps by incorporating predictive analytics and real-time data to align resources with current and projected health needs.

By adopting these theoretical perspectives, this study seeks to advance health budgeting approaches by integrating priority-setting principles and evidence-based resource allocation. It builds on these theories to develop a data-driven methodology that aims to address the unique financial and operational challenges faced by healthcare systems, particularly in economically constrained environments like Greece. The theoretical basis informs the proposed framework's structure and guides the use of analytics and predictive tools for optimized health budgeting.

8.1.3 IDENTIFYING GAPS IN THE EXISTING RESEARCH

Despite the critical importance of health budgeting, existing literature primarily focuses on traditional approaches, such as incremental or historical budgeting, which often fail to fully incorporate modern predictive analytics or data-driven methodologies. Moreover, much of the research is concentrated on high-income countries, where healthcare systems benefit from advanced technological infrastructure and financial resources. Limited attention has been given to the unique challenges of health budgeting in countries with economic constraints or complex healthcare systems, such as Greece, where accurate forecasting and efficient resource allocation are crucial for maintaining healthcare services amid financial pressures.

This study seeks to address this gap by developing a novel data-driven framework for health budgeting, offering a more dynamic approach to resource allocation through predictive modelling and value-based care principles. By focusing on Greece, the study aims to demonstrate how data-driven tools can be adapted to specific national contexts, thus offering practical solutions for improving the sustainability of healthcare systems in similar countries.

8.1.4 COMPARATIVE CONTEXT: GREECE AND OTHER COUNTRIES

While several countries, such as the United States and the United Kingdom, have made significant progress in incorporating data analytics and predictive modelling into their health budgeting processes, Greece presents a unique case study. The country's economic challenges and its reliance on public healthcare make accurate budgeting particularly critical. Unlike wealthier nations with greater technological and financial flexibility, Greece must ensure that healthcare resources are allocated efficiently to avoid disruptions in care.

This study not only highlights the distinctive characteristics of Greece's healthcare system but also provides a broader comparative context. By analyzing how other countries have tackled similar issues with data-driven budgeting, this paper emphasizes the need for adaptable frameworks that can be applied across diverse healthcare landscapes. The findings of this study, therefore, aim to contribute to both the Greek context and the broader international discourse on improving healthcare sustainability through innovative budgeting approaches.

8.1.5 STRUCTURE OF THE STUDY

The rest of the paper is organized as follows. In Section 8.2 the challenges in health budgeting, traditional methods, modern tools for sustainability, and the shift towards data-driven approaches for optimizing healthcare resource allocation are discussed. In Section 8.3, we provide a comprehensive review of traditional methods employed in health budgeting, highlighting their limitations and the factors driving the adoption of data analytics for budgeting purposes. It explores how predictive modelling can be utilized to forecast healthcare expenditure and demand, discussing the benefits, challenges, ethical considerations, and limitations associated with data analytics in health budgeting. In Section 8.4, we introduce a novel data-driven framework for health budgeting, delineating its key components, methodology, and advantages over traditional methodologies. We discuss how this framework integrates quantitative and epidemiological data, predictive modelling techniques, and principles of value-based care to inform budget allocations and optimize resource utilization within healthcare systems. In Section 8.5, we apply the proposed data-driven framework to the context of Greece, demonstrating its practical relevance and effectiveness in informing health budgeting decisions within a specific healthcare system. Through case study analysis and empirical evidence, we illustrate how the framework can be tailored to address the unique challenges and priorities of the Greek healthcare landscape. In the final section, we provide a thorough discussion of the findings, drawing conclusions regarding the implications of the proposed data-driven framework for health budgeting. We reflect on the strengths and limitations of the framework, identify areas for improvement, and offer recommendations for future research directions. By synthesizing the key insights gleaned from the literature review, framework application, and discussion, we aim to provide valuable insights into the evolving landscape of health budgeting and the potential of data analytics to drive transformative change within healthcare systems. Moreover, some concluding remarks are presented.

8.2 INNOVATIVE HEALTH BUDGETING APPROACHES: CHALLENGES, METHODS, AND DATA-DRIVEN SOLUTIONS

8.2.1 CHALLENGES AND LIMITATIONS IN HEALTH BUDGETING: NECESSITY FOR NEW APPROACHES

Health budgeting is essential for ensuring the stability and efficiency of healthcare systems; however, it faces numerous challenges that limit the effectiveness of traditional methods. This section discusses these challenges, emphasizing the necessity for new approaches and drawing on existing literature to highlight key issues in current budgeting practices.

Economic Fluctuations and Funding Uncertainty

One of the primary challenges in health budgeting is the uncertainty of funding streams, which are often affected by economic fluctuations and shifting government priorities (White, 2014). Budgeting in healthcare is inherently vulnerable to macroeconomic factors, making long-term planning difficult and causing inconsistencies in resource allocation Resource allocation. This inconsistency often disrupts the provision of essential services, leading to potential gaps in care that compromise patient outcomes. Traditional budgeting models, such as incremental budgeting, are not well-suited to address these fluctuations as they rely on minor adjustments to previous budgets, often ignoring larger economic shifts and leading to inefficiencies in resource use.

Rising Healthcare Costs

The rising cost of healthcare services, pharmaceuticals, and medical technology is another major challenge for traditional budgeting methods (Paul *et al.*, 2020). The complexity and rapid advancement of medical technologies make it difficult for static budgeting systems, such as fixed or top-down budgeting, to keep pace. As a result, healthcare systems often experience budget shortfalls, service cutbacks, and, at times, inefficient resource allocations that strain both the fiscal stability and responsiveness of the system. This economic reality necessitates flexible budgeting models that can adapt to changing costs and allocate resources based on actual needs rather than historical expenditures.

Demographic Shifts and Aging Populations

Demographic changes, particularly the aging population, are reshaping healthcare needs globally. Older populations generally require more healthcare resources due to a higher prevalence of chronic diseases, which traditional budgeting approaches may not accommodate adequately (Paul *et al.*, 2020). This growing demand for long-term care and chronic disease management challenges incremental and fixed budgeting models, which lack the flexibility to adjust for significant shifts in healthcare demands. Consequently, these models often fail to provide equitable access to care for aging populations, underscoring the need for more adaptable, patient-centred budgeting frameworks.

Fragmentation in Health System Governance

The fragmentation of healthcare governance further complicates budget coordination, as health systems often involve multiple government levels, departments, and external agencies (White, 2014). This fragmentation can lead to siloed budgets and a lack of coordination in resource allocation, resulting in inefficiencies and suboptimal use of resources. Traditional top-down budgeting models may exacerbate these challenges, as centralized decision-making might overlook regional or local healthcare needs, reducing system responsiveness. Activity-based budgeting offers some advantages by linking funding to specific services or procedures, but it may still struggle with resource allocation at a system-wide level, where collaboration across entities is crucial.

Short-Term Prioritization Over Long-Term Health Outcomes

Traditional budgeting methods often emphasize short-term financial goals, potentially overlooking long-term health outcomes and preventive care needs (Paul *et al.*, 2020). By focusing primarily on immediate costs rather than on outcomes, traditional methods may underfund preventive health measures and infrastructure that yield future benefits, such as early disease detection programs or public health initiatives. This short-sightedness limits investment in essential services that could reduce costs in the long term, demonstrating the need for budgeting approaches that prioritize long-term health system resilience and population health outcomes.

Limitations of Traditional Budgeting Methods

Several traditional budgeting methods are widely used in healthcare systems worldwide, each with its strengths and limitations:

Incremental Budgeting: Although incremental budgeting offers simplicity, it may fail to accommodate evolving healthcare needs, as budget adjustments are typically based on historical allocations without considering current healthcare demands (White, 2014).

Fixed Budgeting: Fixed budgeting provides stability but can restrict flexibility and responsiveness to new healthcare challenges, such as unexpected public health emergencies.

Top-Down Budgeting: While this approach aligns with overarching policy goals, it may overlook local healthcare needs, potentially leading to disparities in resource distribution.

Activity-Based Budgeting: Linking budgets to specific services can promote transparency and accountability, but it may struggle to address system-wide needs and lacks adaptability to sudden demographic changes or cost surges.

These methods, though widely adopted, are increasingly inadequate for the complex needs of modern healthcare systems. A data-driven, flexible approach is necessary to meet the current challenges of funding uncertainty, rising costs, and demographic shifts.

The Necessity for Data-Driven, Adaptive Budgeting Frameworks Given the limitations of traditional budgeting approaches, there is a growing need for adaptable, data-driven budgeting frameworks that utilize predictive analytics and real-time data to inform resource allocation. The integration of modern data analytics with priority-setting and flexible resource management allows healthcare systems to allocate resources based on current demand, future projections, and evidence-based priority setting (Paul *et al.*, 2020). By providing insights into patient needs, population trends, and cost drivers, data-driven models offer a sustainable and equitable solution for optimizing resource allocation, ensuring that healthcare systems are better prepared to meet both current and future demands.

Ultimately, modern, data-driven frameworks that leverage predictive modelling and patient-centred principles are essential for addressing the challenges faced by traditional budgeting systems. These approaches support a more efficient, resilient, and adaptable healthcare budgeting process, empowering healthcare systems to respond to demographic changes, cost escalations, and funding uncertainties effectively.

8.2.2 CONCURRENT TOOLS FOR GUARANTEEING THE SUSTAINABILITY OF THE HEALTHCARE SYSTEMS

While the challenges in health budgeting, such as uncertain funding streams and rising healthcare costs, are significant, healthcare systems have developed various tools to navigate these complexities. Moving beyond traditional methods, policymakers and administrators increasingly rely on innovative financial mechanisms (Kutzin, 2013). Capped health budgets, clawbacks, and rebates are integral components of healthcare financing and reimbursement systems, playing critical roles in ensuring fiscal sustainability and efficiency:

1. **Capped Health Budgets:** A capped health budget imposes a financial constraint on the total amount of funds allocated to healthcare within a specific timeframe, such as annually or quarterly. The primary aim of implementing a capped budget is to exercise control over healthcare expenditures, mitigate budgetary pressures, and promote fiscal responsibility.

By setting predetermined limits on healthcare spending, policymakers and administrators seek to manage costs while striving to maintain the quality and accessibility of healthcare services.

2. **Clawbacks:** Clawback mechanisms are employed by payers, typically government health agencies or insurers, to recover excess payments disbursed to healthcare providers. Clawbacks may occur when providers receive payments exceeding the agreed-upon reimbursement rates, often due to billing errors, overcharging, or changes in reimbursement policies. In such instances, the surplus funds are "clawed back" by deducting them from future payments to the healthcare provider. Clawbacks serve as a corrective measure to prevent financial losses and ensure that payments to providers align with established reimbursement guidelines.
3. **Rebates:** Rebates in healthcare refer to discounts or refunds extended by pharmaceutical companies or healthcare service providers to payers, such as health insurance plans or government health agencies. Rebates are commonly negotiated as part of pricing agreements for pharmaceutical products, aiming to optimize market access and affordability. The determination of rebate amounts may be influenced by factors such as product volume, market share, or inclusion in formularies. By leveraging rebates, payers can negotiate favourable terms and manage pharmaceutical expenditures more effectively, thereby enhancing cost-containment efforts within healthcare systems.

Overall, capped health budgets, clawbacks, and rebates constitute essential tools in the complex landscape of healthcare financing and reimbursement. These mechanisms enable stakeholders to strike a balance between cost containment objectives and the delivery of accessible healthcare services, thereby contributing to the sustainability and efficiency of healthcare systems.

8.2.3 LIMITATIONS AND SHORTCOMINGS OF TRADITIONAL BUDGETING METHODS

Although capped budgets, clawbacks, and rebates provide effective mechanisms to manage healthcare costs and promote fiscal responsibility, they are not without limitations. Traditional budgeting methods indeed have inherent limitations that can impede effective resource allocation and responsiveness to evolving healthcare needs. These limitations include:

1. **Rigidity:** Traditional budgeting methods are often rigid and inflexible, relying heavily on historical data and predetermined allocations. This rigidity can hinder adaptability to changing circumstances and emerging healthcare priorities.
2. **Focus on Inputs:** Traditional budgeting methods typically concentrate on input-based metrics, such as expenditure on personnel, equipment, and supplies, rather than emphasizing outcomes or performance metrics. This input-focused approach may overlook the broader impacts of healthcare interventions and fail to incentivize efficiency and innovation.
3. **Fragmentation:** Traditional budgeting methods may result in fragmented funding streams, with allocations distributed across various departments, programs, and initiatives without a cohesive strategic vision. This fragmentation can undermine coordination and integration efforts within healthcare systems, leading to inefficiencies and gaps in service delivery.
4. **Short-Term Orientation:** Traditional budgeting methods often exhibit a short-term orientation, focusing on immediate financial targets and annual budget cycles rather than long-term sustainability and strategic planning. This short-termism can hinder investments in preventive care, population health initiatives, and infrastructure development.

In contrast, concepts such as capped health budgets, clawbacks, and rebates represent innovative approaches aimed at addressing some of the shortcomings of traditional budgeting methods and enhancing the efficiency and effectiveness of healthcare financing and reimbursement systems. While capped health budgets provide logical constraints on expenditure, clawbacks, and rebates offer mechanisms for managing costs and incentivizing value-based care.

However, it's important to recognize that rebate and clawback mechanisms are not without challenges. These mechanisms can introduce complexities and potential drawbacks, including:

1. **Lack of Transparency:** Rebate and clawback agreements between payers and healthcare providers or pharmaceutical companies may lack transparency, making it difficult for stakeholders to assess the true cost of medications or healthcare services.
2. **Administrative Burden:** Managing rebate and clawback arrangements can impose administrative burdens on healthcare organizations, requiring robust tracking, reporting, and reconciliation processes to ensure compliance and accuracy.
3. **Potential for Disputes:** Disputes may arise between payers and healthcare providers or pharmaceutical companies regarding the calculation, timing, or validity of rebates and clawbacks, leading to conflicts and delays in payment reconciliation.

Addressing these challenges requires ongoing collaboration, transparency, and accountability among stakeholders involved in healthcare financing and reimbursement. By leveraging innovative approaches and embracing principles of value-based care, healthcare systems can strive to overcome the limitations of traditional budgeting methods and achieve more sustainable, equitable, and patient-centred outcomes.

8.2.4 EFFECTIVE HEALTH BUDGETING USING DATA-DRIVEN APPROACHES

Given the significant limitations of traditional and modern health budgeting methods, healthcare organizations and Social Security Funds are exploring alternative data-driven budgeting approaches based on population needs and outcomes to address these limitations and enhance overall efficiency and effectiveness (Kaplan, Porter, *et al.*, 2011). Recent examples highlight how data-driven approaches are transforming health budgeting in diverse healthcare systems across the world.

For instance, the National Health Service (NHS) in England has integrated population health analytics into its budgeting process, leading to better resource allocation by focusing on the most pressing healthcare needs of regional populations. This shift has helped reduce hospital admissions by targeting preventive measures at high-risk populations through early interventions. Similarly, the Centers for Medicare and Medicaid Services (CMS) in the U.S. has employed predictive analytics to improve budget planning for future healthcare needs. By using historical claims data, CMS has been able to forecast potential health outcomes and adjust its budgeting priorities, accordingly, allowing for more efficient allocation of funds towards high-need areas like chronic disease management. These examples illustrate the practical value of data-driven budgeting in addressing real-world healthcare challenges.

Data-driven health budgeting refers to the use of comprehensive and accurate data to inform the allocation of financial resources in the healthcare sector. This approach involves leveraging data analytics, statistical models, and evidence-based insights to make informed decisions about budget priorities, resource allocation, and strategic planning. The goal is to optimize the use of financial resources, improve healthcare outcomes, and enhance the overall efficiency of healthcare systems.

This approach requires the extended application of health information systems (HIS), including electronic health records (EHRs). These tools help payers extract meaningful insights from large datasets, such as analyzing patient demographics, health trends, disease prevalence, and treatment outcomes to inform budgetary decisions. Moreover, it requires the extended use of analytics such as population health analytics, which enables healthcare organizations to identify key health determinants, prioritize interventions, and allocate resources based on the specific needs of different demographic groups. In addition, it requires in-depth cost analyses for various healthcare services and interventions which allows decision-makers to understand the financial implications of different programs as well as to optimize resource allocation.

Data-driven benchmarking against industry standards and using performance metrics help assess the effectiveness and efficiency of healthcare programs. Data-driven insights from performance evaluations inform future budget decisions. Data-driven health budgeting helps identify areas of inefficiency, waste, or unnecessary spending, enabling healthcare organizations to optimize resource allocation and reduce costs.

The extended use of EHRs offers the possibility of applying outcome-based budgeting, which focuses on outcomes rather than inputs by aligning budget priorities with the expected impact on health outcomes. This approach ensures that financial resources are directed toward interventions that deliver the most significant benefits. Incorporating patient-centric data, such as patient preferences, satisfaction levels, and experiences, allows for a more holistic understanding of healthcare needs. This information helps tailor services to meet patient expectations.

Predictive modelling techniques enable healthcare organizations to anticipate future healthcare needs, identify potential risk factors, and allocate resources proactively to address emerging health challenges. Establishing mechanisms for continuous monitoring and evaluation ensures that budgetary decisions are regularly reviewed and adjusted based on the latest data. This iterative process allows for ongoing improvement.

8.2.5 ETHICAL CHALLENGES IN DATA-DRIVEN HEALTH BUDGETING

While data-driven budgeting offers significant advantages, it also raises a number of ethical concerns that need careful consideration. One of the primary ethical challenges is data privacy. As healthcare systems increasingly rely on large volumes of patient data, safeguarding this information becomes crucial. Breaches of sensitive data, particularly when used to inform financial decisions, can lead to a loss of patient trust, legal liabilities, and potential discrimination (Zhang *et al.*, 2023). For example, if predictive models incorrectly categorize patients as high-risk, these individuals may face higher insurance premiums or reduced access to services based on faulty predictions.

Another ethical concern is algorithmic bias. Predictive models and machine learning tools rely on historical data, which may reflect existing inequalities in healthcare access or treatment outcomes. This can lead to biased budgeting decisions that inadvertently reinforce disparities, particularly for underserved communities. To mitigate this, healthcare organizations must ensure that data-driven tools are carefully monitored for bias and adjusted to ensure equitable outcomes across all demographic groups. Furthermore, transparency and accountability are critical when using data analytics to make budgetary decisions. The use of complex algorithms can make it difficult for stakeholders to understand how certain decisions are made, raising concerns about the fairness and legitimacy of budget allocations. It is essential for healthcare organizations to adopt transparent processes that allow for public scrutiny of the data sources and models used in budget planning. Finally, the increasing reliance on data analytics in health budgeting may shift the focus too heavily towards cost-efficiency, potentially compromising patient-centred care. There is a risk that budget decisions driven primarily by data may undervalue qualitative aspects of healthcare, such as patient-provider relationships, patient satisfaction, and holistic well-being, which are harder to quantify but critical to delivering high-quality care.

In summary, by embracing data-driven health budgeting, healthcare organizations can enhance their ability to allocate resources effectively, improve patient care, and respond more efficiently to evolving healthcare needs. Recent case studies demonstrate the practical advantages of data-driven approaches, while ongoing improvements in predictive modelling and real-time adaptability continue to advance this field. However, as healthcare systems adopt these tools, they must also navigate significant ethical challenges, including data privacy, algorithmic bias, and ensuring transparency. A balanced approach that combines data-driven insights with ethical safeguards is essential for optimizing healthcare budgeting in an equitable and patient-centred manner

8.3 REVIEW OF THE LITERATURE

8.3.1 A COMPREHENSIVE EXPLORATION OF TRADITIONAL HEALTH BUDGETING METHODS WORLDWIDE AND IN EUROPEAN HEALTHCARE SYSTEMS

The intricate landscape of healthcare budgeting globally and within the European context is a multifaceted domain, intricately linked to the overarching goals of resource optimization, equitable distribution, and improved healthcare outcomes (Hollingsworth, 2016). The traditional methods employed in healthcare budgeting represent a spectrum of approaches, each with its strengths and limitations, shaping the financial architecture of healthcare systems worldwide (Zhang & Bohlen, 2023).

Incremental budgeting, deeply entrenched in historical spending patterns, often mirrors a cautious and conservative approach (Hartman *et al.*, 2006). The method's simplicity facilitates implementation, yet its propensity to perpetuate existing resource allocations raises questions about adaptability to the dynamic and evolving nature of modern healthcare needs. More specifically, incremental budgeting is a widely utilized traditional method in health systems globally. Rooted in the concept of incrementalism, this approach involves making minor adjustments to the previous year's budget, primarily based on historical spending patterns. While valued for its simplicity and ease of implementation, incremental budgeting has faced criticism for perpetuating existing resource allocations, potentially hindering innovation, and adaptation to changing healthcare needs. The method assumes that historical expenditure patterns accurately reflect future requirements, raising questions about its effectiveness in the dynamic landscape of modern healthcare.

Line-item budgeting, on the other hand, offers transparency by breaking down expenditures into distinct categories, providing a comprehensive view of resource allocation (Rajan *et al.*, 2016). This approach provides a detailed breakdown of expenditures, offering transparency in budgetary allocations. However, critics argue that line-item budgeting may lack the flexibility needed to address emerging health challenges or evolving priorities. As a result, the method has been associated with potential inefficiencies and difficulties in adapting to dynamic healthcare landscapes.

Historical budgeting relies on past expenditures as a benchmark for future resource allocations, serving as a foundational pillar when data or sophisticated analytical tools are lacking (Hilton & Joyce, 2007). Yet, the approach's potential limitations in accommodating evolving healthcare needs necessitate critical scrutiny. **Formula-based budgeting**, often employed to ensure equitable resource distribution, may fall short in capturing the nuanced and intricate healthcare requirements, potentially oversimplifying the complex dynamics at play. **Activity-based budgeting**, while offering a detailed understanding of costs associated with specific services, necessitates careful consideration of cost assignment challenges (Rajan *et al.*, 2016). This method is particularly relevant in healthcare settings where a granular analysis of expenses is crucial. By aligning budgeting with actual services delivered, ABB aims to enhance transparency and accountability. However, challenges may arise in accurately assigning costs to activities and maintaining flexibility in the face of changing healthcare demands.

Within the European context, **global budgeting** emerges as a cornerstone, providing financial stability by setting overall budgets for healthcare systems, regions, or facilities (Wolfe & Moran, 1993). This approach, while offering predictability, faces challenges in adapting to the diverse and dynamic healthcare needs of populations. **Decentralized budgeting**, prevalent in European countries with decentralized healthcare systems, seeks to balance local autonomy with a unified standard of care. **Per capita allocation** is a widely adopted budgeting method in several European countries, where budgets are determined based on the population served by a particular healthcare entity. This approach aims to ensure a fair distribution of resources, considering the demographic variations. However, challenges may arise in accurately estimating population health needs and addressing disparities among different demographic groups (Downes *et al.*, 2017).

Performance-based budgeting introduces a paradigm shift by linking financial allocations to predefined performance metrics, promoting efficiency and accountability (Downes *et al.*,

2017). However, the approach requires a careful definition of relevant indicators to align financial incentives with overarching healthcare goals. Finally, **budget negotiation**, a collaborative decision-making process involving diverse stakeholders, ensures inclusivity but may grapple with challenges in consensus-building and maintaining transparency.

8.3.2 THE EMERGING ROLE OF DATA ANALYTICS IN HEALTHCARE BUDGETING AND COMPARISON WITH TRADITIONAL HEALTH BUDGETING METHODS

In the contemporary landscape of healthcare, the integration of data analytics has become imperative, fundamentally altering the traditional paradigms of budgeting. The ever-expanding volume and complexity of healthcare data necessitate advanced analytical techniques to glean actionable insights for informed decision-making (Raghupathi & Raghupathi, 2014). With healthcare budgets under constant pressure to optimize resource allocation and enhance efficiency, the integration of data analytics emerges as a pivotal strategy to address the multifaceted challenges inherent in healthcare budgeting (Cozzoli *et al.*, 2022). In particular, the allocation and prioritization of budgets are pivotal elements in the management of practices, particularly amid periods of uncertainty and change (Kunnathuvalappil Hariharan, 2020). Nevertheless, making well-informed and strategic decisions regarding resource allocation and activity prioritization can pose challenges in the absence of dependable and pertinent data (Seixas *et al.*, 2021). The utilization of data analytics has the potential to enhance the processes of budgeting and prioritization for practice managers, offering valuable insights, guidance, and feedback (Bergmann *et al.*, 2020).

Healthcare systems generate vast amounts of data, ranging from patient records and treatment outcomes to operational efficiency metrics. The application of data analytics facilitates a comprehensive understanding of these datasets, unveiling patterns, trends, and correlations that might remain hidden in traditional budgeting approaches (Akinrinmade *et al.*, 2023). Predictive analytics, a key component of data analytics, enables healthcare providers and policymakers to anticipate future healthcare needs, thereby enhancing the accuracy of budget projections (Chao *et al.*, 2023). By leveraging historical data, predictive models can forecast patient admission rates, disease prevalence, and other critical factors, enabling proactive resource planning.

When juxtaposed with traditional healthcare budgeting methods, data analytics techniques offer several distinct advantages. Unlike incremental budgeting, which relies on historical spending patterns without necessarily considering emerging trends, data analytics allows for real-time monitoring and adaptation to evolving healthcare needs (Batko & Slkezak, 2022; Paganelli *et al.*, 2022). Line-item budgeting, while providing detailed expenditure breakdowns, may lack the agility to respond to dynamic healthcare environments, whereas data analytics offers a dynamic, adaptive approach by continuously assessing patterns and adjusting resource allocations accordingly (Jia *et al.*, 2022; Zhao *et al.*, 2021).

Historical budgeting, a cornerstone of traditional methods, may not fully capture the intricacies of evolving healthcare landscapes. Data analytics, on the other hand, thrives on historical data but extends its capabilities to predictive modelling, scenario analysis, and identification of cost-saving opportunities (Ahmed *et al.*, 2022). Formula-based budgeting, while striving for equitable distribution, might oversimplify complex healthcare needs, whereas data analytics techniques, such as machine learning algorithms, can identify nuanced patterns and account for diverse factors influencing resource requirements (Ramirez *et al.*, 2024).

Moreover, activity-based budgeting, though detailed in understanding costs associated with specific services, may face challenges in accurate cost assignment. Data analytics techniques enhance precision by employing sophisticated algorithms to allocate costs more accurately, contributing to a more granular and transparent understanding of healthcare expenditures (Raghupathi & Raghupathi, 2014). The integration of data analytics also aligns with the goals of global budgeting, decentralized budgeting, and per capita allocation in Europe, allowing for a more nuanced and responsive approach to diverse healthcare needs (Larson & Chang, 2016).

8.3.3 APPLICATIONS OF DATA ANALYTICS AND MACHINE LEARNING ALGORITHMS IN OPTIMIZING HEALTH BUDGET ALLOCATION

In the past few years, there has been a growing preoccupation among policymakers, national public health managers, and academics regarding the financing of public health. Moreover, the intricate and unpredictable nature of systems is exacerbated by the elevated data dimensionality arising from numerous risk factors influencing patient expenditures, including demographics, diagnoses, comorbidities, and more (Li *et al.*, 2013). Most real-world financial systems demonstrate nonlinear and uncertain behaviours that can evolve over time. Consequently, there is an increasing demand for addressing highly nonlinear and time-variant challenges, leading to the growing popularity of artificial intelligence (AI) (Bahrammirzaee, 2010). In essence, these emerging applications are supplanting older ones that were limited to estimating short-term flow through the calculation of pertinent variables and the establishment of current and potential future relationships (Wang, 2020).

Research has explored diverse implications and features related to the application of AI and advanced data management tools (Gupta & Tripathi, 2016) in the governance of health financing. On a broader scale, AI can be instrumental in examining the social determinants of health (SDH) that might contribute to higher costs within specific patient populations (Davis *et al.*, 2021). An investigation introduced a machine learning (ML) framework designed to automate the classification of health projects into different global common goods for health (CGH) categories, showcasing the feasibility and efficiency of monitoring financing for global CGH (Dixit *et al.*, 2022). Another study utilized various datasets to analyze the extremes of micro and macro variables, along with the flow of funds at the country level. The research aimed to construct a general equilibrium model that encompasses regional variations in financial friction, calibrating it to measure fluctuations in regional aggregates (Moll *et al.*, 2017). Furthermore, a study has introduced a streamlined simulation system framework aimed at enhancing the efficiency of emergency material preparation and dispatch (Du *et al.*, 2020). In a separate research endeavor, scholars created an intricate decision support system referred to as the "Cost Calculation Tool" for the primary Austrian social occupational insurance institution (AUVA). This tool delves into the comprehensive assessment of total direct and primary indirect costs associated with individual injury claims, spanning from the incident occurrence to the affected individuals' demise (Rauner *et al.*, 2015). Moreover, the examination of a subset of healthcare cost data is poised to facilitate cost audits and enable artificial sampling audits (Ji, 2016).

The advent of big data technology introduces both new foundations and challenges to tax management, contingent upon the establishment of a big data-oriented mindset (Wang, 2020). Achieving an optimal tax necessitates the precise collection of data, or at the very least, the capture of the appropriate magnitude reflecting the sum of negative externalities. Governments, whether at the national, regional, local, or any other political level, should leverage big data to decipher underlying issues and explore viable pathways forward (Chloupkova *et al.*, 2018). Another study emphasizes that determining the optimal size of a meat tax requires a comprehensive analysis integrating foresight and big data. This strategic approach offers a dual advantage: optimizing meat production levels while concurrently generating tax revenue. The resulting tax revenue should be strategically invested through international negotiations, with potential areas including development aid, disaster funds, climate change initiatives, research, new technology, and more (Chloupkova *et al.*, 2018).

In comparison to traditional health budgeting methodologies, the utilization of data analytics can yield significant advantages across various dimensions:

- **Precision and Accuracy:** Traditional health budgeting often relies on historical patterns and basic extrapolations, potentially lacking precision in forecasting future healthcare needs. Conversely, advanced analytics, integrating machine learning and predictive modelling, provide a more nuanced understanding of healthcare demand. By analysing extensive datasets, these methodologies enhance the accuracy of budget predictions, facilitating more refined resource allocation.

- **Real-time Adaptability:** Conventional approaches may struggle to promptly adjust to dynamic shifts in healthcare environments, relying on static assumptions and historical averages. In contrast, the real-time data processing and analysis capabilities of advanced analytics empower health budgeting to dynamically respond to evolving situations. This immediate adaptability is crucial for effectively addressing unforeseen changes, such as disease outbreaks or fluctuations in population health.
- **Identification of Cost Drivers:** Pinpointing precise factors influencing healthcare costs can be challenging with traditional methods, limiting the capability to address pivotal cost drivers. In contrast, advanced analytics, employing techniques like regression analysis and machine learning algorithms, excel in revealing intricate relationships within healthcare data. This proficiency facilitates the identification of specific cost drivers, providing decision-makers with tools to devise targeted strategies for effective cost containment.
- **Optimized Resource Allocation:** Conventional approaches may result in suboptimal resource allocation, lacking a comprehensive consideration of factors shaping healthcare demands. Conversely, the use of optimization algorithms and simulation modelling within advanced analytics enhances the effectiveness of resource allocation. This ensures resources are directed towards areas with the most substantial needs, optimizing the overall impact of the budget.
- **Enhanced Decision Support:** Traditional approaches often fall short in delivering comprehensive decision support, posing a challenge for healthcare administrators in making well-informed choices. In contrast, advanced analytics provide robust decision support tools, including interactive dashboards and scenario modelling. This empowerment equips decision-makers with the necessary insights to adeptly navigate complex budgeting decisions.
- **Cost-Benefit Analysis and Return on Investment (ROI):** Traditional approaches may face challenges in conducting intricate cost-benefit analyses, limiting the capacity to evaluate the return on investment for various healthcare interventions. In contrast, advanced analytics, utilizing cost-benefit analysis techniques and predictive modelling, enable a more comprehensive assessment of potential returns on investment. This facilitates the prioritization of interventions with the most significant impact within budget constraints.

8.4 THE PROPOSED FRAMEWORK

The methodological framework of this study revolves around the development of a data-driven approach aimed at estimating, forecasting, and long-term monitoring of healthcare costs necessary for covering treatment and care needs associated with specific ICD10 codes. This approach integrates quantitative and epidemiological data with the principles governing the appropriate utilization of healthcare products and services.

Unlike prevailing budget management methods, which typically employ a top-down strategy, this study advocates for a bottom-up approach. Paramount importance is placed on understanding the genuine healthcare expenditure requirements of the population before delineating essential budget allocations across disease categories.

This shift in approach acknowledges the inherent complexities within healthcare expenditures. By leveraging quantitative data and epidemiological insights, alongside adherence to usage protocols, the proposed methodology endeavours to provide a more accurate, comprehensive, and tailored framework for budget allocation. Such an approach holds the potential to enhance resource allocation efficiency, improve patient outcomes, and ensure better alignment between budgetary

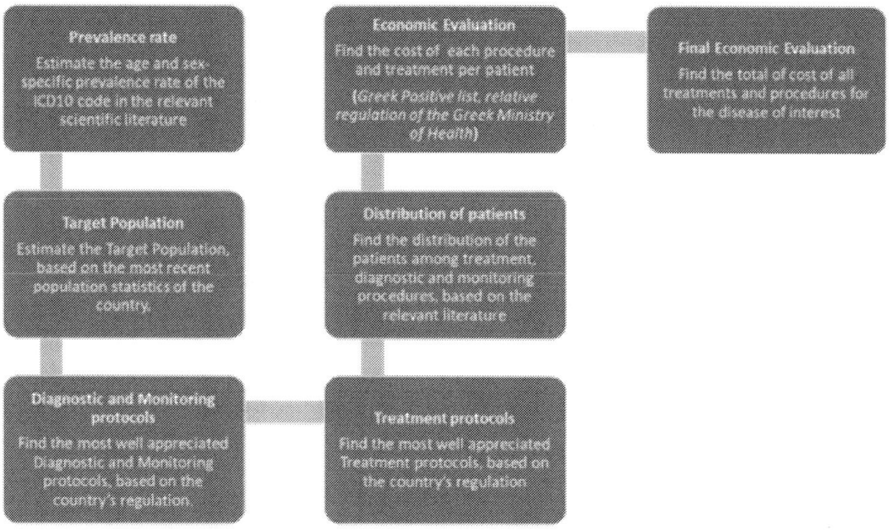

Figure 8.1 The proposed framework for each ICD10 code

allocations and actual healthcare needs at the community level.

8.4.1 THE OUTLINE OF THE PROPOSED FRAMEWORK

The outline of the proposed methodological framework is presented in Figure 8.1. The proposed framework consists of the following steps:

Step 1: Estimate related **ICD10 prevalence rate**, by using an appropriate sampling procedure or by meta-analyzing appropriate literature-acquired data and estimating the target population to be treated.

Step 2: Estimate the **target population** using either census data or any other validated source.

Step 3: Find the most appreciated **diagnostic and monitoring protocols** based on each country's regulation.

Step 4: Find the most appreciated **therapeutic protocols** based on the country's regulation.

Step 5: Based on the diagnostic and treatment protocols identified in previous steps, find the **distribution of patients among diagnostic, treatment, and monitoring procedures**, based on the relevant literature. The distribution of patients between the different treatments and treatment lines can be found through a literature search of PubMed and Google Scholar for randomized clinical trials.

Step 6: Performance of **economic evaluation** by calculating the relevant cost of each procedure and treatment per patient.

Step 7: Finally, the **total economic evaluation** emerges, by calculating the total cost of all treatments and procedures for the disease of interest.

In more detail, in the first step of the model, we estimate the prevalence p of a specific disease. Next, we estimate the number of people suffering from this disease, as $T = N \cdot p$, where N is the total population. Next, we find, through the Therapeutic Protocols, the most prevalent treatments. Suppose, for example, that there are 2 predominant treatments. Next, we estimate the subpopulations per treatment $T_1 = N \cdot p \cdot d_1$ and $T_2 = N \cdot p \cdot d_2$, where d_i is the total number of people receiving i-th treatment. Next, we look for the cost of the first and second treatments, let C_1 and C_2, respectively.

Finally, we calculate the total cost (TC) as:

$$TC = TC_1 + TC_2 = N \cdot p \cdot d_1 \cdot C_1 + N \cdot p \cdot d_2 \cdot C_2.$$

8.4.2 INPUTS OF THE METHOD

In case we would like to estimate the healthcare costs for a specific disease category [Disease X] the essential inputs that would be required in order to make the final estimation would be:

- Identify the prevalent drugs prescribed for the treatment and management of [Disease X]. This can be obtained from official treatment guidelines or health ministries. Consider multiple lines of treatment, including first-line, second-line, etc.
- Determine the diagnostic tests recommended for monitoring the health status and disease progression of [Disease X] patients. Rely on established protocols from authoritative.
- Estimate the expenses related to general practitioners' and specialists' consultations for [Disease X] patients. Define the frequency of visits based on treatment objectives or disease severity. Use guidelines or recommendations from health authorities.
- Identify medical procedures related to the treatment of complications arising from [disease X]. Utilize relevant systems like the Diagnosis-Related Group (DRG) system to associate costs with specific complications.

The required inputs for the aforementioned steps are summarized below:

Input 1: Disease Prevalence – Gather data on the prevalence of [Disease X] from reputable sources, such as national health databases, epidemiological studies, or international health organizations, and utilize meta-analysis of the literature to estimate the prevalence of [Disease X]. More specifically:

Source	**Prevalence** (p; 95% Confidence Interval)
1	$p_1\ (p_{11}, p_{12})$
2	$p_2\ (p_{21}, p_{22})$
3	$p_3\ (p_{31}, p_{32})$
4	$p_4\ (p_{41}, p_{42})$
Meta-analysis result	$p\ (p_1, p_2)$

Input 2: Population Data – Acquire demographic information on the target population from official sources like national statistical agencies or population databases. More specifically, suppose that based on the Greek census database, the Greek population equals x, then the target population is estimated as:

$$Target\ population\ (with\ the\ [Disease\ X]) = p \cdot x$$

Input 3: Treatment Protocols – Refer to official treatment protocols for [Disease X] published by health ministries or recognized medical organizations. Identify the hierarchy of treatment lines and associated drugs. More specifically, let us assume a [Disease X] with 2 treatment lines:

Treatment line and drugs	**Monthly cost per drug** (D)	**Proportion of patients receiving each drug** (q)
First treatment line		
Drug 1	D_1	q_1
Drug 2	D_2	q_2
Second treatment line		
Drug 1 +	D_1+	
Drug 3	D_3	q_3
Drug 4	D_4	q_4
Drug 2 +	D_2+	
Drug 3	D_3	q_5
Drug 4	D_4	q_6

Then the total monthly cost per line of treatment will be:

$$\begin{aligned} \textit{Monthly Cost of first-line} =& (q_1 \cdot \textit{Target population (with the [Disease X]))} \cdot d_1 \\ &+ (q_2 \cdot \textit{Target population (with the [Disease X]))} \cdot d_2 \\ \textit{Monthly Cost of second-line} =& (q_3 \cdot \textit{Target population (with the [Disease X]))} \cdot (d_1 + d_3) \\ &+ (q_4 \cdot \textit{Target population (with the [Disease X]))} \cdot (d_1 + d_4) \\ &+ (q_5 \cdot \textit{Target population (with the [Disease X]))} \cdot (d_2 + d_3) \\ &+ (q_6 \cdot \textit{Target population (with the [Disease X]))} \cdot (d_2 + d_4) \end{aligned}$$

Based on these equations, the total annual cost for all treatment lines for the [Disease X] will be:

$$\begin{aligned} &\textit{Total Annual cost for all treatment lines for the [Disease X] for} \\ &\textit{the Target population (with the [Disease X])} \\ &= (\textit{Monthly Cost of first-line} + \textit{Monthly Cost of second-line}) \cdot 12 \end{aligned}$$

It should be noted that the proportion of patients receiving each drug is gathered from reputable sources, such as national health databases, epidemiological studies, or international health organizations, and then meta-analysis is utilized in order to estimate the final proportion in each drug and in each line of treatment. In the case that for the disease of interest, there are more treatment lines, and the aforementioned equations should be populated accordingly.

Input 4: Medical Procedures and Costs – Obtain information on medical procedures and their costs related to [Disease X] from healthcare databases or relevant medical societies. Utilize systems like DRG for cost allocation. More specifically, let us assume a [Disease X] with 3 disease-related complications. Then the total annual cost for these disease-related complications is estimated with the following equation:

$$\begin{aligned} &\textit{Total annual cost of disease related complications} \\ &= (k \cdot \textit{Target population (with the [Disease X]))} \cdot (c_1 + c_2 + c_3), \end{aligned}$$

where k is the proportion of patients experiencing disease- related complications, c_1 is the total cost of 1st complication, c_2 is the total cost of 2nd complication, and c_3 is the total cost of 3rd complication.

It should be noted that the proportion of patients experiencing disease-related complications is gathered from reputable sources, such as national health databases, epidemiological studies, or international health organizations, and then meta-analysis is utilized in order to estimate the final proportion of patients experiencing at least one complication.

Input 5: Diagnostic Tests and Costs – Refer to established protocols for diagnostic tests in [Disease X]. Determine the frequency of tests and associated costs according to the relevant regulations. Use official regulations or health authority guidelines for cost data. More specifically, in this step of the proposed framework firstly the proportion of patients (m) who monitor their health status on a frequent basis should be gathered from reputable sources, such as national health databases, epidemiological studies, or international health organizations.

Let us assume that there are 4 different medical examinations which are being conducted in order to monitor the status of the patients with Disease X:

Medical examination	Frequency with which patients should be submitted to the relevant medical examinations for monitoring	Cost / medical examination / time	Annual Cost / medical examination
Med_1	n_1 times/year	c_1	$c_1 \cdot n_1$
Med_2	n_2 times/year	c_2	$c_2 \cdot n_2$
Med_3	n_3 times/year	c_3	$c_3 \cdot n_3$
Med_4	n_4 times/year	c_4	$c_4 \cdot n_4$

Then the total annual cost for these disease-related medical examinations is estimated with the following equation:

$$\begin{aligned}&\textit{Total annual cost of disease related medical examinations}\\&= (m \cdot \textit{Target population (with the [Disease X])}) \cdot (c_1 \cdot n_1 + c_2 \cdot n_2 + c_3 \cdot n_3),\end{aligned}$$

Input 6: Physician Visit Data – Determine the frequency of physician visits for [Disease X] patients based on established guidelines or recommendations. Estimate the associated costs per visit. In this step of the process, information about how often patients typically see their physician (let us assume f times/year) should be gathered from official databases and articles, as well as the cost of each visit (let us assume c/visit).

Then the total annual cost for these disease-related visits to the physician is estimated with the following equation:

$$\begin{aligned}&\textit{Total annual cost of disease related visits to physician}\\&= (\textit{Target population (with the [Disease X])}) \cdot (c \cdot f).\end{aligned}$$

Based on the aforementioned equations, the total disease-related annual cost is estimated as:

$$\begin{aligned}&\textit{Total Annual cost for all treatment lines for the [Disease X] for the Target population}\\&\quad\textit{(with the [Disease X])}\\&+ \textit{Total annual cost of disease related complications}\\&+ \textit{Total annual cost of disease related medical examinations}\\&+ \textit{Total annual cost of disease related visits to physician.}\end{aligned}$$

8.5 THE PROPOSED DATA-DRIVEN FRAMEWORK IN THE CONTEXT OF GREECE

In this section, we will give a practical example of applying the proposed data-driven framework to the Greek healthcare system, showcasing its relevance and effectiveness in guiding health budgeting decisions.

Numerical Inputs of the Methodology in the case of a Disease X.

Input 1: Prevalence of [Disease X]. Data regarding the prevalence of a given disease, let us assume disease X, can be drawn from the Global Burden of Disease (GBD) project, which provides a comprehensive picture of mortality and disability across countries, time, age, and sex. In particular, GBD project, with more than 400 scientific publications since 2010, is being collected and analyzed by a consortium of more than 9,000 researchers in 162 countries and territories, with the data capturing premature death and disability from 370 diseases and injuries in 204 countries and territories, by age and sex, from 1990 to the present (Armocida *et al.*, 2022; Burkart *et al.*, 2022; Collaborators *et al.*, 2022).

Let us assume that the prevalence of disease X has been calculated as 7%.

Input 2: The target population in Greece is the product of multiplying the prevalence rate for the disease X with the total Greek population. This leads to a total of 700.000 people suffering from disease X (based on the data of the Greek census (2011)).

Input 3: Prescription therapeutic protocols: Regarding the treatment of disease X, we can be based on the Prescription Therapeutic Protocols being published by the Greek Ministry of Health.

There are multiple available lines of treatment. In order to distribute patients in each line of treatment, we are based on several studies and the international literature. Let us assume that the distribution of patients among treatment lines for the disease X are three and that the distribution of patients among these three treatment lines is 45%, 35% and 20% respectively. The cost of each treatment line can be retrieved from the Greek Positive list, which is published by the Greek Ministry of Health. This list contains all drugs reimbursed by the Greek Government. Let us assume that the cost of each treatment line is €10 for treatment line 1, €30 for treatment line 2, and €100 for treatment line 3. Assuming the distribution of patients among treatment lines and the cost per line of treatment we conclude that the estimated expenditure for disease X among patients is €24,500,000.00.

Input 4: Medical procedures for disease X related complications. Concerning the medical procedures (based on the Diagnosis Related Groups (DRG) system) that are being applied as well as the costs related to these DRGs, data can be drawn from the DRG-matching application (http://kenicd.e-healthnet.gr), which is offered by the Medical Society of Athens in collaboration with the Onassis Cardiac Surgery Center and the Ministry of Health, in the case of Greece. Let us assume that the cost of hospitalization related to disease X amounts to €38,687,500.00.

Input 5: Diagnostic tests for Monitoring the course of patients. As regards the protocols for monitoring the course of disease X patients, we can be based on those published by the General Secretariat of the Ministry of Health. Concerning their cost, it was retrieved from the relative regulation, as well as on the appropriate relative decision from the Greek Ministry of Health and Social Solidarity.

Let us assume that the annual cost for diagnostic tests for disease X amounts to €23,646,000.00

Input 6: Visits-Consultations with physicians. Information about how often individuals with disease X typically see their doctor can be extracted from the recommendations provided by the Centres for Disease Control and Prevention (CDC). Utilizing this data, the hypothesis was formulated that each patient tends to consult their doctor around four times a year. Finally, as regards their cost, it was hypothesized that each visit costs at least a minimum of €10. Total cost for consultations of patients with physicians accounts for €28,000,000 annually.

The total minimum annual cost of treatment of disease X amounts to €114,833,500.

8.5.1 FINANCIAL AND OPERATIONAL CHALLENGES OF IMPLEMENTING THE FRAMEWORK IN GREECE

Implementing this data-driven budgeting framework in Greece presents several financial and operational challenges, given the unique characteristics of the Greek healthcare system. One notable financial barrier is the limited healthcare budget, which has been impacted by the economic constraints Greece has faced over the past decade. Public healthcare spending in Greece has traditionally been lower than in many EU countries, and the funding required to upgrade data systems, increase staffing, and train personnel could strain the existing budget. Transitioning to a data-driven budgeting system would likely require substantial initial investment to digitize records, maintain updated databases, and enhance IT infrastructure to support data integration and analysis across healthcare sectors.

From an operational perspective, Greece faces challenges related to data collection and standardization. The Greek healthcare system has limited interoperability between its various public and private healthcare providers, which can complicate efforts to collect comprehensive and reliable data for the framework. For instance, although electronic health records (EHRs) are increasingly used, their adoption is not yet uniform, and compatibility issues between healthcare information systems create barriers to efficient data sharing. Additionally, the lack of widespread health information systems (HIS) across rural healthcare facilities further complicates data gathering, posing a significant barrier to the framework's accurate and timely implementation in rural areas.

Another key operational challenge lies in personnel readiness and data literacy. Many healthcare professionals in Greece have limited training in data analytics and digital health technologies, which are essential for successful framework deployment. Bridging this skills gap would require training and capacity-building efforts, adding another layer of cost and complexity to the implementation. Moreover, due to the reliance on traditional budgeting practices, transitioning to a data-driven approach would necessitate a cultural shift within healthcare management, requiring strategic change management initiatives to foster support and buy-in from stakeholders

8.5.2 COMPARATIVE ANALYSIS WITH OTHER EU HEALTHCARE SYSTEMS

In comparison to other EU countries, the Greek healthcare system faces distinct challenges and differences in healthcare funding, data infrastructure, and readiness for a data-driven approach. Unlike Germany or the Netherlands, which have robust health IT infrastructures and high interoperability standards, Greece's healthcare data systems are still in the early stages of digital transformation. Countries like Sweden and Denmark, which have long invested in centralized digital health records and data-sharing frameworks, have advanced infrastructures that enable efficient data-driven budgeting, whereas Greece would need to build or significantly upgrade its systems to achieve similar functionality.

In terms of healthcare funding, Greece allocates a smaller proportion of its GDP to healthcare compared to many EU nations, which can limit the resources available for comprehensive, data-driven budgeting initiatives. For example, countries like France and Germany allocate around 11-12% of GDP to healthcare, while Greece's expenditure is closer to 8%, impacting the availability of funds for large-scale digital and operational enhancements. Moreover, healthcare in Greece relies more heavily on out-of-pocket spending and has fewer resources allocated to preventive and population health initiatives than systems like those in the UK or Nordic countries, where preventive care and data-driven budgeting have long been prioritized.

These comparative differences indicate that while the framework is practical for Greece, adapting it to the current realities and resources of the Greek healthcare system may involve a phased implementation approach. This could start with a focus on high-burden diseases and gradual expansion as data infrastructure, personnel capacity, and funding resources improve. Additionally, Greece could draw lessons from EU counterparts that have successfully implemented data-driven budgeting by adopting modular components of their frameworks, particularly in data standardization, interoperability protocols, and funding mechanisms, to gradually enhance its capacity to support a full data-driven healthcare budgeting system.

8.6 DISCUSSION & CONLCUSIONS

The proposed methodology presents a comprehensive and adaptable approach to estimating healthcare costs associated with Disease X. By integrating various aspects of disease management and leveraging diverse data sources, this framework enhances the accuracy and reliability of cost estimates. This methodology systematically outlines the steps involved in estimating healthcare costs, from defining treatment protocols to assessing expenses for medications, procedures, diagnostic tests, and consultations. By emphasizing the integration of national health databases, epidemiological research, and international health organization data, the methodology ensures the robustness of input parameters for cost estimation.

One of the notable features of this framework is its flexibility. It accommodates multiple treatment lines and considers the proportion of patients receiving each drug, providing a nuanced understanding of real-world scenarios. Additionally, the inclusion of disease-related complications and associated costs, guided by the Diagnosis-Related Group (DRG) system and supported by reputable sources, adds depth to the estimation process.

Furthermore, the methodology addresses costs related to diagnostic tests and physician visits, incorporating the frequency of medical assessments and consultations. By employing meta-analysis techniques to estimate proportions, such as complication rates or drug utilization rates, the methodology enhances the accuracy of estimates derived from diverse data sources. Lastly, the framework acknowledges the importance of adjusting costs based on the target population size. By linking the estimated prevalence of Disease X to the overall population, the methodology enables scalable cost projections, facilitating informed decision-making in resource allocation.

This new analytical bottom-to-top strategy is designed to address the increasing financial burdens encountered when establishing healthcare budgets and offers a robust and sophisticated approach to healthcare cost estimation for any given disease with implications for informing policy decisions and resource allocation strategies in formulating efficient and optimum budgets across healthcare systems.

To further enhance the robustness and applicability of this framework, future research could explore the integration of emerging technologies such as artificial intelligence (AI), machine learning (ML), and real-time data analytics. The incorporation of AI and ML could support the framework by providing predictive capabilities that allow for dynamic adjustments in budgeting based on evolving healthcare needs and trends. For example, AI-driven predictive models could improve the accuracy of prevalence estimates and detect shifts in disease patterns, enabling a more proactive approach to resource allocation. Machine learning techniques could also help refine estimates related to drug utilization, complication rates, and treatment effectiveness by analysing large-scale, real-world data from electronic health records (EHRs) and other sources.

In addition, real-time data analytics could enhance the responsiveness of the framework to new information, such as emerging health crises or seasonal disease patterns. This capacity for real-time data integration would be particularly valuable in healthcare systems where demand fluctuates, and rapid adjustments are needed to optimize resource allocation. Future research could explore methods for integrating continuous data feeds into the framework to enable real-time updates, allowing healthcare providers and policymakers to make more informed decisions.

To strengthen the proposed framework, it is essential to recognize potential limitations and challenges related to data collection and implementation, particularly in healthcare systems with varying levels of data infrastructure.

The effectiveness of this framework relies heavily on the availability and quality of comprehensive epidemiological, demographic, and cost data. However, in many healthcare systems, especially in low- and middle-income countries, data may be incomplete, outdated, or lacking in detail. These gaps can affect the accuracy of disease prevalence estimates, treatment cost assessments, and patient distribution calculations, which may lead to budget misalignments. To address these challenges, alternative approaches to data collection, such as partnerships with international health organizations or the use of regional data proxies, can be considered until national databases are further developed.

The proposed data-driven approach also depends on robust data collection infrastructure, which may not be present in all healthcare settings. Systems lacking standardized health information systems (HIS) and electronic health records (EHRs) will face additional obstacles in collecting, storing, and analyzing the large volumes of data required by this framework. For healthcare systems with limited digital infrastructure, manual data collection methods or phased digitization of health records might be necessary as interim solutions. This gradual approach may require supplementary resources and training to ensure staff are capable of efficiently managing and utilizing the data.

Healthcare systems vary significantly in their treatment protocols, regulatory guidelines, and cost structures, which may complicate efforts to standardize the framework. For example, recommended treatment protocols for a specific ICD-10 category may differ due to national resources, regulatory standards, and policy approaches. This variability highlights the importance of adapting the framework to align with local guidelines. By designing the framework with modularity in mind, healthcare providers can input specific treatment protocols and cost models, ensuring it remains relevant and adaptable to local healthcare settings.

While data-driven approaches enhance decision-making, there is a risk of bias if underlying data are not representative of all demographic groups. For instance, data collection that primarily reflects urban populations may result in the underrepresentation of rural or marginalized groups, leading to skewed budget allocations. To counter this, healthcare systems should aim to collect disaggregated data that reflect a diverse range of demographic and geographic populations. Additionally, incorporating data from community-based studies or using sampling techniques can provide a more balanced view and support more equitable resource allocation.

Another critical concern when using extensive patient data is ensuring data privacy and security. As the framework involves sensitive health and demographic data, strong privacy and security measures are essential for maintaining public trust and regulatory compliance. Systems with limited digital infrastructure may face additional hurdles in safeguarding data against unauthorized access or breaches, underscoring the need for investments in secure data handling practices.

To increase the adaptability of the framework across healthcare systems with varying data capacities, the proposed model includes several strategies. For instance, well-developed systems with existing HIS and EHRs can use these resources to enhance data accuracy and streamline the process, while systems with basic or fragmented infrastructure might rely on data from smaller regional databases or public health studies until more comprehensive systems are implemented. Additionally, partnerships with international health organizations like the World Health Organization (WHO) or the Global Burden of Disease (GBD) initiative could provide supplemental data on disease prevalence, treatment guidelines, and cost estimates, along with technical support and training to build capacity over time.

The modular design of the framework also allows customization for different healthcare contexts. Each step—such as prevalence estimation, treatment cost calculation, and protocol identification—can be adjusted based on the availability of data and regulatory requirements in each country. This flexibility enables healthcare providers to tailor the framework progressively as their data infrastructure improves. In settings with limited digital capacity, building data literacy and technical skills among healthcare professionals will further support implementation. Providing training on data collection, management, and interpretation empowers staff to use the framework accurately and contributes to a sustainable workforce capable of maintaining and enhancing the framework's application over time.

In summary, while the proposed framework offers a structured, bottom-up approach to healthcare budgeting, several limitations need to be considered, including data availability, infrastructure limitations, cost variability, and privacy concerns. By acknowledging these constraints and incorporating customizable modules, the framework can be adapted to various healthcare systems, promoting equitable and efficient budgeting decisions across diverse healthcare environments. This approach not only enhances resource allocation but also aligns budget decisions more closely with the actual healthcare needs of diverse communities.

REFERENCES

Ahmed, S., Alshater, M. M., El Ammari, A., & Hammami, H. (2022). Artificial intelligence and machine learning in finance: A bibliometric review. *Research in International Business and Finance*, *61*, 101646.

Akinrinmade, A. O., Adebile, T. M., Ezuma-Ebong, C., Bolaji, K., Ajufo, A., Adigun, A. O., Mohammad, M., Dike, J. C., & Okobi, O. E. (2023). Artificial intelligence in healthcare: Perception and reality. *Cureus*, *15*(9).

Amiri, M. M., Kazemian, M., Motaghed, Z., & Abdi, Z. (2021). Systematic review of factors determining health care expenditures. *Health Policy and Technology*, *10*(2), 100498.

Armocida, B., Monasta, L., Sawyer, S., Bustreo, F., Segafredo, G., Castelpietra, G., Ronfani, L., Pasovic, M., Hay, S., Sawyer, S. M., *et al.* (2022). Burden of non-communicable diseases among adolescents aged 10–24 years in the eu, 1990–2019: A systematic analysis of the global burden of diseases study 2019. *The Lancet Child & Adolescent Health*, *6*(6), 367–383.

Arrow, K. J. (1978). *Uncertainty and the welfare economics of medical care*. In *Uncertainty in economics* (345–375). Elsevier.

Bahrammirzaee, A. (2010). A comparative survey of artificial intelligence applications in finance: Artificial neural networks, expert system and hybrid intelligent systems. *Neural Computing and Applications*, *19*(8), 1165–1195.

Batko, K., & Slkezak, A. (2022). The use of big data analytics in healthcare. *Journal of big Data*, *9*(1), 3.

Bergmann, M., Brück, C., Knauer, T., & Schwering, A. (2020). Digitization of the budgeting process: Determinants of the use of business analytics and its effect on satisfaction with the budgeting process. *Journal of Management Control*, *31*(1), 25–54.

Burkart, K., Causey, K., Cohen, A. J., Wozniak, S. S., Salvi, D. D., Abbafati, C., Adekanmbi, V., Adsuar, J. C., Ahmadi, K., Alahdab, F., *et al.* (2022). Estimates, trends, and drivers of the global burden of type 2 diabetes attributable to pm2· 5 air pollution, 1990–2019: An analysis of data from the global burden of disease study 2019. *The Lancet Planetary Health*, *6*(7), e586–e600.

Chao, K., Sarker, M. N. I., Ali, I., Firdaus, R. R., Azman, A., & Shaed, M. M. (2023). Big data-driven public health policy making: Potential for the healthcare industry. *Heliyon*, *9*(9).

Chloupkova, J., TINGGAARD SVENDSEN, G., & Zdechovsky, T. (2018). A global meat tax: From big data to a double dividend. *Agricultural Economics/Zemědělská Ekonomika*, *64*(6).

Collaborators, G. 2. A., *et al.* (2022). Global, regional, and national burden of diseases and injuries for adults 70 years and older: Systematic analysis for the global burden of disease 2019 study. *bmj*, *376*.

Cozzoli, N., Salvatore, F. P., Faccilongo, N., & Milone, M. (2022). How can big data analytics be used for healthcare organization management? literary framework and future research from a systematic review. *BMC health services research*, *22*(1), 809.

Davis, S., Nourazari, S., Granovsky, R., & Fard, N. (2021). Predicting a need for financial assistance in emergency department care. *Healthcare*, *9*(5), 556.

Dixit, S., Mao, W., McDade, K. K., Schäferhoff, M., Ogbuoji, O., & Yamey, G. (2022). Tracking financing for global common goods for health: A machine learning approach using natural language processing techniques. *Frontiers in Public Health*, *10*, 1031147.

Downes, R., Moretti, D., & Nicol, S. (2017). Budgeting and performance in the european union: A review by the oecd in the context of eu budget focused on results. *OECD Journal on Budgeting*, *17*(1), 1–60.

Du, L., *et al.* (2020). Medical emergency resource allocation model in large-scale emergencies based on artificial intelligence: Algorithm development. *JMIR medical informatics*, *8*(6), e19202.

Gupta, S., & Tripathi, P. (2016). An emerging trend of big data analytics with health insurance in india. *2016 International conference on innovation and challenges in cyber security (ICICCS-INBUSH)*, 64–69.

Hartman, M., Smith, C., Heffler, S., & Freeland, M. (2006). Monitoring health spending increases: Incremental budget analyses reveal challenging tradeoffs. *Health care financing review*, *28*(1), 41.

Hilton, R. M., & Joyce, P. G. (2007). Performance information and budgeting in historical and comparative perspective. *Handbook of Public Administration: Concise Paperback Edition*, 247.

Hollingsworth, B. (2016). *Health system efficiency: Measurement and policy*. In *Health system efficiency: How to make measurement matter for policy and management [internet]*. European observatory on health systems; policies.

Ji, Z. (2016). Building of a standardized health insurance monitoring model based on data mining. *Int J Simul Syst Sci Technol*, *17*(18), 101–4.

Jia, J., Lai, Y., Yang, Z., & Li, L. (2022). The optimal strategy of enterprise key resource allocation and utilization in collaborative innovation project based on evolutionary game. *Mathematics*, *10*(3), 400.

Kaplan, R. S., Porter, M. E., *et al.* (2011). How to solve the cost crisis in health care. *Harv Bus Rev*, *89*(9), 46–52.

Kunnathuvalappil Hariharan, N. (2020). Rethinking budgeting process in times of uncertainty.

Kutzin, J. (2013). Health financing for universal coverage and health system performance: Concepts and implications for policy. *Bulletin of the World Health Organization*, *91*, 602–611.

Larson, D., & Chang, V. (2016). A review and future direction of agile, business intelligence, analytics and data science. *International Journal of Information Management*, *36*(5), 700–710.

Li, L., Bagheri, S., Goote, H., Hasan, A., & Hazard, G. (2013). Risk adjustment of patient expenditures: A big data analytics approach. *2013 IEEE international conference on big data*, 12–14.

Martin, S., Rice, N., & Smith, P. C. (2008). Does health care spending improve health outcomes? evidence from english programme budgeting data. *Journal of health economics*, *27*(4), 826–842.

Mitton, C., & Donaldson, C. (2001). Twenty-five years of programme budgeting and marginal analysis in the health sector, 1974-1999. *Journal of Health Services Research & Policy*, *6*(4), 239–248.

Moll, B., Townsend, R. M., & Zhorin, V. (2017). Economic development, flow of funds, and the equilibrium interaction of financial frictions. *Proceedings of the National Academy of Sciences*, *114*(24), 6176–6184.

Paganelli, A. I., Mondéjar, A. G., da Silva, A. C., Silva-Calpa, G., Teixeira, M. F., Carvalho, F., Raposo, A., & Endler, M. (2022). Real-time data analysis in health monitoring systems: A comprehensive systematic literature review. *Journal of Biomedical Informatics*, *127*, 104009.

Paul, E., Sambieni, N. E., Wangbe, J.-P., Fecher, F., & Bourgeois, M. (2020). Budgeting challenges on the path towards universal health coverage: The case of benin. *Health Economics Review*, *10*, 1–8.

Raghupathi, W., & Raghupathi, V. (2014). Big data analytics in healthcare: Promise and potential. *Health information science and systems*, *2*, 1–10.

Rajan, D., Barroy, H., & Stenberg, K. (2016). Budgeting for health. *Strategizing National Health in the 21st Century: A Handbook. 1st ed. World Health Organization.*

Ramirez, E., Liu, P., Frankwick, G. L., *et al.* (2024). Evaluating underlying factor structures using novel machine learning algorithms: An empirical and simulation study. *Journal of Business Research*, *173*, 114472.

Rauner, M. S., Schaffhauser-Linzatti, M. M., & Bauerstaetter, J. (2015). Decision support system for social occupational injury insurance institutions: Cost analysis and targeted resource allocation. *Central European Journal of Operations Research*, *23*, 1–29.

Seixas, B. V., Dionne, F., & Mitton, C. (2021). Practices of decision making in priority setting and resource allocation: A scoping review and narrative synthesis of existing frameworks. *Health Economics Review*, *11*, 1–11.

Singh, S. R. (2014). Public health spending and population health: A systematic review. *American Journal of Preventive Medicine*, *47*(5), 634–640.

Wang, W. R. (2020). Research on china's tax policy for responding to public health emergencies under the background of big data. *2020 International Conference on Modern Education and Information Management (ICMEIM)*, 232–235.

White, J. (2014). The challenge of budgeting for healthcare programmes. *OECD journal on budgeting*, *14*(1), 73–107.

Wolfe, P. R., & Moran, D. W. (1993). Global budgeting in the oecd countries. *Health Care Financing Review*, *14*(3), 55.

Zhang, R., & Bohlen, J. (2023). Healthcare business budgeting.

Zhang, W., Cao, G., Wu, F., Wang, Y., Liu, Z., Hu, H., & Xu, K. (2023). Global burden of prostate cancer and association with socioeconomic status, 1990–2019: A systematic analysis from the global burden of disease study. *Journal of Epidemiology and Global Health*, *13*(3), 407–421.

Zhao, J., Xi, X., Wang, S., & Gong, C. (2021). Dynamic analysis of different resource allocations: Implications for resource orchestration management of strategic alliances. *Computers & Industrial Engineering*, *158*, 107393.

9 Statistics and Machine Learning in Industry: Combining Heterogeneous or Multi-Scale Model Outputs

Jean-Michel Poggi
LMO, University Paris-Saclay, Orsay, France
University Paris Cité, Dept SD, IUT Paris-Rives de Seine, Paris, France
E-Mail: jean-michel.poggi@universite-paris-saclay.fr. ORCID: 0000-0002-8222-1653.

9.1 INTRODUCTION

This chapter proposes some applied considerations about the prominent role of data and the necessity to focus on the data produced by the algorithms. The idea of this chapter comes from a discussion after a talk given by Neil D. Lawrence (see Lawrence, 2019 and also Lawrence, 2017) about an important problem: to characterize the process for machine learning systems design by presenting the 3D process (Decomposition, Data, and Deployment). This leads to "a co-evolution of systems, upstream errors can be compensated by downstream corrections and end-to-end training is to be considered". The approach suggests using Machine Learning to assist with Machine Learning. It focuses on data-driven decision-making and proposes to move from a software first paradigm to a data first paradigm.

This is useful to the field of statistics as promoters of a typical Machine Learning (ML) approach to data feeding models or software outputs. Following Feurer *et al.*, 2015, such systems automatically consider past performance on similar datasets and construct ensembles from models evaluated during optimization.

As a statistician, I could focus on a discussion about the quite traditional question of Data Quality which has a fundamentally different meaning than the one implicitly considered by computer scientists and learners. The latter rather insists on data as a flow and focuses on the process of successive transformations of the data. I would like to stress, on the contrary, that such approaches highlighting the prominent role of data and the necessity to consider the data produced by the algorithms as new data, would be very successful for statisticians.

Applied statisticians could benefit greatly from experimenting with some of the methods of machine learning in a more systematic way. Sequential learning proposes to combine optimally, and therefore also to optimally reject, data produced by algorithms, whether or not based on

DOI: 10.1201/9781032664330-9

statistical methods, or by experts. I propose to illustrate this important contribution to applied statistics methodology by considering two examples of forecasting in the industry. These examples use ML algorithms to combine "second-order" data, i.e., data from model outputs coming from heterogeneous experts (first example referred to as "heterogeneous case" in the sequel) and experts related to multiscale aggregates (second example referred to as "multiscale case" in the sequel). We will use the sequential prediction paradigm which allows us to combine several models of internal structure, known or unknown, built on very different assumptions in a unified approach. A very attractive feature is that it does not require any prior knowledge about the internal way used by each expert to generate predictions. The crucial ingredient is to not select the best model or the best resolution but at the contrary by leveraging diversity and optimally organizing the cooperation of the different models or different resolutions.

To get back to the seminal motivation and make clearer the connection with Lawrence, 2019, let us sketch two remarks. The second example will illustrate meta-learning, that is Machine Learning to assist Machine Learning since the models are built internally using procedures supposed to be assessed a priori. The first example makes more complex this view by allowing us to consider in addition predictions made externally, made by others, and of unknown or questionable quality. These two situations address to some extent the "Data First Paradigm" in such context as well as a way of "Combining Data and Systems Design".

The organization of the paper is as follows. The next section presents the two running examples with industrial contexts. We then briefly recall the principle of sequential learning. The third section summarizes the key products of the two applications. A short conclusion is the last section.

9.2 TWO RUNNING EXAMPLES

The "heterogeneous case" comes from a problem of pollution forecasting and deals with combining heterogeneous model outputs. Each day, many forecasts of the daily average concentration of PM10 for the day after, are available by monitoring agencies and come from two families. First, statistical models are constructed using different methods and different set of predictors, and second, deterministic models of physicochemical prediction modelling pollution, weather, and atmosphere. The challenge is to find a way to integrate these sources. Note that Data integration is the third dimension in the information quality framework, Kenett & Shmueli, 2016. Indeed, for each given instant, the forecasts often disagree. The idea is to perform a sequential prediction by combining forecasts to have a single better ensemble prediction coming from "heterogeneous" methods. A solution is provided in Auder *et al.*, 2016.

The "multiscale case" comes from bottom-up forecasting of electricity demand; see Cugliari *et al.*, 2016 and Auder *et al.*, 2018. Typically, starting from the consumption of individual clients, the problem is to forecast the aggregate demand for the next day. Here the challenge is to deal with a sliding scale of data resolution (another dimension in the information quality framework), i.e., to disaggregate the global signal to improve forecasts. The idea is to look for a hierarchical clustering generating groups of customers at various scales, predict the demands of the different groups, and combine it to provide the best aggregate forecast. A solution involving the sequential aggregation of multiscale random forests-based experts is provided in Goehry *et al.*, 2019 (for a recent review of multiscale and multi-granularity analysis tools in the industry, see Reis, 2019).

9.3 KEY TOOL: SEQUENTIAL PREDICTION

The theory of prediction of individual sequences focuses on the rules of aggregation of a set of experts and examines how to weigh and combine these experts. Valuable references are a first survey paper by Clemen, 1989, the book of Cesa-Bianchi & Lugosi, 2006 containing the key material, and a comprehensive general paper by Stoltz, 2010 in French.

The principle is as follows. Let us denote the complete time series from instant 1 to T:

$$y_1, \dots, y_t, \dots, y_T \in [0, B]$$

where B is a known upper bound on the y-values.

The problem: forecasting y_t, given the past values of instant t: $y_1, \dots, y_{t-1}$ using K given experts known only through their instantaneous predictions of y_t denoted by

$$x_t = (x_{1.t}, \dots, x_{K.t}) \in [0, B]^K$$

and the history of their predictions

$$x_1^t = x_1, \dots, x_t$$

The idea is to optimally linearly combine the predictions:

$$\hat{y}_t = p_t.x_t = \sum_{k=1}^{K} p_{k.t} x_{k.t} \tag{9.1}$$

by choosing the weights $p_t = (p_{1.t}, \dots, p_{K.t})$ at each step with respect to the instantaneous losses: the losses at time t of the individual experts

$$L_t(x_{k.t}) = (y_t - x_{k.t})^2 \tag{9.2}$$

and the global one

$$L_t(x) = (y_t - x)^2 \tag{9.3}$$

Observations and predictions are made sequentially and are not subject to any stochastic modelling. The theoretical objective is to predict almost as well as the best expert. It should be noted that this expert is unknown in sequential prediction and can only be determined at the end of the period (at instant T).

Two families of methods with theoretical guarantees are available.

First, the methods adaptively updating the weights at each step like EWA (Exponential Weighted Average):

$$p_{k.t} = c.exp(-\eta \sum_{s=1}^{t-1} L_s(x_{k.s})) \tag{9.4}$$

where $\eta > 0$ is the learning rate which can be simply chosen of the form $\eta = 1/MT$ where M is the loss upper bound, in many practical applications (see Stoltz, 2010 for more details to fine tune this parameter). The constant c is chosen to normalize the weights in such a way that $\sum_{k=1}^{K} p_{k.t} = 1$ leading to a convex combination of experts, suitable for unbiased experts mixing.

Second, the methods optimizing at each step a global criterion on the history of measurements and expert predictions (with or without convexity constraints), like RR (Ridge regression type criterion):

$$p_{..t} = argmin\{\sum_{s=h}^{t-1} (y_s - u.x_s)^2 + \lambda \|u - p_0\|_2^2; u \in \mathbb{R}^K\} \tag{9.5}$$

or the Lasso criterion:

$$p_{..t} = argmin\{\sum_{s=h}^{t-1} (y_s - u.x_s)^2 + \lambda \|u - p_0\|^1; u \in \mathbb{R}^K\}$$

These schemes allow to relax the convexity constraint which is especially relevant when the set of experts contains some biased elements. Of course, again the choice of the regularization parameter λ is crucial and various approaches are proposed in Stoltz, 2010 including simple cross-validation.

In fact, since this type of method does not require information about the internal structure of each expert, the usual results such as consistency of parameter estimation are not relevant. Then some other type of reference is needed to compare the actual performance with. Usually, two oracles are used, based on minimizing the loss over the whole period, which is, of course, unknown to the forecaster making the forecasts in real time. The first oracle is the best expert over the whole period (from time 1 to T) and the second is the best constant (over time) convex combination of experts.

Let us finally mention that these procedures together with various methods to set the learning rate or the regularization parameter are implemented in the R package opera; see Gaillard *et al.*, 2021.

9.4 SEQUENTIAL PREDICTION FOR COMBINING OUTPUTS

Sequential prediction thus allows several models based on very different assumptions to be combined in a unified approach. Secondly, it does not require any prior knowledge of the internal way in which each expert generates predictions. It is therefore especially relevant for combining outputs from models of different natures (see example 1) or to combine multiscale aggregates (see example 2).

9.4.1 COMBINING HETEROGENEOUS EXPERTS

This illustration deals with heterogeneous experts. The problem is to predict the concentration for the next day. In Auder *et al.*, 2016 a dataset from April 2013 to March 2014, is considered. It gives the measures in the monitoring station of Normandy of the daily average concentration of PM10 (including the volatile fraction) and the corresponding forecasts of the day for the day after coming from 10 different prediction models:

- 4 statistical models, built by the regional air quality agency: a mixture of regression models, two specialized linear models: one fitted focusing on slightly polluted days and the other one learning on polluted days) and a generalized additive model (GAM). The variables entering in these models are classical for local pollution modelling: some other pollutants, some meteorological variables (including temperatures, temperature gradients, wind speed and direction, relative humidity, rain), and calendar variables.
- 2 statistical generalized linear (GLM) models, built by national organisms. These models incorporate in addition to the previously considered variables, some large-scale information coming from the deterministic and giving more accurate indications on air mass circulation.
- 3 numerical models at different spatial resolutions coming from meteorological agencies. These models are based on entirely different ideas and are deterministic models of physicochemical prediction modelling pollution, weather, and atmosphere. They provide a multi-scale chemistry-transport model designed to produce daily forecasts of pollutants at different scales. They can be very useful, for example during episodes of heavy pollution or imported pollution, but are very cumbersome to implement and partly opaque.
- 1 classical baseline, the persistence model, predicting for tomorrow the average concentration of today. It is of course the simplest model and also the less accurate.

In Fig. 9.1, the different forecasts (in grey) superimposed with the actual measures of PM10 (in bold) are displayed during one year in a given station in the centre of Rouen, the main city of the Normandy region.

The forecasts often disagree as can be seen in Fig. 9.2 zooming in on 50 days and displaying the forecasts using different light colours. A key question is how to exploit the large variety of models and the diversity of the forecasts to obtain a better prediction.

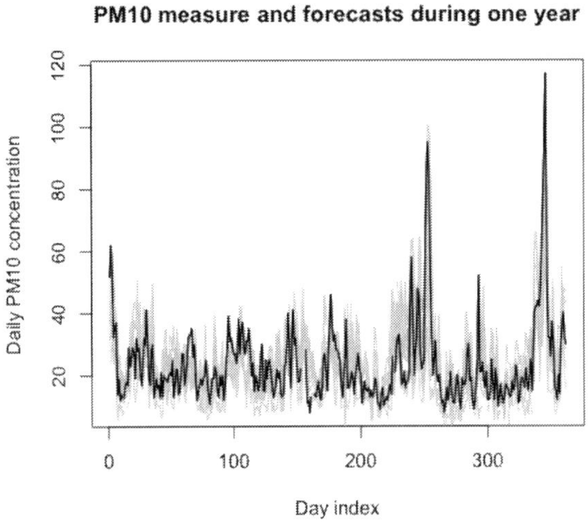

Figure 9.1 For a monitoring station located in the centre of Rouen (France), the different forecasts (in grey) superimposed with the actual measures of PM10 (in bold) are displayed during one year.

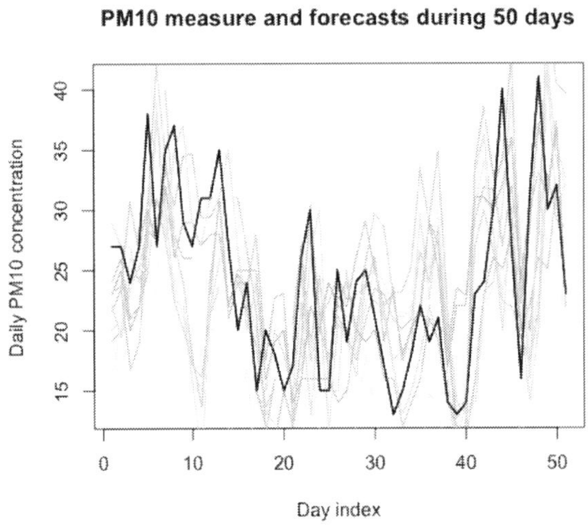

Figure 9.2 For a monitoring station located in the centre of Rouen (France), the different forecasts (displayed using different light colours) superimposed with the actual measures of PM10 (in bold) are displayed during 50 days.

Even if we had some information on the design ingredients for some of these models, we could not take this into account to combine the forecasts properly.

Sequential prediction (see Equation 9.1) allows several models based on very different assumptions to be mixed in a unified approach, and secondly, it does not require any prior knowledge of the internal way each expert uses to generate predictions. It is therefore particularly relevant for combining the results of models of different types (statistical models and physicochemical

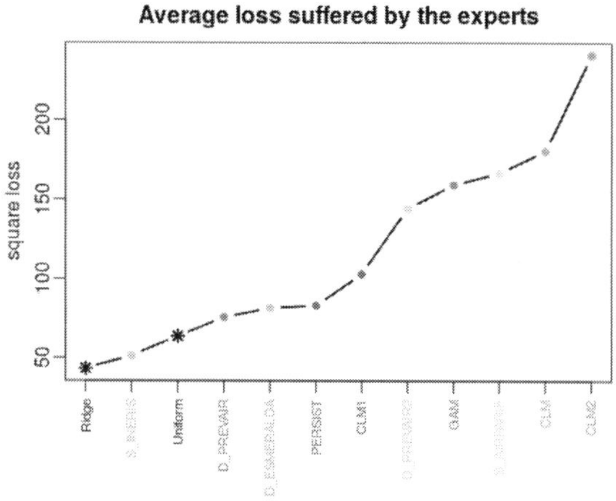

Figure 9.3 For a monitoring station located in the centre of Rouen (France), the mean square prediction error of the different individual experts is compared with the best expert (which is an oracle, unknown in real time) and compared with the average expert (known since it suffices to take all the weight equal to the same value).

deterministic models).

Before analyzing more deeply the performance, let us illustrate the results globally. In Fig. 9.3, the mean square prediction error of the different individual experts is compared with the best expert (which is an oracle, unknown in real time) and compared with the average expert (known since it suffices to take all the weight equal to the same value).

Sequential prediction strategy improves clearly the performance of the best expert both in errors and in alerts as illustrated by the observed-forecasted scatterplot (see Fig. (9.4)). Two well-known indicators are used to assess the quality of the forecast: the RMSE (Root Mean Square Error), which focuses on the average forecast error, and the Threat Score, which is dedicated to alerts and is positive and equal to 1 for a perfect forecast. EWA (Exponential Weighted Average) in the middle and RR (Ridge Regression) on the right. The baseline, given by the best expert performance, is a RMSE of about 7.9 and a TS of about 0.53. In this station, EWA does improve the performance but for RR the contribution of the combination is clearly better since it reaches, thanks to the non-convex weighting strategy RR, the "unbiasedness" is extremely difficult to obtain by classical methods.

The "unbiasedness" and good performance of RR (see Equation 9.5), resulting in a good balance between false alarms and missed alarms, comes from two sources. First, for each day, some methods overestimate and others underestimate, and these subsets of methods change over time. Second, the RR method can take advantage of this situation because it is not constrained by the convexity of the weights. Therefore, some weights of the combination can be negative and large in absolute value as can be seen in Fig. (9.5) displaying the weights of the experts over time obtained with the RR strategy. This allows the forecasts not to shrink and helps to obtain a debiased aggregate.

9.4.2 DISAGGREGATED FORECASTING

In the "heterogeneous case", we have a given set of experts corresponding to outputs of different models and sequential aggregation provides a way to optimally combine them. The "multiscale case" is different and illustrates another aspect. Indeed the strategy starts with the generation of a huge number of models, the experts, and sequential aggregation is used to select appropriate ones and to automatically reject the others.

More precisely, this second illustration deals with disaggregated forecasting, Goehry *et al.*, 2019

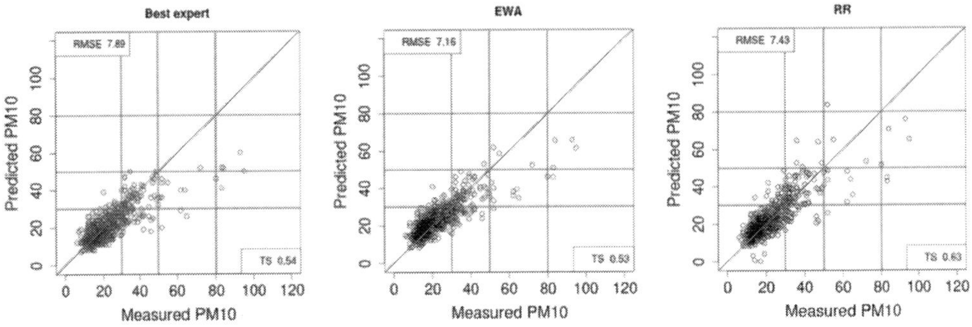

Figure 9.4 Scatterplots predicted-observed of three mixtures of experts for a given station. ideally, all the points should line up on the bisector. On the left, the performance of the best expert is presented which is satisfactory but with the classical bias: the rare large values are underestimated and the small values are slightly overestimated. In the middle, the results obtained using the convex strategy EWA are improved but the bias is not corrected. On the right, using RR and relaxing the convex constraint, the expected debiased shape is obtained while the global performance is only slightly less good. (This figure comes from the paper Auder *et al.*, 2016 and is reproduced courtesy of the journal Atmospheric Pollution Research).

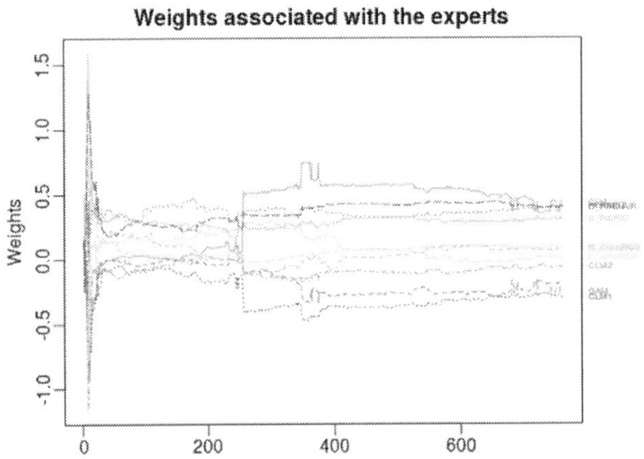

Figure 9.5 Weights of the individual experts over time obtained with the RR strategy for the same given station. This figure comes from classical outputs of the R package opera; see Gaillard *et al.*, 2021. The weights are extremely variable during a few initial iterations and then oscillate more smoothly before the final stabilization. It should be noted that some weights are positive while others are negative.

consider N individual customers and the problem is to disaggregate the global signal to improve the forecast of the global demand Z. The bottom-up forecasting strategy consists of grouping individuals into clusters, fitting forecasting models to each cluster, and aggregating the forecasts to predict the total. The intuition is that the population could be divided into sub-populations with different consumption habits, requiring different models.

Indeed, disaggregation intuitively introduces more sources of error, but it may be useful to disaggregate the global signal in order to significantly improve the forecast. One idea is to find a good trade-off between homogeneity within clusters and the quality of model estimation. Homogeneity increases as the number of clusters increases, while quality decreases as the number of clusters increases (see Misiti *et al.*, 2010), although it should be noted that in the electrical context, homogeneous clusters open up the possibility of making better use of covariates (weather, calendar, etc.).

Moreover, in the bottom-up forecasting literature, customer grouping is achieved by clustering methods, and the final forecast is obtained by a simple aggregation of the group forecasts. In Goehry *et al.*, 2019, we propose an aggregation performed with time-varying weights derived from robust online expert aggregation methods to take advantage of a very diverse and potentially large number of forecasts.

The approach is to build experts using random forests trained on some subsets of customers, then normalize their predictions and aggregate them using a convex expert aggregation algorithm to predict system load.

More precisely, the strategy is in three steps:

- First, to disaggregate the global signal by generating using hierarchical clustering, K groups of customers, with possible overlaps, with $K << N$. Then the demand $Z_{t,k}$ of group k at time t, is to be predicted involving to model the electricity demand.
- The second step generates many models from clusters at different scales using random forests, a flexible nonparametric forecasting method introduced by Breiman Breiman, 2001, allowing to easily consider exogenous variables.
- The last step is to combine these different model outputs to produce a forecast of Z_t the aggregate demand at instant t.

Applying these ideas on the well-known public dataset Irish data, the new aggregation method is compared with two strategies for building subsets of customers: hierarchical clustering based on based on survey data and/or load characteristics and the baseline: random clustering strategy. We observed that disaggregation leads to a huge gain (but not more than random) of about 25% in terms of errors. Random forests provide useful predictors for all aggregation scales, but also a lot of irrelevant ones, and decisive additional gains are obtained thanks to the sequential combination of the group's predictors.

9.5 CONCLUSION

This chapter has proposed, through two industrial use cases, some considerations about the prominent role of data and the need to focus on the data produced by the algorithms. Applied statisticians could benefit from experimenting more systematically with some machine learning methods. Sequential learning proposes to optimally combine, and therefore to optimally reject, the data produced by algorithms, whether based on statistical methods or not, or by experts.

It promotes the bridge between machine learning and applied statistics, but also looks at the problem of data quality produced by algorithms from a different perspective, more operational and indirect. In fact, it focuses more on how the data can be useful, rather than just considering its quality as a measure, which is often impossible in the context of new data in the era of big data.

Finally, the problem of obtaining a prediction interval is important from a statistical point of view. It remains unsolved until recently since this problem is not easy to solve with this kind of model-free method, especially in the time series context. But, let us mention that some very recent advances in conformal prediction in the non-exchangeable case open up the possibility of computing such an interval (see, for example, Barber *et al.*, 2023, Zaffran *et al.*, 2022).

ACKNOWLEDGEMENTS

We would like to thank the reviewers for their very constructive feedback and suggestions.

REFERENCES

Auder, B., Bobbia, M., Poggi, J.-M., & Portier, B. (2016). Sequential aggregation of heterogeneous experts for pm10 forecasting. *Atmospheric Pollution Research*, *7*(6), 1101–1109.

Auder, B., Cugliari, J., Goude, Y., & Poggi, J.-M. (2018). Scalable clustering of individual electrical curves for profiling and bottom-up forecasting. *Energies*, *11*(7), 1893.

Barber, R. F., Candes, E. J., Ramdas, A., & Tibshirani, R. J. (2023). Conformal prediction beyond exchangeability. *The Annals of Statistics*, *51*(2), 816–845.

Breiman, L. (2001). Random forests. *Machine learning*, *45*, 5–32.

Cesa-Bianchi, N., & Lugosi, G. (2006). *Prediction, learning, and games*. Cambridge university press.

Clemen, R. T. (1989). Combining forecasts: A review and annotated bibliography. *International journal of forecasting*, *5*(4), 559–583.

Cugliari, J., Goude, Y., & Poggi, J.-M. (2016). Disaggregated electricity forecasting using wavelet-based clustering of individual consumers. *2016 IEEE International Energy Conference (ENERGYCON)*, 1–6.

Feurer, M., Klein, A., Eggensperger, K., Springenberg, J., Blum, M., & Hutter, F. (2015). Efficient and robust automated machine learning. *Advances in neural information processing systems*, *28*.

Gaillard, P., Goude, Y., Plagne, L., Dubois, T., & Thieurmel, B. (2021). *Opera: Online prediction by expert aggregation* [R package version 1.2.0]. https://CRAN.R-project.org/package=opera.

Goehry, B., Goude, Y., Massart, P., & Poggi, J.-M. (2019). Aggregation of multi-scale experts for bottom-up load forecasting. *IEEE Transactions on Smart Grid*, *11*(3), 1895–1904.

Kenett, R. S., & Shmueli, G. (2016). *Information quality: The potential of data and analytics to generate knowledge*. John Wiley & Sons.

Lawrence, N. D. (2017). Data readiness levels. *arXiv preprint arXiv:1705.02245*.

Lawrence, N. D. (2019). Data science and digital systems: The 3ds of machine learning systems design. *arXiv preprint arXiv:1903.11241*.

Misiti, M., Misiti, Y., Oppenheim, G., & Poggi, J.-M. (2010). Optimized clusters for disaggregated electricity load forecasting. *REVSTAT-Statistical Journal*, *8*(2), 105–124.

Reis, M. S. (2019). Multiscale and multi-granularity process analytics: A review. *Processes*, *7*(2), 61.

Stoltz, G. (2010). Agrégation séquentielle de prédicteurs: Méthodologie générale et applications à la prévision de la qualité de l'air et à celle de la consommation électrique. *Journal de la Société française de Statistique*, *151*(2), 66–106.

Zaffran, M., Féron, O., Goude, Y., Josse, J., & Dieuleveut, A. (2022). Adaptive conformal predictions for time series. *International Conference on Machine Learning*, 25834–25866.

10 The Value of Data Exploitation in Sports Industry: An Overview

Christos Spyropoulos
Department of Business Administration, University of Piraeus, Greece

Polychronis Economou
Department of Civil Engineering, University of Patras, Greece
E-Mail: peconom@upatras.gr. ORCID: 0000-0001-6452-5920.

Sotiris Bersimis
Department of Business Administration, University of Piraeus, Greece
E-Mail: sbersim@unipi.gr. ORCID: 0000-0002-7448-1172.

Abbreviations	
Convolutional Neural Network	CNN
Deep Learning	DL
General Data Protection Regulation	GDPR
K Nearest Neighbour	KNN
Machine Learning	ML
National Football League	NFL
Principal Component Analysis	PCA
Reinforcement Learning	RL
Statistical Learning	SL

10.1 INTRODUCTION

Sports analytics has a history that goes back to the 1861 Baseball guide, The Beadle's Dime Baseball Player (Irwin P. Beadle & Co., 1860). The book states that analyzing a player's play on the field is necessary to assess his abilities. Well, that's what sports analysis is all about. As the stakes got more significant, it became evident that the practice required a digital push to become more of an art and science combo and to provide predicted impact. And we currently live in that era (Singh, 2020).

In the 21st century, sports data is able to leverage advanced data collection and analysis tactics, from real-time video data collection to powerful algorithms (Cintia *et al.*, 2015) in order to develop player plans and game simulations for practice sessions and major games. The use of analytics in sports is not a new concept. In fact, it has been around for decades (Jayal *et al.*, 2018). However, with the increasing availability of data and the development of new technologies, analytics in sports have become more sophisticated and widespread. Nowadays, analytics are used in a variety of sports, including football, basketball, baseball, hockey, and soccer (Jayal *et al.*, 2018).

Sports data preprocessing is the process of collecting and analyzing data to inform and improve

DOI: 10.1201/9781032664330-10

decision-making in sports (Jayal *et al.*, 2018). Currently, analytics has become an essential part of sports management, as it can help coaches, athletes, and owners to make better decisions and gain a competitive advantage (Singh, 2020). One of the main advantages of sports analytics is that it allows teams to find hidden patterns and insights in their data that might not be visible to the naked eye (Horvat *et al.*, 2019; Russell *et al.*, 2021). For example, analytics can help coaches identify which players are the most effective in certain situations or which strategies work best against specific opponents (Kniffin *et al.*, 2017; Wang *et al.*, 2018). This information can be used to create more efficient game plans and give teams a competitive edge (Jayal *et al.*, 2018).

Another benefit of analytics is that it can help to reduce injury rates (Combs *et al.*, 2019; Myer *et al.*, 2012; Sharma & Sharma, 2023) and improve player performance (Bharathan *et al.*, 2015; Horvat *et al.*, 2019). By analyzing data on player movement and exertion, teams can identify early signs of fatigue and make adjustments to training programs or playing time to prevent injuries (Fleisig *et al.*, 2018; Myer *et al.*, 2012; Sharma & Sharma, 2023). Analytics can also help coaches and trainers optimize player training regimes, leading to improved performance on the field (Jayal *et al.*, 2018).

Analytics is an essential part of modern sports management. It provides teams with valuable insights and data that can help them to make better decisions, and gain a competitive advantage (Jensen *et al.*, 2016; Solntsev, 2014). The sports sector has a significant economic impact both on and off the pitch, making high-value business decisions. Today, every team depends on revenue from game events, merchandise, sponsorships, and media rights (Arslan *et al.*, 2022; Naraine *et al.*, 2022). To make the most of these, use off-field analysis (Arslan *et al.*, 2022). Sports analytics may assist in optimizing income streams for everything from determining the ideal time to sell tickets to forecasting attendance at an event (Arslan *et al.*, 2022). Athletic metrics is also an excellent tool for tracking social media trends and data related to player or team popularity, which may be utilized to launch TV series, merchandising, and other items (Arslan *et al.*, 2022; Jensen *et al.*, 2016; Naraine *et al.*, 2022; Solntsev, 2014).

The introduction and widespread application of data analysis in sports are the topics of this paper. The paper begins with a mention of the inception of the use of data analysis in sports, followed by references to the regions it influences, the sports it can have a beneficial influence on, and the manner in which it can do so (Bharathan *et al.*, 2015; Singh, 2020; Worsey *et al.*, 2020). In the following section, it will be discussed how data is gathered, how it is organized and classified, and how it is stored (Bai & Bai, 2021; Siddiqa *et al.*, 2016). It's vital to note the difficulties that can arise in some sports when there is no centralized method of data management and no set taxonomy to adhere to (Bai & Bai, 2021; Russell *et al.*, 2021; Siddiqa *et al.*, 2016; Singh, 2020).

Additionally, there will be numerous references to techniques commonly utilized in sports (Jayal *et al.*, 2018). It will start by discussing the methods and the primary types of problems each method is used to solve (Jayal *et al.*, 2018). Examples of well-known techniques that greatly increased the value of data that had previously existed but had not been put to good use will be provided (Giles *et al.*, 2020; Gong & Wang, 2021; Jayal *et al.*, 2018). In-depth references will be made and analyzes of some of the methods that are most commonly employed (Jayal *et al.*, 2018). The quantity of publications on each topic and the use of methodology is a crucial consideration. To determine which nations, which industries, and how quickly data analytics in sports has developed, even the volume of publications every year is crucial. The use of analytical techniques in sports will be covered in detail in the following section. The participation and improvement contributed to their perception of the sport (coaches, gymnasts, athletes) would first be divided into groups based on the sport they were interested in. The division of approaches according to the field of interest will be covered in the following sections.

For instance, a topic of interest might be a player's unique skill or the impact of fans on a team's success (Aloufi *et al.*, 2018; Lacey & Kennett-Hensel, 2016; Mueller, 2020). This section will be analyzed in terms of the field being examined, giving it a uniform expression regardless of the sport being discussed. The conclusions that may be derived from the recent usage of data analysis in sports are then presented in the next two chapters. It will be questioned whether there are any areas where the current research could advance. We'll also list certain items that have become a staple of the sport in recent years and for which our prior efforts have clearly paid off. Questions and recommendations for more data-driven sports analysis will be raised by this analysis.

10.2 SPORTS DATA

The primary goals of sports big data management are to use data management strategies, tools, and platforms to store, preprocess, process, and secure sports big data (Siddiqa *et al.*, 2016). However, the unstructured and variable nature of data sources makes big data management a challenging undertaking (Siddiqa *et al.*, 2016). The administration of sports big data is critical to the prosperity of the national sports industry, teams, and people (Liu, 2019a; Liu, 2019b; Liu, 2020). Increasing data quality and accessibility for decision-making, as well as exploring the potential value of sports big data, are the main objectives of sports big data management (Liu, 2019a). This segment discusses gathering big sports data, labeling big sports data, enhancing existing data, data protection, and data management methods.

10.2.1 GATHERING DATA

In the sphere of sports, there are now more methods than ever before for gathering data (Siddiqa *et al.*, 2016). The demand for better data management has arisen as a result of both the growth and diversity of data (Cao, 2012; Morgulev *et al.*, 2018; Siddiqa *et al.*, 2016). During games, practices, and training sessions, sophisticated technologies including sensors, wearable gadgets, and high-definition cameras are used to collect a wide range of data (Schumaker *et al.*, 2010; Veal & Darcy, 2014). This information, which includes player biometrics, video footage, and statistics, is rigorously organized and kept in safe databases that are frequently hosted on the cloud for quick access and analysis (Cao, 2012; Veal & Darcy, 2014). This is how teams store the information about their players because it is common to have access to the information about players who are affiliated with them yet play for another team. Data from leagues or events is still centrally maintained and is available on big data platforms that either have free access to them or charge a fee to view the data (Schumaker *et al.*, 2010). This information is used by teams for player measurements, training, injury prevention, and occasionally tactical considerations (Meehan I. I. I. *et al.*, 2011; Schumaker *et al.*, 2010).

The diversity of sports big data is a crucial component. Data types can be extremely complex, and data sources might originate from a wide range of sources (Bai & Bai, 2021). The number of sports data has increased as a result of the growth of the Internet, the Internet of Things, and the sports sector (Bai & Bai, 2021). The gathering of network sports big data is especially difficult since network data are varied, complicated in composition, and have distinct usage methods and utilization values for a variety of reasons (Bai & Bai, 2021; Liu, 2019b; Liu, 2020). Web crawlers are widely used to collect copious amounts of data about online sports.

To aid researchers when analyzing the growth of the smart sports industry in light of big data, several freely downloadable datasets are easily accessible on the website (Bai & Bai, 2021; Liu, 2019b). Google Dataset Search can be used to look up hard-to-locate data online. Google Dataset Search was recently added to help users locate linked online dataset sources (Bai & Bai, 2021). Dataset providers must define their datasets using a range of information (such as author, publication date, data content, and terms for using the data) in order to make them more searchable. The social network can also be utilized to gather extensive sports data (Bai & Bai, 2021). For instance, Stats Perform (Opta) can be accessed through the website https://www.statsperform.com/opta/. Data Performance Opta is a business that gathers and examines thorough statistics from various athletic events. This Company specializes in sports data analytics. It is frequently utilized by sports teams, media organizations, and fans to improve their comprehension of the games. Opta has a global presence and a reputation for offering thorough data insights.

To capture video of live games, practices, and training sessions, cameras must be placed carefully (Yu & Farin, 2005). The motions of the players and key moments during games are captured using top-notch cameras with specialized settings (Barris & Button, 2008). Video is captured and archived for analysis, coaching input, and scouting. Instant replays and official decisions both depend on it (Barris & Button, 2008). Highlight reels enhance scouting and recruitment, while video analysis helps coaches and athletes improve performance (Xu *et al.*, 2006). Through broadcasts and social media, the footage increases fan involvement (Xu *et al.*, 2006). In conclusion, video recording in sports offers insightful information, supports judgment, and improves the whole experience for players, coaches, spectators, and authorities (Barris & Button, 2008; Xu *et al.*, 2006; Yu & Farin, 2005).

10.2.2 PROTECTION OF DATA

Despite the widespread use of big data applications today, difficulties and security concerns still exist (Liu *et al.*, 2018). Processing of the data requires extensive sports data collection. For all kinds of data applications that can use the results of big data analysis, secure sports big data gathering is a necessary initial step (Liu *et al.*, 2018). Secure sports big data collection approach is essential for many data applications because dubious data sources enable data collection to investigate numerous harmful assaults and acts (Liu *et al.*, 2018). Researchers have achieved some progress in this area, such as offering energy-efficient data gathering and data security in a distributed setting (Bai & Bai, 2021), to enable secure massive data collecting. It is advised to use blockchain technology and Reinforcement Learning (RL) to solve these issues (Liu *et al.*, 2018).

Protecting private data on players, teams, and athletic events is part of data protection in sports (Harriss & Atkinson, 2009). It involves protecting private information, such as contact information and medical records, from identity theft and unauthorized access. To ensure fair competition and make informed decisions, performance data must be kept accurate and complete (Synrem, 2019). Compliance with rules, such as the GDPR and other data privacy laws, is essential to avoid legal repercussions (Synrem, 2019). To stop cyberattacks that target priceless sports data, it's essential to use strong cybersecurity measures like encryption and firewalls (Socolow & Jolly, 2017). To allow authorized parties to access data while preventing leaks or breaches, secure data exchange mechanisms must be in place (Harriss & Atkinson, 2009; Socolow & Jolly, 2017; Synrem, 2019). Additionally, sustaining trust and respecting privacy depends on safeguarding fan data, including personal information and purchasing history. Building trust with athletes, teams, and fans requires open disclosure of data protection rules (Socolow & Jolly, 2017). Data is held responsibly, privacy concerns are reduced, and a quick reaction to potential breaches is made possible by establishing data retention regulations and an incident response strategy (Harriss & Atkinson, 2009). Data protection is essential in the digital age of sports, where technology and analytics are at the forefront, to maintain integrity and privacy while maximizing the potential of data-driven insights (Harriss & Atkinson, 2009; Socolow & Jolly, 2017; Synrem, 2019).

10.2.3 LABELING OF DATA

When enough sports data have been gathered, the next phase involves labeling (Bai & Bai, 2021). For example, umpire gesture recognition for automated labeling of sports video (Chambers *et al.*, 2004). It uses a hierarchical hidden Markov model in combination with a filler model to handle the problems of automated segmentation and robust gesture classification (Chambers *et al.*, 2004). Data labeling and data gathering are typically completed concurrently (Bai & Bai, 2021). One of three categories can be made from the data labels (existing labels, crowd-based labels, weak labels) (Bai & Bai, 2021). One might utilize the currently used labels as a learning resource to forecast the remaining labels (Bai & Bai, 2021). Although producing accurate labels all the time would be ideal, this implementation procedure might be highly expensive (Bai & Bai, 2021).

The best approach for labeling occurrences is by hand (Bai & Bai, 2021). In the Machine Learning (ML) community, active learning has long been considered an essential technique for cost-effectively selecting real samples to classify (Ehrenberg *et al.*, 2016). Task assignment strategies are even more crucial now that crowdsourcing techniques have been suggested for the labeling process to ensure high-quality labels (Roh *et al.*, 2019). The accuracy and usability of the data programming model have improved (Ratner *et al.*, 2017). Compared to training with fewer manual labels, training a discriminative model with a large number of weak labels may result in higher accuracy (Zhang, 2015). Numerous systems, including DeepDive, DDLite, and Snorkel, have been developed for data programming (Ehrenberg *et al.*, 2016; Ratner *et al.*, 2017; Zhang, 2015).

ML technology can be used to handle noisy data and inaccurate labeling. To enhance data quality, a great deal of scholarly research has been done (Wang *et al.*, 2019). A representative cleaning system called HoloClean uses quality standards, value correlations, and reference data to develop a probabilistic model to explain how the data were produced (Rekatsinas *et al.*, 2017). More technologies for data cleaning have also been created to enhance the format of raw data for study in the future. To increase accuracy, some cleaning models have been developed. To increase the effectiveness of data cleaning, the ActiveClean model employs a kind of stochastic gradient descent for both cleaning and training (Rekatsinas *et al.*, 2017).

Using oracles, TARS can be applied to the problem of cleaning crowdsourced labels. TARS bears the name of an intelligent robot from Interstellar that can offer wise counsel to humans (Dolatshah, 2018). TARS is a label-cleaning advisor that may provide data scientists with two crucial pieces of advice when they need to train or test a model using noisy labels (Dolatshah, 2018). First, given a test dataset with noisy labels and a classification model, TARS may use the test data to forecast how well the model will perform ground-truth labels during the model testing stage. Second, given a training dataset with noisy labels and a classification method, TARS can determine which label should be sent to an oracle for cleaning during the model training stage to optimize the model potential for improvement. TARS provides two recommendations. However, given training data with noisy labels, TARS can select which instances to send to an oracle to optimize the projected model, which can improve each noise accuracy (Dolatshah, 2018). To obtain high-quality data labels, it is possible to enhance the quality of the current labels (Sheng *et al.*, 2008). It is noticed whether (or not) recurrent labeling improves data quality and concentrates on enhancing training labels for supervised induction (Sheng *et al.*, 2008). The trial outcome demonstrates the following benefits: the label and model can become better with repeated labeling (Sheng *et al.*, 2008). Repeated labeling can improve noisy labels, and a dependable method for raising label quality is also offered (Dolatshah, 2018).

10.3 AN OVERVIEW OF THE APPLICATION OF ANALYTICS IN SPORTS

Numerous publications will be examined on various levels in this section. First, a Scopus summary of the number of publications in the domain of sports analytics during the past ten years will be given. Also, it is important to examine the fields that research has mostly focused on in the sections that follow. The number of nations having sports data analysis activity and the extent of that activity are crucial additional factors. In this section, will be made a separation by Sport and Topic.

The world of sports has undergone a change because of sports analytics (Ghosh *et al.*, 2023). It gave rise to "Moneyball" techniques in baseball, where roster decisions and player performance are made using data-driven insights (Singh, 2020). Basketball teams use analytics to evaluate individual effectiveness, shot selection, and defensive tactics, enhancing team performance as a whole (Barris & Button, 2008; Singh, 2020).

For player scouting, performance evaluation, and injury prevention in soccer, analytics are becoming crucial (Ghosh *et al.*, 2023; Singh, 2020). To keep track of players' whereabouts and make tactical modifications, teams use GPS trackers and video analysis (Ghosh *et al.*, 2023; Maulud & Abdulazeez, 2020).

10.3.1 THE EVOLUTION OF SPORTS ANALYTICS

Sports analytics until 2000 was not used extensively, mostly concentrating on conventional statistical metrics like field goal percentages, rebounds, and points scored. The landscape was gradually shaped by the arrival of technology, but in comparison to modern standards, the depth and breadth of data available were restricted. Teams used a more instinctive grasp of the game to guide decision-making processes, relying on basic statistical analysis for player evaluations and strategic choices (Singh, 2020).

A discernible change occurred with the introduction of sophisticated statistics and computerized tracking systems in the late 1990s (Singh, 2020). Even if they weren't as advanced as modern sports analytics, these early advancements set the stage for the upcoming data-driven revolution. A more thorough understanding of player performance, team chemistry, and game strategy was made possible by analysts' experimentation with new measurements and models (Cintia *et al.*, 2015).

Sports analytics underwent a transformation between 2000 and 2010, characterized by an increase in the amount of data available and the use of cutting-edge technologies (Singh, 2020). The introduction of high-speed cameras, GPS technology, and player-tracking devices changed the game and made a lot of detailed data available for study. Teams and analysts investigated player efficiency ratings, plus-minus statistics, spatial analysis, and other sophisticated measures in addition to more conventional statistics. The Moneyball craze Lewis's book highlighted the tactical benefits of making decisions based on statistics. Sports analytics have become widely used in a variety of sports this decade, impacting fan interaction, game strategy, and player recruiting while laying the groundwork for further developments in the years to come (Cintia *et al.*, 2015).

Sports analytics saw a significant rise after 2014, driven by wearable technologies, machine learning, and artificial intelligence developments. Teams adopted a comprehensive strategy, using real-time performance analysis, player workload management, and predictive modelling to prevent injuries. Sports strategy was transformed by the combination of big data and state-of-the-art analytics tools, which increased the significance and complexity of data-driven decision-making (Singh, 2020).

This study was performed using the Scopus database. 1637 papers in the field of sports data analysis were discovered in Scopus during the last ten years (from 2014 to 2023). Generally speaking, Scopus has been home to sports analytics literature since 2000. There are just 45 articles overall from 2000 to 2013. Sports and analytics were the keywords utilized for this particular Scopus research project. There has been a marked growth in the number of academic articles on sports analytics (Barris & Button, 2008; Bengio *et al.*, 2013; Ghosh *et al.*, 2023; Maulud & Abdulazeez, 2020; Singh, 2020). This growth has several causes. The number of sports analytics papers published each year is on the rise, as seen in Figure 10.1.

In Figure 10.1, it is obvious that only in 2014 there was almost the same number of papers (40) as in previous years (from 2000 to 2013 there were only 45 papers in Scopus). Furthermore, it is noticed that in the last 5 years, there have been approximately 250 papers or more every year for sports analytics in Scopus 10.1.

In American football, analytics aid clubs in selecting players, calling plays, and making draft choices based on data (Ghosh *et al.*, 2023). They assist in preventing injuries by keeping track of player health information (Maulud & Abdulazeez, 2020). The most popular categories of papers in sports analytics from Scopus over the past ten years are shown in Figure 10.2. In previous years, seems there is an interest in Data Mining and ML.

Sports analytics is a young field that has become very popular all around the world. According to variances in research priorities, resources, and sports cultures, different nations have different numbers of research publications on sports analytics (Barris & Button, 2008; Bengio *et al.*, 2013;

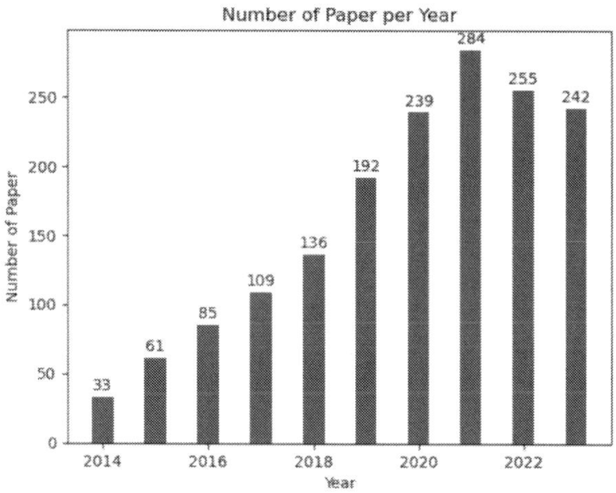

Figure 10.1 Number of Sports Analytics Papers publication per Year created 10/12/2023

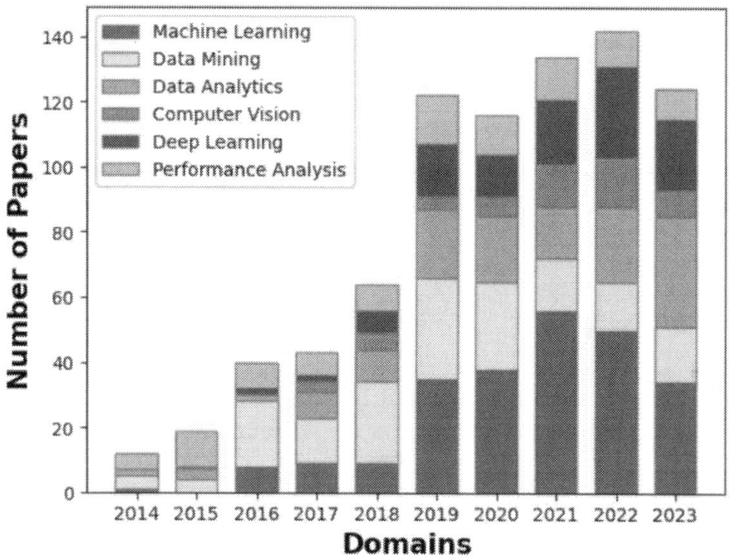

Figure 10.2 Number of Sports Analytics Papers publication in different Domains of Analytics per Year created 10/12/2023

Ghosh *et al.*, 2023; Maulud & Abdulazeez, 2020; Singh, 2020). There is a summary of the involvement and contributions of different nations in sports analytics research below.

The US has significantly advanced the field of sports analytics (Cervone *et al.*, 2014; Gudmundsson & Horton, 2017). This is due to the highly embedded sports culture in the country, the abundance of professional sports leagues, and the powerhouse intellectual institutions (Cervone *et al.*, 2014; Gudmundsson & Horton, 2017). The analysis of player performance, audience participation, and injuries are frequently the focus of American research (Cervone *et al.*, 2014; Gudmundsson & Horton, 2017).

Thanks in part to the research of institutions like MIT sports analytics have evolved substantially (Davenport, 2014; Goldman & Rao, 2012; Macdonald, 2012). As it mentioned in Figure 10.3, the USA is the country with the biggest interest in sports analytics. In the subject of sports analytics,

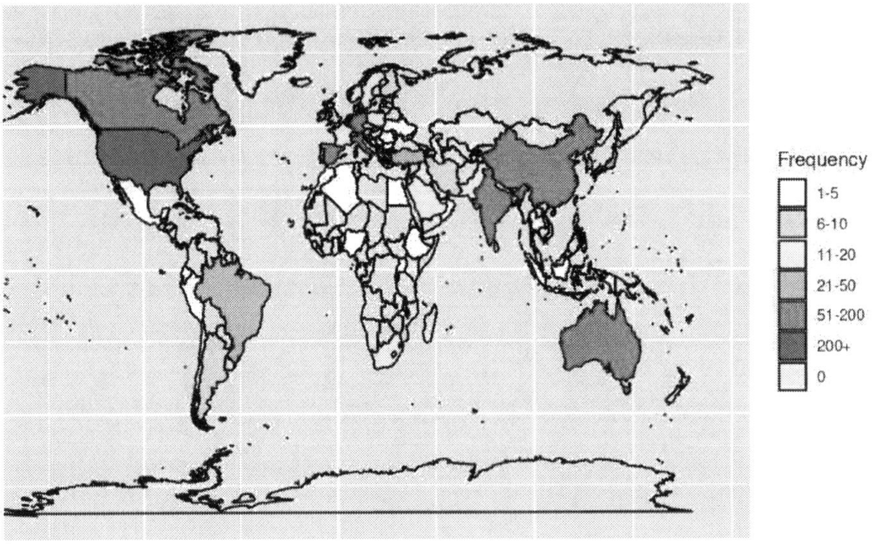

Figure 10.3 Number of Sports Analytics Papers publication per Country created 10/12/2023

the UK is well-represented, with a focus on cricket, rugby, and football (soccer) (Zhang, 2015). Premier League clubs have embraced analytics for player procurement and performance evaluation (Gomez-Lopez *et al.*, 2019). A substantial number of articles have been authored with a focus on data-driven sports decision-making by British universities and research institutions (Gomez-Lopez *et al.*, 2019; Zhang, 2015).

The love of sports in Australia, particularly cricket, rugby, and Australian rules football, has had an impact on sports analytics research (Teune *et al.*, 2021). The country places a great emphasis on sports science and sports-related research, with universities like the Australian Institute of Sport contributing to the discipline (Teune *et al.*, 2021). Canada's vibrant ice hockey culture has allowed for significant advancements in sports analytics (Thornton *et al.*, 2016). Popular study areas in Canada include hockey analytics, player development, and injury prevention. The NHL adoption of analytics for decision-making has sparked an increase in research across the country (Thornton *et al.*, 2016). The Netherlands has embraced sports analytics and is renowned for its understanding of football (soccer) (Berrar, 2018). Contributions from Dutch researchers have been made in a variety of domains, such as player monitoring, tactical analysis, and injury prevention (Berrar, 2018).

Football has been played for a very long time in Spain, and several of its football teams, like Barcelona and Real Madrid, have started using analytics (Núñez-Pomar *et al.*, 2020). In Spain, study on football analytics, including player placement, passing networks, and match strategy, is common (Núñez-Pomar *et al.*, 2020). Football analytics is also gaining popularity in Germany since teams in Bundesliga employ data analytics for player procurement and match analysis (Sheng *et al.*, 2008). Research on topics linked to sports analytics, such as injury analysis and performance projection, has also been done at universities in Germany (Sheng *et al.*, 2008).

Interest in sports analytics has increased in Japan, especially in baseball and soccer (Manzenreiter & Horne, 2013). Japanese researchers have looked into both player performance analysis and injury prevention. Japanese baseball groups have embraced data analytics to help their players develop (Manzenreiter & Horne, 2013). Investing in soccer-specific sports analytics research has just lately started in China (Liu, 2019a). Due to the nation's growing sports industry and the staging of key international sporting events in China, there is now increasing interest in analytics for player development and team success (Liu, 2019a).

10.3.2 SPORTS-SPECIFIC APPLICATION OF ANALYTICS

Overall, sports analytics is becoming increasingly important in all team sports, as it allows teams to make data-driven decisions that can improve their performance on the field (Joash Fernandes *et al.*, 2020; Reyers & Swartz, 2023; Tian *et al.*, 2019). Several team sports can benefit from sports analytics (Joash Fernandes *et al.*, 2020; Reyers & Swartz, 2023; Tian *et al.*, 2019).

10.3.2.1 Basketball

There are many studies about basketball analytics (Cao, 2012; Cervone *et al.*, 2014; Chen *et al.*, 2021; Giasemidis, 2020; Goldman & Rao, 2012; Gómez-Carmona *et al.*, 2021; Horvat *et al.*, 2019; Ji, 2020; Jia *et al.*, 2020; Lacey & Kennett-Hensel, 2016; Liu, 2019b; Liu, 2020; Russell *et al.*, 2021; Sarlis & Tjortjis, 2020; Tian *et al.*, 2019; Wang *et al.*, 2018). In the NBA, teams use analytics to track player performance, shot charts, and lineups to determine which combinations work best together (Wang *et al.*, 2018). Teams also use data to create game plans and predict opponent tendencies (Lacey & Kennett-Hensel, 2016; Wang *et al.*, 2018). The identification of basketball strategies is a very important topic (Tian *et al.*, 2019). Lately, there has been a great deal of curiosity around academia and the international sports industry involving the use of spatiotemporal tracking data and ML to identify and classify offensive and defensive tactics in team sports (Cervone *et al.*, 2014). The learning of basketball defensive strategies using data is the main topic of this research (Ji, 2020; Liu, 2019b; Sarlis & Tjortjis, 2020). Basketball strategy learning has been mostly examined from an offensive standpoint, ignoring the other players' roles and concentrating on the relationship between the major on-ball defender and the on-ball player (Goldman & Rao, 2012). Furthermore, most sports analytics programs that provide play-by-play information are skewed greatly in favour of offensive measures like passes, dribbles, and shoots (Cervone *et al.*, 2014; Goldman & Rao, 2012). Basketball game outcomes can be predicted using analytics by looking at team statistics, player performance, and historical data to project possible outcomes, especially in NBA or Euroleague (Chen *et al.*, 2021; Giasemidis, 2020).

Ji, 2020, has the purpose of classifying, using ML, the several defensive strategies that basketball players employ when they change up their original defensive approach. Researchers and the global sports industry often have many discussions about how offensive and defensive strategies in team sports can be recognized and categorized using spatiotemporal monitoring data. The primary focus of this research is on teaching basketball defensive methods utilizing data. The interplay between the on-ball player and the major on-ball defender has generally been the focus of basketball strategy learning research from an offensive angle, ignoring the other players' roles. Furthermore, the majority of play-by-play data analytics for sports heavily favour offensive actions like passes, dribbles, and shots. The goal of this research was to use ML to classify basketball players' different defensive techniques when they deviate from their original defensive plan.

10.3.2.2 Football

National Football League (NFL) teams use analytics to decide on player personnel, game planning, and play calling (Joash Fernandes *et al.*, 2020; Muraoka *et al.*, 2021; Reyers & Swartz, 2023). Coaches and scouts use data to evaluate players physical abilities, athleticism, and effectiveness on the field. One of the most important topics in the NFL is predicting plays (Joash Fernandes *et al.*, 2020). This research aims to create an interpretable ML model to predict National Football League plays (pass versus rush) that coaches and players may find value in the present (Joash Fernandes *et al.*, 2020). Utilizing information from 2013-2014 to 2016-2017 to determine the highest feasible prediction accuracy, they first create and contrast several ML models using data from the 2017 NFL regular seasons, which includes 1034 games and 130,344 pass/rush plays (Joash Fernandes *et al.*, 2020). The most effective model, a neural network, can compete with cutting-edge methods used for other datasets (Joash Fernandes *et al.*, 2020). An important study about player performance is (Reyers & Swartz, 2023). This study is about using monitoring data, the NFL evaluates quarterbacks. The NFL quarterback performance is evaluated (Reyers & Swartz, 2023).

Player monitoring data makes it possible to assess the various options quarterbacks have at their disposal and the expected points that each one will yield (Reyers & Swartz, 2023). A quarterback's options are predicated on taking into account numerous frames during a play, therefore one option may change into another throughout the play. They also take the likelihood of quarterback running options into account. Tracking data makes it possible to measure the openness and placement of receivers on the field, two factors that are crucial when evaluating quarterback alternatives. Next, the estimated expected points obtained from the alternatives and the passing options likelihood of success are calculated using ML algorithms. The method of estimating also takes into consideration what can happen after a reception. The quarterback's actual performance is then evaluated in comparison to the best available choice (Reyers & Swartz, 2023). In another domain, an important study about Chronic Traumatic Encephalopathy for NFL players (Muraoka *et al.*, 2021). A neurodegenerative tauopathy known as chronic traumatic encephalopathy (CTE) is linked to repeated head traumas experienced by athletes participating in contact or collision sports, such as boxers, American tackle football, soccer, rugby, and ice hockey players (Muraoka *et al.*, 2021).

10.3.2.3 Soccer

In soccer, teams use analytics to track players movements, passes, and shots to identify patterns and weaknesses in the opposing team defense (Aloufi *et al.*, 2018; Arbues-Sanguesa *et al.*, 2020; Cintia *et al.*, 2015; Combs *et al.*, 2019; Eryarsoy & Delen, 2019; Hossain *et al.*, 2017; Hostrup & Bangsbo, 2023; Kumar, 2013; Majumdar *et al.*, 2022; Pantzalis & Tjortjis, 2020; Ribeiro *et al.*, 2022; Solntsev, 2014; Tuyls *et al.*, 2021; Yiğit *et al.*, 2020). It is common to use data to measure player performance and to make strategic decisions on substitutions and formations (Eryarsoy & Delen, 2019; Hostrup & Bangsbo, 2023; Pantzalis & Tjortjis, 2020; Ribeiro *et al.*, 2022; Tuyls *et al.*, 2021; Yiğit *et al.*, 2020). There are many papers with these applications.

Hostrup & Bangsbo, 2023 is a study about highly intensive internal training. High-intensity interval training is an exercise that effectively improves the physical conditioning of athletes at intensities that are close to or above the capacity of aerobic energy systems. However because highly intensive internal training places heavy demands on the body, it raises the possibility of an overload-related drop in match performance and harm. Because of this, some coaches are more likely to use highly intensive internal training both during and before the season. Since these phases can be crucial, the difficulty is figuring out how to maximize and dosage high intensive internal training during them (Hostrup & Bangsbo, 2023). Studies have shown how useful it is to carry out periods of intense training to reduce the risk of overload while also improving performance. Intensity is enhanced by increasing the amount of highly intensive internal training, for example, during intensified training periods, which are normally a few weeks long (Hostrup & Bangsbo, 2023).

Another work is about player performance and wages (Ribeiro *et al.*, 2022). Football players' salaries make up the largest portion of club spending and are one of the main sources of inspiration for athletes. What motivates football players to continually improve their play, and how do strong performances affect pay? Are wage differentials between clubs being put up to encourage players, much as in a rank-order competition? They examine the correlation between player performance and earnings collecting precise player salary data from the Portuguese First League Football. They show that players who play more minutes are paid more, and the top players face a convex relationship that suggests football teams should hold a tournament where players are ranked based on their minutes played, games played, and performance statistics throughout the season (Ribeiro *et al.*, 2022).

10.3.2.4 Baseball

In Major League Baseball, analytics has become a crucial part of team strategy (Fleisig *et al.*, 2018; Hamilton *et al.*, 2014; Mueller, 2020; Mun *et al.*, 2023). Teams use data to calculate hitter and pitcher tendencies, track player development, and create effective lineups. The sports business has become increasingly interested in sports game prediction utilizing ML techniques (Fleisig *et al.*, 2018; Hamilton *et al.*, 2014; Mueller, 2020; Mun *et al.*, 2023). Especially, sports-related industries such as sports marketing, sports betting, in-game strategy development, etc. need this activity. It is vital to take into account a variety of game aspects, including player characteristics and historical match results, to increase prediction accuracy. Win-lose outcome prediction in baseball games is defined in this article as a binary classification problem (Mun *et al.*, 2023). They begin by gathering numerous team and player attributes from Korean Baseball Organization data for the years 2012 through 2021 (Mun *et al.*, 2023). Then, utilizing Siamese neural networks, they provide a brand-new prediction model called CompeteNet to discover the competitive dynamics between a player for two clubs pitcher and hitter (Mun *et al.*, 2023). Despite its simplicity, CompeteNet beats other ML models, according to substantial experimental results (Mun *et al.*, 2023).

An important research about Baseball Pitch Prediction is Hamilton *et al.*, 2014. One of the most well-known sports leagues in North America is Major Organization Baseball, a professional baseball organization that operates in both the US and Canada. Baseball has undergone extensive statistical and mathematical research, in part due to its popularity and the accessibility of game data. Pitch analysis is especially crucial for helping a team better understand the pitch behaviour it may encounter during a game so that the team may plan a batting strategy to counter that predicted pitch behaviour. They classify pitches by type using a variety of popular machine-learning classification techniques on PITCH f/x data. They then use features that are only known before the pitch is thrown to extend the classification challenge to prediction (Hamilton *et al.*, 2014).

10.3.2.5 Volleyball

In the field of volleyball, sports analytics has taken on more significance as it offers teams and coaches insightful information to enhance performance (Almujahed *et al.*, 2013; Chellatamilan *et al.*, 2015; Hoege *et al.*, 2020; Hsu *et al.*, 2014; Tracy *et al.*, 2023; Van Haaren *et al.*, 2016). Modern statistical methods and technology have completely changed how volleyball is studied and played (Almujahed *et al.*, 2013; Tracy *et al.*, 2023; Van Haaren *et al.*, 2016). The evaluation of a player's performance is a crucial component in volleyball analytics (Hsu *et al.*, 2014). Teams can evaluate player location, speed, and accuracy by utilizing cameras and sensors to track player movements and ball trajectories (Almujahed *et al.*, 2013; Hoege *et al.*, 2020; Van Haaren *et al.*, 2016). Coaches can use this information to make educated judgments on replacements, rotations, and tactical changes (Chellatamilan *et al.*, 2015; Tracy *et al.*, 2023). Opponent analysis is another crucial aspect (Hsu *et al.*, 2014). Teams employ analytics to research the tactics, playing styles, and weaknesses of their competitors (Almujahed *et al.*, 2013; Tracy *et al.*, 2023; Van Haaren *et al.*, 2016). The creation of counter strategies to achieve a competitive advantage is made possible by this information (Hoege *et al.*, 2020; Hsu *et al.*, 2014; Tracy *et al.*, 2023).

Analytics also assists in preventing injuries by keeping track of player effort and weariness (Chellatamilan *et al.*, 2015). Teams can lower the likelihood of overuse injuries and improve training regimens by monitoring player movements and physical data (Almujahed *et al.*, 2013; Chellatamilan *et al.*, 2015). Additionally, by giving real-time information and visualizations during games, analytics improves spectator engagement and makes sports more entertaining and interactive (Hoege *et al.*, 2020; Tracy *et al.*, 2023). Through enhancing player performance, strategy, and fan engagement, volleyball analytics is revolutionizing the sport (Tracy *et al.*, 2023; Van Haaren *et al.*, 2016). It has developed into a crucial weapon for teams trying to succeed in this fiercely competitive sport (Almujahed *et al.*, 2013; Hoege *et al.*, 2020; Tracy *et al.*, 2023).

10.3.2.6 Other sports

While traditionally associated with team sports like basketball, soccer, and football (Ji, 2020; Joash Fernandes *et al.*, 2020; Tian *et al.*, 2019), sports analytics can also be applied to individual sports like boxing, ice skating, fencing, golf, wrestling, tennis, martial arts, skiing, and rodeo events. Boxing and wrestling are two combat sports that can benefit from sports analytics (Bigsby & Ohlmann, 2017; Kasiri *et al.*, 2017; Kasiri-Bidhendi *et al.*, 2015; Okagbue *et al.*, 2018; Worsey *et al.*, 2020). Coaches and trainers may pinpoint problem areas and create more efficient training regimens by monitoring variables like punch volume, accuracy, and power. Additionally, data on an opponent's fighting style and tendencies can inform game plans and strategies (Kasiri *et al.*, 2017; Kasiri-Bidhendi *et al.*, 2015). ML is an effective technology for categorizing data and has been used to categorize movement data captured during everyday activities and sports using wearable inertial sensors (Worsey *et al.*, 2020). In combat sports like boxing, inertial sensors can provide valuable biofeedback; yet, their use has not been widely accepted (Kasiri-Bidhendi *et al.*, 2015; Okagbue *et al.*, 2018). The foundation for combat sports players' automatic workload tracking is laid if simple combinations of inertial sensors can be employed to automatically determine strike type and save time-consuming processes like video labeling (Bigsby & Ohlmann, 2017; Kasiri *et al.*, 2017; Kasiri-Bidhendi *et al.*, 2015). The researchers of this study evaluate the classification performance of six different supervised ML models (tuned and untuned) using two simple inertial sensor setups (configuration 1: inertial sensor worn on both wrists; configuration 2: inertial sensor worn on both wrists and the ankle)(Worsey *et al.*, 2020).

Golf is another sport that relies heavily on data analysis (Bacic *et al.*, 2017; Drappi & Co Ting Keh, 2019; Elmore & Urbaczewski, 2021; Fry & Ohlmann, 2012). By tracking swing mechanics, shot distances, and putting accuracy, golfers can identify strengths and weaknesses in their game and adjust their training accordingly (Fry & Ohlmann, 2012). Similarly, course data such as slope, wind, and temperature can be used to inform club selection and shot strategy (Bacic *et al.*, 2017; Elmore & Urbaczewski, 2021). In article (Drappi & Co Ting Keh, 2019), a model is presented that can accurately forecast the discrete probability distribution of a golfer's score for every tournament hole, shot by shot. They first generalized Broadie's score-based skill estimate technique to allow a golfer's skill (such as scoring average, driving spray, iron play, and putting) to alter continuously (Drappi & Co Ting Keh, 2019).

Fencing and martial arts are both highly technical sports that can benefit from sports analytics in similar ways (Leman & House, 2012; Smolkin, 2019). By tracking metrics like footwork speed, blade work accuracy, and reaction time, athletes can fine-tune their technique and develop more efficient training programs (Leman & House, 2012; Smolkin, 2019). Additionally, data on an opponent's style and tendencies can inform tactical decisions during a match (Leman & House, 2012; Smolkin, 2019).

Tennis is another individual sport that can benefit from sports analytics (Giles *et al.*, 2020; Gong & Wang, 2021; Nieto & Sánchez, 2013; Polk *et al.*, 2019; Rodenberg & Feustel, 2014; Rodenberg *et al.*, 2016; Vinyes Mora & Knottenbelt, 2017). By tracking metrics like serve speed, spin rate, and shot placement, players can identify areas for improvement and refine their training regimens (Polk *et al.*, 2019; Rodenberg & Feustel, 2014). Similarly, data on an opponent's style and tendencies can be used to inform game plans and strategies (Nieto & Sánchez, 2013; Rodenberg *et al.*, 2016). Using tracking data, it is possible to identify and classify change of direction (COD) movements in professional tennis; the study of Giles *et al.*, 2020 set out to develop an automated technique. Nineteen professional tennis players, including 9 males and 10 females from the Australian Open Grand Slam, had their match play films classified by three specialists in sports science, strength, and fitness for moderate-to-high intensity COD. A total of 1,494 modifications were identified and synchronized with the 2D player position captured at 25 Hz based on camera tracking data. Several machine-learning classifiers were trained and tested using a collection of 1,128 time-motion features. The optimal performance outside of the sample was determined to be the random forest algorithm, which, with an F1-score of 0.729, categorized changes in medium and high intensity. This work offers a novel and useful approach to automatically detect and categorize COD movements in professional tennis using player tracking data and ML techniques (Giles *et al.*, 2020).

Internet video data is growing at a rapid rate, and finding relevant information has become a crucial problem that needs to be resolved. Gong & Wang, 2021 examine video categorization and content-based retrieval methods with an emphasis on sporting events. Its goals are to create an index and mark for video material, also to promote user acquisition by computer processing, interpretation, and analysis of the video material (Gong & Wang, 2021). Classification of video tennis has significant scholarly and practical implications (Gong & Wang, 2021).

Statistics demonstrate that tennis spectators rank fourth, even though the number of participants cannot crack the top ten (Gong & Wang, 2021; Polk *et al.*, 2019; Rodenberg *et al.*, 2016; Vinyes Mora & Knottenbelt, 2017). Each year, the masters and crown tennis events are conducted in four different countries, watching major international events. Tennis tournaments become a mainstay of many people's leisure and holiday time (Gong & Wang, 2021). Tennis matches can run anywhere between two and four hours, and there are innumerable big and little matches played worldwide every year, thus the amount of tennis records set is astounding. Additionally, just 5% of tennis matches in the sports world use artificial intelligence, compared to 50% of football matches. As a result, they must quickly and effectively find a way to process videos while handling enormous volumes of data (Gong & Wang, 2021).

Ice skating and skiing are two winter sports that can benefit from data analysis (Brunauer *et al.*, 2020; Delibašić *et al.*, 2018; Kniffin *et al.*, 2017; Knobbe *et al.*, 2017; Vleugels *et al.*, 2021). By tracking metrics like speed, acceleration, and turn radius, skaters and skiers can identify areas for improvement and develop more effective training programs (Brunauer *et al.*, 2020; Knobbe *et al.*, 2017). Additionally, weather and course data can be used to inform equipment selection and race strategy (Brunauer *et al.*, 2020; Vleugels *et al.*, 2021). Sports analytics is best known as the field of research that focuses on finding small but significant improvements within competitions, even though broader sets of athlete and team-level data from external competitions (such as strength and conditioning metrics) have typically been left out of such analyzes (Delibašić *et al.*, 2018; Vleugels *et al.*, 2021).

Given that strength and conditioning routines are arguably the most common way that individuals expect extra-competitive progress to convert into within-competition performance, it is clear that strength and conditioning metrics deserve deeper analytical attention. A research conducted by the National Collegiate Athletic Association (NCAA) on Division 1 Men's Ice Hockey players uses both (a) strength and conditioning data and (b) in-game performance assessments to illustrate this methodology (Kniffin *et al.*, 2017). Although multivariate studies indicate favourable associations between conditioning and strength measurements and playing time as the more significant conclusion, bivariate analysis demonstrates a substantial positive link between bench press performance, points scored (r = 0.15). The fundamental strategy they develop emphasizes the significance of taking into account extra-competitive variables like strength and conditioning metrics to comprehend coaching decisions on playing time and performance within the competition, even though within-competition data are becoming more readily available for analytics research. They also talk about how players, coaches, and team managers could all benefit from the integrated approach they provide, as well as experts in strength and conditioning (Kniffin *et al.*, 2017).

10.3.3 TOPIC-SPECIFIC APPLICATION OF ANALYTICS IN SPORTS

This systematic review aims to explore different domains in sports analytics, including player performance analysis (Arbues-Sanguesa *et al.*, 2020; Ashford *et al.*, 2021; Bharathan *et al.*, 2015; Cintia *et al.*, 2015; Cotterill, 2010; Ghosh *et al.*, 2022; Giles *et al.*, 2020; Gómez-Carmona *et al.*, 2021; Gomez-Lopez *et al.*, 2019; Gudmundsson & Horton, 2017; Liu & Schulte, 2018; Luo, 2020; Macdonald, 2012; Muraoka *et al.*, 2021; Pantzalis & Tjortjis, 2020; Ribeiro *et al.*, 2022; Russell *et al.*, 2021; Sarlis & Tjortjis, 2020; Tian *et al.*, 2019; Vats *et al.*, 2022; Yiğit *et al.*, 2020), game strategy optimization (Ashford *et al.*, 2021; Horvat *et al.*, 2019; Kumar, 2013; Sharma & Sharma, 2023; Tian *et al.*, 2019; Wang *et al.*, 2018), injury prevention (Combs *et al.*, 2019; Delibašić *et al.*, 2018; Fleisig *et al.*, 2018; Myer *et al.*, 2012; Pyne *et al.*, 2014; Wäsche *et al.*, 2017), fan engagement (Aloufi *et al.*, 2018; Brown & Billings, 2013; Karg *et al.*, 2021; Lacey & Kennett-Hensel, 2016; Mueller, 2020) and business operations (Arslan *et al.*, 2022; Jensen *et al.*, 2016; Mondello & Kamke, 2014; Naraine *et al.*, 2022; Solntsev, 2014).

10.3.3.1 Player Performance

Player Performance analysis is one of the most popular and widely researched domains in sports analytics (Bharathan *et al.*, 2015; Cintia *et al.*, 2015; Giles *et al.*, 2020; Luo, 2020; Russell *et al.*, 2021; Sarlis & Tjortjis, 2020). It involves using data to evaluate a player skills, strengths, weaknesses, and potential for improvement (Bharathan *et al.*, 2015; Russell *et al.*, 2021). This domain encompasses various sub-domains such as player evaluation, scouting, talent identification, and development, which are crucial for building competitive teams (Bharathan *et al.*, 2015; Russell *et al.*, 2021; Yiğit *et al.*, 2020). Invasion based on teams sports like football, basketball, and hockey are comparable in that players are free to move about the field of play, and player and team performance cannot be adequately analyzed without taking into account the movements and interactions of all players collectively (Yiğit *et al.*, 2020). State-of-the-art object tracking systems currently generate high-definition and spatiotemporal traces of player trajectories at high frequency, which has enabled numerous research initiatives spanning many fields to glean knowledge from the trajectory data. Gudmundsson & Horton, 2017 review current studies that make use of spatiotemporal data from team sports are used as input and need complex computation. The research endeavors are categorized in this article in a coherent framework and a list of open research questions is provided in Gudmundsson & Horton, 2017.

Basketball is a game that is constantly changing, and this is an important factor that affects player health (Sarlis & Tjortjis, 2020). To assess and determine the physical fitness level and profiles of basketball players according to sex, Gómez-Carmona *et al.*, 2021 set out to do just that 26 semi-pro basketball players (13 men and 13 women) were evaluated using inertial devices during various fitness tests that had already been scientifically approved (Gómez-Carmona *et al.*, 2021). To examine sex-related differences and pinpoint physical fitness profiles, they employed PCA and the t-test for independent data (Gómez-Carmona *et al.*, 2021).

Football statistics had an extraordinary metamorphosis as a result of the startling spread of sensing technologies that offer high-fidelity data streams gathered from every game (Bharathan *et al.*, 2015; Cintia *et al.*, 2015; Gómez-Carmona *et al.*, 2021; Muraoka *et al.*, 2021; Ribeiro *et al.*, 2022; Russell *et al.*, 2021). Football clubs, coaches, and leagues are now receiving data from professional statistical analysis companies like ProZone and Opta. These organizations are starting to analyze this data to keep track of their players and develop team plans (Gómez-Carmona *et al.*, 2021). Standard methods for assessing and forecasting team performance are based on historical variables, such as previous wins or losses, performance in qualification games, and margin of victory in previous contests (Gómez-Carmona *et al.*, 2021; Ribeiro *et al.*, 2022; Russell *et al.*, 2021). In contrast to conventional models, (Cintia *et al.*, 2015) suggests a model in this study that is based on the observation of players' on-field conduct. To demonstrate the benefits of their simple network measures, they model a team game as a network (Cintia *et al.*, 2015).

The creation of a pre-performance routine is still a regularly used and intuitively appealing method to improve sporting performance preparation (Arbues-Sanguesa *et al.*, 2020; Cotterill, 2010). The degree of actual comprehension, however, is not quite sufficient to support the creation of a pre-performance routine with assurance (Arbues-Sanguesa *et al.*, 2020). Fundamentally speaking, it is still unclear what purpose routines serve, what they ought to include, or how to teach them (Arbues-Sanguesa *et al.*, 2020; Cotterill, 2010). The purpose of the review of Cotterill, 2010 is to clarify existing knowledge and identify areas in need of more study. Greater comprehension is needed to comprehend both the psychological elements of routines and the method through which a routine is created or taught.

Basketball physical preparation models can be improved along with athlete health and performance by measuring the physical work and acute psychobiological reactions that ensue (Gómez-Carmona *et al.*, 2021; Russell *et al.*, 2021; Sarlis & Tjortjis, 2020). Although the amount of literature documenting the physical demands of basketball has increased recently and training load monitoring technologies have also advanced, as of right now, there are no evaluations that aggregate all of the research done on basketball (Gómez-Carmona *et al.*, 2021; Sarlis & Tjortjis, 2020). The existing literature also lacks a comprehensive analysis of the load surveillance techniques and metrics used in basketball (Gómez-Carmona *et al.*, 2021; Russell *et al.*, 2021; Sarlis & Tjortjis, 2020). To better understand the physical demands of different sports, reliable comparisons between research would be made possible by this kind of critical analysis (Russell *et al.*, 2021).

This systematic review goal was to analyze the approaches and tools available for keeping track of the physical demands placed on competitive basketball players (Russell *et al.*, 2021). The physical requirements of training and playing video games were referred to as the "training load" because it was anticipated that the latter would also serve as a training stimulus. This study set out to identify methodological flaws in the literature and provide basketball-specific operational definitions and recommendations for enhancing basketball training load monitoring (Russell *et al.*, 2021).

10.3.3.2 Game strategy optimization

Game strategy optimization involves using data-driven techniques to develop winning strategies that take advantage of opponents' weaknesses and minimize one's team weaknesses (Ashford *et al.*, 2021; Horvat *et al.*, 2019; Kumar, 2013; Sharma & Sharma, 2023; Tian *et al.*, 2019; Wang *et al.*, 2018). This domain includes sub-domains such as game simulation, predictive modelling, play calling, and game plan development (Ashford *et al.*, 2021; Tian *et al.*, 2019). Game strategy optimization is critical for coaches and teams seeking a competitive edge in today's highly competitive sports landscape (Kumar, 2013; Sharma & Sharma, 2023; Wang *et al.*, 2018).

There is a publication for insights on ecological dynamics, recognition-primed decision-making, and information processing were used to explain decision-making in team sports, which caused conceptual conflict and practical ambiguity (Ashford *et al.*, 2021; Horvat *et al.*, 2019; Kumar, 2013; Sharma & Sharma, 2023; Tian *et al.*, 2019; Wang *et al.*, 2018). The purpose of authors in this study (Ashford *et al.*, 2021) was to investigate empirical data to understand how decisions are made in team sports and to pinpoint the characteristics of competent decision-making in a team sports environment. Nine internet databases PORTDiscus, PsycINFO, PsycArticles, PsycTests, PubMed, SAGE journals online, Web of Knowledge, Academic Search Complete, and Web of Science—were searched until the final return in March 2021 (Ashford *et al.*, 2021). 53 pieces that met the requirements for inclusion underwent thematic analysis and narrative synthesis. Findings show that characteristics including complexity, typicality, and other variables influence whether mental representations are relatively absent or present during the decision-making process. Before proposing implications for practitioners, they recommend that to better understand decision-making in team sports, future studies incorporate ideas and methods shared by each perspective (Ashford *et al.*, 2021).

Sports like baseball and basketball have successfully used sports analytics (Bharathan *et al.*, 2015). Its use in soccer, however, has been quite modest. There is little research on soccer analytics using ML approaches, and forecasts are the main use (Bharathan *et al.*, 2015). It is necessary to ascertain whether using ML in soccer analytics may produce better and more informative outcomes. In this work (Kumar, 2013), they analyze player and match performance data using both descriptive and predictive methods. Evaluating a soccer player's performance often involves using expert ratings as a standard practice (Kumar, 2013). However, experts typically do not disclose the specific criteria they use for evaluating players. Instead, they strive to identify the key performance factors that influence their assessments (Kumar, 2013).

Sports are using statistical analysis, data mining, and ML more and more (Bharathan *et al.*, 2015). The initial information system or data-driven Web application named Basketball Coach Assistant (later BCA) for sports statistics and analysis is described in this paper along with its new and enhanced version (Horvat *et al.*, 2019). This study also provides a summary of the commercial information systems in use at the moment for in-game analysis. The purpose of BCA is to offer crucial data for making decisions regarding the training process and managing basketball teams. The player progress indicators and statistical analysis based on data mining techniques are given particular attention in addition to statistical analysis to identify the various classes of played game points. The data from the BCA information system, displayed in tables, were beneficial for planning training activities and making strategies (Horvat *et al.*, 2019).

When defending against the Cavaliers in the 2017 NBA Playoffs, Celtics coach Brad Stevens was presented with a dilemma: "Do you double and risk giving up easy shots, or stay at home and do the best you can?" Finding a sound defensive plan that successfully uses in the NBA, double may make all the difference, yet it's a tough call. The trade-off between risk and return is quantified as they examine double teaming in the NBA (Cervone *et al.*, 2014; Gómez-Carmona *et al.*, 2021; Russell *et al.*, 2021; Sarlis & Tjortjis, 2020). They determined when the ball handler was double-teamed using player trajectory information gathered from over 643,000 possessions (Wang *et al.*, 2018). With these data and the corresponding result (i.e., was the defense successful?), they used deep RL to assess the efficacy of the defensive tactics (Wang *et al.*, 2018). They give qualitative and quantitative findings that summarize the defensive approach they have learned to use. They demonstrate that the points per possession and win percentage predicted by their policy value calculations are accurate (Wang *et al.*, 2018). The proposed framework, taken as a whole, offers a step towards a deeper comprehension of NBA defensive strategies (Wang *et al.*, 2018).

Sports researchers have only lately started to pay attention to network analysis, even though it is an important analytical tool in many sectors of the social and natural sciences (Sharma & Sharma, 2023; Wäsche *et al.*, 2017). The usefulness of network analysis for analyzing athletic events is assessed in Sharma & Sharma, 2023. First, some background on social network analysis (SNA) and related concepts. To look into research topics and methodology, they conduct a systematic review of the empirical literature on SNA and its application to sport. Based on this review, they provide a six-dimensional conceptual typology of SNA applications in sports, which includes intra-organizational networks, competitive networks, affiliation networks, and interaction networks. Future directions for this exciting field of athletic research are explored (Sharma & Sharma, 2023).

10.3.3.3 Injury prevention

Injury prevention is a growing domain in sports analytics, as teams look to mitigate the risk of injuries to their players. This domain includes sub-domains such as player tracking, fatigue monitoring, injury prediction, and rehabilitation planning (Combs *et al.*, 2019; Delibašić *et al.*, 2018; Fleisig *et al.*, 2018; Myer *et al.*, 2012; Pyne *et al.*, 2014; Wäsche *et al.*, 2017). By analyzing data on player movements, biometrics, and other factors, teams can identify potential injury risks and take proactive steps to prevent them.

Basketball is an outdoor sport that takes a lot of practice and physical stamina from the player (Wäsche *et al.*, 2017). Even a minor gesture mistake might cause harm and cause the team to lose the game (Combs *et al.*, 2019; Delibašić *et al.*, 2018; Fleisig *et al.*, 2018; Myer *et al.*, 2012; Pyne *et al.*, 2014; Wäsche *et al.*, 2017). An environment for injury-free sports training will be activated by the suggested ML-based gesture recognition and recommender model (Wäsche *et al.*, 2017). "The free injury stadiums" will give the athlete receiving training real-time feedback (Wäsche *et al.*, 2017). It would take a close inspection (Wäsche *et al.*, 2017). The use of dribble maneuvers, quick bends, and motions at the locomotive controls, and produce a simulation for an improved response from the basketball trainee (Wäsche *et al.*, 2017).

Using a treatment appropriately for the demographic, it is intended for is a crucial first step in the treatment of a specific injury etiology (Delibašić *et al.*, 2018; Fleisig *et al.*, 2018; Myer *et al.*, 2012; Pyne *et al.*, 2014; Wäsche *et al.*, 2017). An algorithm for determining athletes with a high risk of anterior cruciate ligament (ACL) injury landing tactics has been developed, and it uses field-based methods rather than a motion analysis systems based in a lab (Myer *et al.*, 2012). The proposed prediction algorithm and the resulting field-based assessment methods make it possible to identify athletes with a lower chance of suffering an ACL injury. The field-based tool ease of use and accuracy make it useful for pinpointing elements that could put athletic women susceptible to harm (Myer *et al.*, 2012).

The researchers goal in Myer *et al.*, 2012 is to show how new algorithm techniques can be used with video analysis software, which is frequently used by coaches, strength and conditioning experts, and athletic trainers, to accurately capture and analyze measurements of knee valgus motion, knee flexion range of motion, body mass, tibia length, and quadriceps to hamstrings ratio. A prediction system was developed to identify those who may be at risk of suffering a noncontact ACL injury and who could gain the most from neuromuscular training. The algorithm combined measurements from the field with software analyzes (Myer *et al.*, 2012).

The goal of Fleisig *et al.*, 2018 was to ascertain how frequently pitching mechanics errors discovered through biomechanical analysis are fixed. 46 Pitchers in baseball had their biomechanics evaluated twice, with an average of 12 months (range: 2–48 months) passing between each evaluation. At the time of both evaluations, pitchers who were participating in minor league, major league, college, or high school levels were in good health. After warming up, each competitor threw ten fastballs with maximum effort. Using computerized three-dimensional motion analysis, eight kinematic parameters were calculated and compared with a database of elite professional pitchers. Each participant or coach received an explanation of any flaws—defined as departures from the elite range—after his initial assessment. According to data from the second review, 44% of the problems had been fixed (Fleisig *et al.*, 2018).

Errors in the following areas seemed to be repaired more frequently than others: stride length, front foot position, shoulder external rotation, shoulder abduction, and elbow flexion, they included knee extension and shoulder abduction, which were observed less frequently. Future studies may establish the level of athletes or the best training regimens for improving defects (Fleisig *et al.*, 2018).

The prevalence of sickness and injury among aquatic sportsmen during training and major international competitions are covered in detail, along with the key principles for injury and illness prevention in aquatic sports, and the pertinent scientific studies on the role of nutrition in reducing disease and injury risk is also covered in the review of Pyne *et al.*, 2014. A carefully thought-out diet that provides adequate calories, macronutrients (particularly protein and carbs), and micronutrients (particularly zinc, iron, and the vitamins A, D, E, B6, and B12) are advised for aquatic athletes to preserve health and performance. During lengthy training sessions, it is advantageous to consume carbohydrates through sports drinks, gels or sports meals to sustain energy availability. Studies on certain strains of probiotic species and foods or supplements containing plant polyphenols have shown promise, but more research is necessary (Pyne *et al.*, 2014).

Between 1.1 and 1.9 million concussions in children under the age of 18 in the US happen annually due to athletics and leisure activities (Combs *et al.*, 2019). Football has a higher incidence of concussions than other child sports. This is especially troubling in light of recent research demonstrating that concussions sustained during athletic competition are linked to cognitive and psychological impairment among former players in their later years. Unfortunately, there aren't many effective therapies available right now to stop concussions in football. This study's goal was to assess a mentored intervention program that was created to teach safe play methods and, as a result, lessen the frequency and the intensity of brain trauma among football players in high school (Combs *et al.*, 2019).

In a clustered quasi-experimental study design, the player education intervention program was made available to two out of every three teams in year one, two out of every four teams in year two, and all four teams in year three (Combs *et al.*, 2019). All players wore in-helmet accelerometers to measure head hits. To determine who on intervention teams is "high-risk" for targeted mentorship, game footage, and head impact data were used. Based on three factors, players were chosen as intervention candidates: A player had a concussion in more than 7% of head impacts that exceeded 60 g in peak linear acceleration and more than 20% of head impacts that struck the highest level of the head. Over three weeks, high-risk athletes participated in three sessions of one-on-one or small-group coaching with a mentor who used game video analysis to impart safer playing methods (Combs *et al.*, 2019).

10.3.3.4 Fan engagement

Fan engagement is a critical domain in sports analytics, as it involves understanding the preferences and behaviours of fans to enhance their experience of the game (Aloufi *et al.*, 2018; Brown & Billings, 2013; Karg *et al.*, 2021; Lacey & Kennett-Hensel, 2016; Mueller, 2020). This domain includes sub-domains such as fan profiling, social media analysis, and fan feedback (Aloufi *et al.*, 2018; Brown & Billings, 2013; Karg *et al.*, 2021; Lacey & Kennett-Hensel, 2016; Mueller, 2020). By analyzing data on fan preferences, purchasing habits, and social media interactions, teams can develop personalized engagement strategies that resonate with their audience (Aloufi *et al.*, 2018; Brown & Billings, 2013; Karg *et al.*, 2021; Lacey & Kennett-Hensel, 2016; Mueller, 2020).

The authors in Mueller, 2020 examine how Major League Baseball game ticket sales can be predicted and use random forests, which are produced using conditional inference trees and classification and regression trees (CART), to pinpoint significant attendance predictors. In contrast to earlier research that forecasts sports demand, I take into account various predicting periods and utilize just information that is readily available to the general public before a game or season. To anticipate the regular season of 2015, training data for the models comes from the years 2013 and 2014. A dynamic method for month-ahead forecasting is used in conjunction with the static within-season method. Individual team out-of-sample performance is assessed and tested using various regression models with dummy variables in the least squares and an ignorantly projected lagged attendance (Mueller, 2020).

Models and forecasting horizons both state, that the results indicate a wide range in team-specific prediction accuracy (Mueller, 2020). On average, the predicted performance of linear and tree-ensemble models is similar; however, least-squares regression, despite taking into account censoring attendance predictions and using team-fixed effects to match stadium capacities, is unable to take into account several team-specific quirks (Mueller, 2020).

The abundance of soccer-related web content produced has been widely employed in sentiment analysis research by supporters and teams (Aloufi *et al.*, 2018). Aloufi *et al.*, 2018 suggests a domain-specific method for deciphering the emotion displayed in soccer-related conversations. They present a vocabulary unique to soccer, which they use to create a sentiment analysis model tuned to the game. Their findings demonstrate the viability of the suggested method for identifying spectator emotion during soccer matches (Aloufi *et al.*, 2018).

Corporate social responsibility (CSR) initiatives have the potential to have an impact on professional sports clubs' brand and image, and ultimately, how well such franchises can connect with their particular fan bases (Lacey & Kennett-Hensel, 2016). Particularly, the NBA incorporates CSR into its marketing and public relations plan. By analyzing how fans' expectations and perceptions of CSR initiatives affect the quality of their relationship with the team, this study contributes to their understanding of CSR. It is based on field survey data that an NBA franchise in the Southeastern United States helped collect throughout three seasons. The findings show that when fans believe their team is socially responsible and that belief is in line with the team's perceived CSR practices, the combined effect raises the bar for the fan experience (Lacey & Kennett-Hensel, 2016).

Professional clubs rely heavily on season ticket holders as a source of income, yet retention is a constant problem (Karg *et al.*, 2021). Prior studies have mainly relied on survey data, which has attenuated issues and has concentrated on broad characteristics like relationship tenure, frequency of game attendance, and renewal intention. Karg *et al.*, 2021 examine and forecast retention throughout a three-year term as a loyalty consequence for a single professional club to advance this significant research issue. To be more precise, the authors use a variety of loyalty metrics and team performance to forecast retention and make use of cutting-edge data mining techniques to boost prediction accuracy (Karg *et al.*, 2021).

Previous studies have examined how companies publicly react to crises (Brown & Billings, 2013). The researchers in this study use the NCAA investigation at the University of Miami as a case study to investigate how Twitter users might act as stand-ins for organizations during times of crisis. An examination of 75 well-known Miami supporters with over 42,000 Twitter followers using Coombs, 2007 reputation-repair strategy reveals that as main coping mechanisms, the most common actions used by fans were to criticize the accuser, remind, ingratiate themselves, and deflect attention. The examination also uncovered fresh techniques for reputation restoration, having consequences for companies, educational establishments, sports studies, and crisis communication theory mentioned (Brown & Billings, 2013).

10.3.3.5 Business operation

Business operations are another important domain in sports analytics, as they involve using data to optimize revenue generation and cost management (Arslan *et al.*, 2022; Jensen *et al.*, 2016; Mondello & Kamke, 2014; Naraine *et al.*, 2022; Solntsev, 2014). This domain includes sub-domains such as ticket pricing, sponsorship management, fan analytics, and venue optimization (Arslan *et al.*, 2022; Jensen *et al.*, 2016; Mondello & Kamke, 2014; Naraine *et al.*, 2022; Solntsev, 2014). By analyzing data on fan behaviour, sponsor interactions, and venue operations, teams can improve their profitability while enhancing the fan experience (Arslan *et al.*, 2022; Jensen *et al.*, 2016; Mondello & Kamke, 2014; Naraine *et al.*, 2022; Solntsev, 2014).

In a market for sports tickets with multiple avenues for sales, they create a framework to analyze the purchasing patterns of various sectors of diverse clientele and to maximize costs for different insurance (Arslan *et al.*, 2022). It can be difficult for sports teams to retain or grow their ticket sales. However, they lack cutting-edge technologies for demand forecasting and pricing optimization that account for the variety of seat section valuations and opponent matchups. They intend to use price-driven-data approaches to increase gate earnings (Arslan *et al.*, 2022).

They work with a collegiate football team to develop a data-driven pricing tool that splits clients into two channels of sales, analyzes customer decision-making processes using the Mixed Multinomial Logit and Multinomial Logit frameworks, and calculates optimal or nearly optimal prices while accounting for certain administrative restrictions set by the team leader (Arslan *et al.*, 2022). Additionally, their approach considers the capacity restrictions of seat classifications as well as the order in which clients arrive (Arslan *et al.*, 2022).

Social media has emerged as a crucial new frontier in the sports sponsorship paradigm (Naraine *et al.*, 2022). It provides marketers with a potent tool to increase customer involvement. Even so, it's unclear to what extent social media content utilized to promote a sports sponsorship can influence customer engagement patterns. The goal of Naraine *et al.*, 2022 sought to ascertain the impact of sponsors on fan participation in sports organizations' social media posts. 13,542 Instagram posts were generated by four professional sports teams between 2017 and 2019. Sponsored material has a negative impact on engagement levels, according to a regression analysis. Consequently, companies need to be more conscious of the possibility that their relationship may not achieve the desired outcomes if they just finance content in an unnatural, coercive way (Naraine *et al.*, 2022).

Because expenses are proprietary, there is a dearth of academic research analyzing the variables affecting their costs for sports marketing activities, like sponsorship (Jensen *et al.*, 2016). The purpose of Jensen *et al.*, 2016 is to provide a methodology for market intelligence analysis that will assist managers in more precisely anticipating and estimating costs in the current dynamic sports marketing environment. The sports apparel industry is highly competitive and dynamic, this context was selected to investigate the influence of four distinct elements on sponsorship prices: market-specific factors, on-field performance, property-specific considerations, and market-specific factors. A predictive empirical model was created using a structured, hierarchical process, and predicted values were then generated for each property (Jensen *et al.*, 2016).

The chapter addresses real-world issues surrounding football club valuation (Solntsev, 2014). As part of the income approach, these issues are handled utilizing the existing discount cash flow model. The article goes into great detail about football clubs' revenue mix, spending habits, and methods for predicting future cash flows. Additionally, it takes into account the procedures for determining the discount rate for the FC's valuation as well as the risks connected to its operations. There are statistics and a few illustrations of several business models from the European football market (Solntsev, 2014).

The competitive and difficult subject of sports analytics has grown as professional sports organizations continue to search for ways to enhance their on-field performance(Mondello & Kamke, 2014). Despite these latest advancements and the accessibility of data, organizations, researchers, and practitioners frequently exchange limited and anecdotal information. In this article (Mondello & Kamke, 2014), they aimed to give a succinct review of analytics before sharing numerous best practice examples of how Tampa Bay Lightning, an NHL team, incorporates analytical methods into a number of its main business organizations. An organizational focus on customer relationship management (CRM) gives management useful information about their clients' recent buying patterns and possibly even helps forecast future purchases (Mondello & Kamke, 2014).

10.4 TECHNIQUES FOR ANALYZING SPORTS DATA

A wide range of analytical techniques are used in the quickly developing discipline of sports analytics to convert raw data into insightful information. As a fundamental strategy, descriptive analytics summarizes historical data to provide a thorough understanding of prior performances, player statistics, and team patterns (Rodgers *et al.*, 2023; Singh, 2020). With the help of predictive analytics, teams may use this past data to build models that predict future events, such as game outcomes or player performance (Ghosh *et al.*, 2023; Rodgers *et al.*, 2023). Sports analytics relies heavily on ML techniques, which provide strong tools for finding patterns and connections in huge datasets. Players are grouped using clustering algorithms based on shared traits, which helps with player profiling and position categorization (Ghosh *et al.*, 2023).

Insights from passing networks and player cooperation are uncovered via network analysis, which probes the complex interactions between players and teams (Ghosh *et al.*, 2023). In order to understand player development and team dynamics, time series analysis, which is essential for game performance data, finds trends, patterns, and seasonal fluctuations (Ghosh *et al.*, 2023; Rodgers *et al.*, 2023). Marketing tactics and fan engagement activities are shaped by sentiment analysis, which is frequently applied to social media data, fan sentiment, and interaction (Sun *et al.*, 2020). Techniques for optimization improve choices like choosing the best lineup or efficiently allocating resources (Ghosh *et al.*, 2023). Last but not least, text mining retrieves useful data from unstructured data sources like sports publications and interviews, offering perceptions into the general public and player sentiments. Each of these analytical techniques plays a special part in releasing the potential of sports data, facilitating decision-making, and providing teams with a competitive edge in the sports sector (Ghosh *et al.*, 2023; Rodgers *et al.*, 2023).

Because of its capacity to extract meaningful and rich information from the raw data, deep learning (DL) has become incredibly popular among other algorithms (Sun *et al.*, 2020). DL architecture has a high time and memory overhead complexity, they need a lot of processing power

(Sun *et al.*, 2020; Vinyes Mora & Knottenbelt, 2017). To further improve performance and optimize the architecture, another task is to modify the hyper-parameters. Feature engineering and selection could create another difficulty because they are crucial to improving the feature space representation of the raw data (Vinyes Mora & Knottenbelt, 2017). Domain-specific expertise is needed for feature engineering and selection to enhance the semantic inference of the raw data (Rodgers *et al.*, 2023; Sun *et al.*, 2020; Vinyes Mora & Knottenbelt, 2017).

In addition, class imbalance, data size constraints, label datasets, and other problems plague DL (Vinyes Mora & Knottenbelt, 2017). Additionally, the same difficulties are frequently encountered when creating scalable algorithms (Sun *et al.*, 2020; Vinyes Mora & Knottenbelt, 2017). Other approaches, to solve such issues, strategies including zero and few-shot learning, active learning, transfer learning, and self-supervised learning can be applied (Ghosh *et al.*, 2023). RL does, however, assist in the resolution of more difficult issues in a variety of applications, including navigation, video games, control sets for augmented and virtual reality (AR/VR), and more (Luo, 2020). One of the biggest challenges to the real-time application of RL is the requirement for large computational resources and a large amount of data to learn from (Ghosh *et al.*, 2023; Luo, 2020).

The dimensionality of the feature space presents another difficulty; RL has its own limitations, which makes it difficult to employ in practical settings (Luo, 2020; Sun *et al.*, 2020). The reward function structure, which aids agents in environment learning, presents another difficulty (Luo, 2020). In contrast, RL results in the agent learning the appropriate behaviour for that environment and for optimum performance (Luo, 2020). Consequently, the RL strategy ought to be used when they require agents to investigate and exploit, then collect information and incentives, and lastly engage with the surroundings (Luo, 2020). Furthermore, techniques such as gradient, deep Q neural network, policy optimization, imitation/inverse RL, and distributed RL with quantile regression, and others are becoming more and more popular in recent years to address the few issues mentioned above (Ghosh *et al.*, 2023). Many of the aforementioned methods employed in the world of sports will be discussed in this section of the chapter. The study will be broken into sections based on the categories that each technique or algorithm falls within.

10.4.1 STATISTICAL LEARNING

Statistical learning (SL) in sports analytics entails using sophisticated statistical methods and ML algorithms to sports data (Hopkins *et al.*, 2009). It aims to extract meaningful insights, predict outcomes, and improve player performance (Albert *et al.*, 2005). This interdisciplinary approach enables teams, coaches, and analysts to obtain a competitive edge, optimize methods, and make data-driven decisions (Albert *et al.*, 2005). By analyzing player statistics, game events, and other relevant data, SL uncovers patterns, identifies key performance indicators, and enhances understanding of sports dynamics (O'Donoghue, 2013). As the field continues to grow, it is poised to revolutionize the future of sports analytics is being shaped by how sports are played, and coached (Albert *et al.*, 2005; Pan, 2019).

Various ML techniques used in the sports analytics field are listed and discussed in this section. It is common to go into great detail about SL, DL, and RL methods. It is observed that, for many sports analytics applications, the researchers primarily used the following algorithms. Interestingly, they found that SL is renowned among coaches for its straightforward and simple-to-interpret qualities despite advances in DL and RL algorithms (Albert *et al.*, 2005; Pan, 2019). Because of SL, new avenues for research have opened up in the big data sports domains, including sports analytics (Albert *et al.*, 2005; Pan, 2019).

To highlight important traits that aid team managers and other individuals in evaluating player performance, statistical measures like variance, min-max values, standard deviation, and others are utilized (Hopkins *et al.*, 2009). To assess and estimate the characteristics (body posture, style, and speed) of individuals who are most likely to make mistakes during training sessions, for example, and to identify those characteristics in those individuals, variance and standard deviation offer fascinating insights (Albert *et al.*, 2005). The gap between analytical tools and player performance is filled by SL. The most often used statistical analysis techniques in sports include sample size calculations, hypothesis testing, mean, median, and mode (O'Donoghue, 2013). Sports data analysts and coaches have recently tended to favour employing conventional statistical methods to ascertain or analyze the player performance (Pan, 2019).

Recently, researchers have concentrated on developing statistics and models that are data-driven to understand the player performance. Maszczyk *et al.*, 2014, for instance, conducted a comparison analysis for predicting javelin throwers and the outcome of the match. Four predictors (specific power of the arms and trunk, grip power, specific power of the abdominal muscles, and cross step) effect the javelin throwers performance, as shown by a correlation matrix and regression analysis.

Basketball is a game of choices; a player can decide to pass, dribble, or shoot at any time to alter the nature of possession. The great majority of player decisions, however, cannot currently be evaluated statistically because most metrics are determined by events that take place at the end of possession or very close to it, such as points, turnovers, and assists (Cervone *et al.*, 2014). By estimating the number of points the offense should score at the end of the possession, or expected possession value (EPV), they provide a mechanism for using player monitoring data to give each possession moment a point value. Any basketball related decision, such as whether to pass, dribble, or shoot, may be evaluated by analysts using EPV. This allows numerous new metrics and analyzes of basketball to be conducted, which quantify value in terms of points. Using player-tracking data from the 2012 – 13 season, they computed EPV and presented the results in this study. They also provide various examples of EPV-derived metrics that address practical basketball-related concerns (Cervone *et al.*, 2014).

To find associations between variables and create predictions, sports analytics frequently uses the fundamental statistical approach known as linear regression (Hong & Calder, 2023). It offers analysts and organizations in the sports world useful insights into a range of topics, from player performance to game outcomes. The recovery of an athlete following an injury is the focus of this particular paper (Hong & Calder, 2023). Based on previous data, linear regression models can be used to forecast player performance in sports. For instance, analysts can utilize historical data on goals, assists, and shots on goals to forecast a player performance in the future (Sarlis & Tjortjis, 2020). Recently, researchers and professionals have developed a keen interest in sports groups entrepreneurial mindset and social impact (Núñez-Pomar *et al.*, 2020). Núñez-Pomar *et al.*, 2020 aims to investigate the relationship between the social performance of sports clubs and their entrepreneurial orientation, as well as the possible impact of management variables such as the level of competition (national vs. regional) and the source of funding (public vs. private) (Núñez-Pomar *et al.*, 2020). Two complimentary techniques have been used to compare the outcomes: models based on qualitative comparative analysis and linear models with moderation analysis (Núñez-Pomar *et al.*, 2020).

Teams can use this information to make educated decisions on player acquisition, contract negotiations, and starting lineup decisions (Cervone *et al.*, 2014). In addition, linear regression can be used to find key performance indicators (KPIs) that have a big impact on how games turn out (Cervone *et al.*, 2014). Analysts can identify the variables, like as possession time, pass completion rates, or shots on goal, that are most closely connected with winning by studying big datasets of historical games (Cervone *et al.*, 2014). This knowledge aids teams and coaches in creating winning game plans and adjusting their tactics as needed. A study uses linear regression to anticipate the situation when sports organizations are faced with crisis conditions, sports entrepreneurship has been viewed as a crucial component (Escamilla-Fajardo *et al.*, 2020). The purpose of this study is to ascertain whether there are any disparities between the predictions of entrepreneurship on the standard of service in nonprofit sports organizations and the crises resulting from COVID-19 on sports entrepreneurship (Escamilla-Fajardo *et al.*, 2020). Another study with a linear regression approach is (Sarlis & Tjortjis, 2020). The data from this study can be used by basketball performance analytics to evaluate team and player performance more accurately. These statistics can also be used to assess how future estimates might come true, evaluate team composition, and track an athlete career development (Sarlis & Tjortjis, 2020).

There is also an important study about team performance (Manasis & Ntzoufras, 2014). In this study, there is a use of different indexes to measure the final position of a team in a league. An empirical study that examines the value and trajectory of all relevant indices further clarifies their key characteristics using data from the English Premier League (Manasis & Ntzoufras, 2014). For example, the G index not only considers the promotion and relegation rule, but it enables this possibility to compare leagues and/or seasons with different numbers of teams (Manasis & Ntzoufras, 2014). Additionally, it calculates how frequently the top K positions should change in comparison to a perfectly balanced league where every game outcome is entirely random. The equation of the G index is the following (Buzzacchi *et al.*, 2003):

$$G(T) = \frac{\sum_{t=1}^{T} y^L(K,t) - \sum_{t=1}^{T} y^L_\alpha(K,t)}{\sum_{t=1}^{T} y^L(K,t)}$$

where T stands for the years under consideration and $y^L_\alpha(K,t)$ stands for the observed number of teams entering at least once in the top K positions in the highest league. The lower bound of G is well defined as it equals zero and it is obtained in the case of perfect balance (Buzzacchi *et al.*, 2003).

Linear regression is a statistical technique used to understand the relationship between a dependent variable and one or more independent variables. It does this by fitting a straight line to observed data points (Hope, 2020). It enables us to comprehend how a change in one or more variables can forecast or justify changes in other variables in more straightforward terms (Maulud & Abdulazeez, 2020).

Not all applications of linear regression are commercial in nature. It is also highly valued in sports (Sarlis & Tjortjis, 2020). For example, one could be curious about the relationship between a basketball team average amount of points scored per game and the total number of games won in a season (Sarlis & Tjortjis, 2020). These factors may have a linear relation, which can be checked by a scatterplot. The number of victories and the average total point score by the opponent is also linearly connected (Maulud & Abdulazeez, 2020). As the number of games won increases, the opponent average score decreases. With a solid model, it is feasible to predict the number of games that each team will win (Maulud & Abdulazeez, 2020).

10.4.2 UNSUPERVISED LEARNING

Unsupervised learning in sports analytics has emerged as a powerful tool for gaining insights from vast and complex datasets without the need for labeled training data (Chmait & Westerbeek, 2021; Kitani *et al.*, 2011). By leveraging unsupervised algorithms, sports analysts can uncover patterns, relationships, and hidden structures that may not be evident through manual examination (Chmait & Westerbeek, 2021; Kitani *et al.*, 2011). This approach is particularly valuable in sports, where the abundance of data, including player statistics, game outcomes, and performance metrics, poses challenges for traditional analysis methods (Chmait & Westerbeek, 2021).

One of the primary applications of unsupervised learning in sports analytics is player profiling (Kitani *et al.*, 2011). Analysts can categorize players based on comparable playing styles, physical characteristics, or performance indicators by employing clustering techniques like K-means or hierarchical clustering (Sanakoyeu *et al.*, 2018). These clusters can provide useful information, like the identification of talented individuals with comparable skill sets, a comparison of various play styles within a team, or the identification of possible areas for player development (Chmait & Westerbeek, 2021; Sanakoyeu *et al.*, 2018).

Anomaly detection is a further area where unsupervised learning thrives. Sports datasets can contain noise, mistakes, and outliers, which can affect the analysis accuracy (Horvat & Job, 2020).

Autoencoders and one-class SVM are two examples of anomaly detection algorithms that can assist in identifying unexpected events or occurrences in data, allowing analysts to explore potential abnormalities, such as doping, match-fixing, or injuries that might have gone unnoticed otherwise (Chmait & Westerbeek, 2021).

Unsupervised learning is useful not only for player-specific analysis but also for assessing team performance (Kitani *et al.*, 2011). Analysts can minimize the dimensionality of data by using techniques like principal component analysis (PCA) or independent component analysis, discovering the most important performance elements that influence a team's success. With this strategy, coaches and strategists can make data-driven judgments to raise team performance while also optimizing game strategies (Horvat & Job, 2020). Furthermore, game strategy analysis can greatly benefit from unsupervised learning (Horvat & Job, 2020). Coaches can identify opponent trends and create methods to effectively counter them by grouping game scenarios and plays (Horvat & Job, 2020). This strategy also aids in spotting novel moves or strategies that rivals might not have come across, giving the team a competitive edge (Chazan-Pantzalis, 2020).

Even with all the advantages, there are still difficulties when using unsupervised learning in sports analytics . Complex tasks can include feature engineering, data preprocessing, and choosing the right algorithms. Additionally, interpretability can be problematic because some unsupervised models are referred to as "black boxes" and may be difficult to understand (Chazan-Pantzalis, 2020; Horvat & Job, 2020). Unsupervised learning is a useful method in sports analytics since it gives analysts the ability to gain important insights, profile players, spot anomalies, assess team performance, and come up with winning game plans. The effect of unsupervised learning in sports analytics will continue to grow as the discipline develops thanks to improvements in algorithms and data collection (Chazan-Pantzalis, 2020; Rentzmann & Wuthrich, 2019).

Using the statistical approach of cluster analysis, related data points are grouped to reveal patterns or structures within a dataset (Cartigny *et al.*, 2021; Gomez-Lopez *et al.*, 2019). To compare differences in implicit beliefs about sports ability, motivational orientation, and intention to be physically active, the study set out to identify the various motivational climate profiles in a large sample of high-performance handball players in Spain. Additionally, the study examined the relationship between gender and motivational climate profiles (Gomez-Lopez *et al.*, 2019). High-performance handball players totaling 444 (233 men, 211 women) decided to take part. The surveys they completed included the validated Spanish versions of the Conceptions on the Nature of Athletic Ability Questionnaire-2, the Perceived Motivational Climate in Sport Questionnaire, the Sport Motivation Scale, and the Intention to be Physically Active Questionnaire (Gomez-Lopez *et al.*, 2019). Two distinct motivational climate profiles, "mastery climate" and "performance climate", were discovered by hierarchical cluster analysis and verified by a K-Means cluster analysis (Gomez-Lopez *et al.*, 2019).

A parallel dual career pathway, which places equal emphasis on the athletic and vocational careers, an educational/vocational pathway, which prioritizes the vocational career, and a sporting pathway, which prioritizes the athletic career are some of the approaches to a dual career that have been identified by research in the past (Cartigny *et al.*, 2021). However, more investigation is required to completely comprehend the identification and self-efficacy aspects of these profiles. To address this, this research aims to collect survey responses from 111 athletes with multiple careers. A cluster analysis was used to examine the survey results, which measured characteristics of self-efficacy, athletic identity, and career identity.

A non-parametric statistical method called kernel density estimation (KDE) is used to estimate the probability density function of a random variable (Chen, 2017; Terrell & Scott, 1992; Zambom & Ronaldo, 2013). The fundamental principle of KDE is to construct a smooth estimate of the underlying probability distribution by placing a kernel, a smooth, symmetric function at each data point, and adding up these kernels (Chen, 2017). Integration and convolution are used in the mathematical formulation. Furthermore, the kernel function integrates to 1 and is symmetric and non-negative. Two examples are the Gaussian (normal) kernel and the Epanechnikov kernel (Chen, 2017; Terrell & Scott, 1992; Zambom & Ronaldo, 2013).

There is a study (Jlassi *et al.*, 2010) that introduces a new method that combines three key colour levels found through histogram analysis with the Singular Value Decomposition (SVD) of each image to enhance the compactness of density distribution. In reality, the suggested method uses colour information and SVD to effectively delete background pixels that correspond to shadow pixels. This technique has been used on colour football video footage. On the other side, coaches and leaders must have the most technical-tactical information possible to gather statistics about players throughout a game (football, handball, volleyball, rugby, etc.) and to fine-tune their strategy. For this reason, it is essential to develop an algorithm for automatically identifying colour zones (players) of interest and resolving the issue of background and foreground confusion in a succession of photos (Jlassi *et al.*, 2010).

PCA is often used in sports analytics to simplify complex datasets and find important patterns in the data (Shlens, 2014). By identifying the most significant components, PCA helps analysts gain valuable insights into player performance, team dynamics, and game strategies. Here are summaries of two research papers that demonstrate the applications of PCA in sports analytics (Shlens, 2014).

The main objective of PCA is to convert a high-dimensional dataset into a lower-dimensional space while retaining as much of the variability present in the original data as feasible. This reduction simplifies data visualization, computation, and analysis, making it easier to identify meaningful patterns and relationships within the data (Mishra *et al.*, 2017; Shlens, 2014). Mathematically, PCA is a linear transformation that involves finding the principal components (eigenvectors) of the data covariance matrix (Mishra *et al.*, 2017). The covariance matrix shows the connections between the dataset various dimensions or attributes and offers details on the distribution and direction of the data (Mishra *et al.*, 2017).

10.4.3 SUPERVISED LEARNING

In sports analytics, supervised learning is a potent method that involves training algorithms using labeled data to create predictions or categorizations (Vats *et al.*, 2022). This method is frequently used in sports to address a variety of problems, from predicting player performance to analyzing game results.

One of the primary applications of supervised learning in sports is the prediction of player performance. Analysts can forecast how a player will perform in upcoming games by instruction for prediction models like as support vector machines (SVM), decision trees, and logistic regression using previous data on player statistics and performance. Making wise decisions about team selection, player moves, or contract talks can be aided by these predictions (Kasiri-Bidhendi *et al.*, 2015).

Injury prevention and prediction is a crucial area of application (Myer *et al.*, 2012; Sharma & Sharma, 2023). Supervised learning models can be created to identify prospective injury risks by gathering information on athletes physical characteristics, training regimens, and injury histories. With the help of this data, sports organizations can customize their training plans, rest policies, and injury recovery plans to reduce injury risk and maintain peak performance levels among their athletes (Sharma & Sharma, 2023). Game outcome analysis also makes extensive use of supervised learning (Vats *et al.*, 2022). Analysts can forecast the likelihood of a team winning a specific match based on multiple criteria such as squad makeup, past performance, weather conditions, and home-field advantage by entering historical game data into models, such as logistic regression or random forests (Kasiri-Bidhendi *et al.*, 2015). Coaches can use this predictive skill to develop winning plans for forthcoming games and make game-time decisions (Vats *et al.*, 2022).

Although supervised learning in sports analytics has many benefits, it also has certain drawbacks (Vinyes Mora & Knottenbelt, 2017). The effectiveness of the models can be influenced by the type and volume of labeled data (Johnson *et al.*, 2018; Vinyes Mora & Knottenbelt, 2017). Additionally, because sports are dynamic, it is not always possible to predict future results with complete accuracy, particularly in high-variance sports like basketball or soccer (Johnson *et al.*, 2018). The ability to accurately anticipate player performance, prevent injuries, analyze game outcomes, and recognize video events are all made possible by supervised learning, which is a useful tool in sports analytics (Johnson *et al.*, 2018; Sarlis & Tjortjis, 2020). Sports teams and analysts can get useful insights to enhance team performance, optimize game strategy, and make data-driven decisions for a competitive advantage by utilizing the power of labeled data and sophisticated ML algorithms (Sarlis & Tjortjis, 2020).

The family of generalized linear classifiers includes SVM methods (Premalatha *et al.*, 2013). Hyperplanes are used by SVM algorithms to divide the problem area into various subdomains of the same class of solutions. The separating hyperplane is a line that divides the plane into two portions, each of which contains the observations of one class when considering a binary classification issue in a 2D space (Sarlis & Tjortjis, 2020). Each observation is represented by a point in a K-dimensional space by SVM algorithms, where k is the number of issue attributes, and each dimension corresponds to one of those qualities (Premalatha *et al.*, 2013).

Each observation has a support vector attached to it, which is the hyperplane that passes through the point and is parallel to the separating hyperplane (Premalatha *et al.*, 2013). Someone should be aware that there are support vectors for each class in issues with many classes. The program then examines the support vectors for the various classes, using the SVs of the data points that are closest to the corresponding hyperplanes as a starting point (Premalatha *et al.*, 2013). These data points are each class outliers, and these support vectors mostly determine where the hyperplanes are located (Sarlis & Tjortjis, 2020).

Due to the significance of the outliers, SVM are a wise choice to prevent overfitting (Bishop, 2006; Yu *et al.*, 2003). However, when the data set contains a lot of noise, this trait turns out to be a significant flaw in the approach (Bishop, 2006). One advantage of this algorithm is that it performs well when the number of features is greater than the number of data points (Yu *et al.*, 2003). The SVM algorithm aims to find a hyperplane that separates different classes.

In binary classification, the chance of a result occurring (often 0 or 1) is predicted using the statistical technique of logistic regression using one or more independent factors (Apostolou & Tjortjis, 2019; Kapadia *et al.*, 2020; Miller, 2015). The logistic function, which converts linear combinations of predictors into a probability score between 0 and 1, is used to model the relationship (Apostolou & Tjortjis, 2019; Kapadia *et al.*, 2020; Miller, 2015). (Kapadia *et al.*, 2020) uses logistic regression for the Indian Premier League. Decision trees are an ML technique used for both regression and classification issues. They construct a model of choices and their effects that resembles a tree (Kapadia *et al.*, 2020). They began with a root node and divided the data into branches depending on feature values, creating decision nodes, which in turn produced leaf nodes that represented the outcomes or predictions at the end (Eryarsoy & Delen, 2019; Jayalath, 2018; Kapadia *et al.*, 2020).

A decision tree will be represented by T. Every internal node in the tree represents a decision rule that takes into account a particular feature, and every leaf node represents an anticipated result. Recursive definitions allow us to mathematically represent a decision tree (Suthaharan, 2016). Random Forest is an ensemble ML technique used for regression and classification (Jayalath, 2018). It constructs many decision trees during training and merges their forecasts to get predictions that are more trustworthy and precise (Jayalath, 2018). Random Forest lessens overfitting and improves model generalization by combining the findings from various trees (Eryarsoy & Delen, 2019; Jayalath, 2018; Kapadia *et al.*, 2020).

In the Random Forest algorithm N is the number of trees and T_i represents the ith decision tree. For each tree T_i, a subset of the training data is randomly selected with replacement (bootstrap sample). Additionally, a subset of features is randomly chosen at each split, introducing randomness and diversity among the trees. A potent ensemble ML technique for classification and regression is gradient boosting (Jayalath, 2018; Kapadia *et al.*, 2020). It sequentially creates several decision trees, each one addressing the flaws of the preceding. Gradient Boosting increases model accuracy over iterations by giving more weight to data items that had previously been misclassified (Jayalath, 2018; Kapadia *et al.*, 2020).

Gradient Boosting is an ensemble learning technique that combines the predictions of multiple weak learners (typically decision trees) to create a strong predictive model. It builds trees sequentially, with each tree correcting the errors of the previous ones (Jayalath, 2018). Popular implementations with distinctive advantages in various applications include AdaBoost, XGBoost, and LightGBM (Chen *et al.*, 2021; Eryarsoy & Delen, 2019; Yiğit *et al.*, 2020). A straightforward and efficient ML approach for classification and regression problems is K-Nearest Neighbours (KNN) (Jayanth *et al.*, 2018; Kapadia *et al.*, 2020). It operates by locating the K data points that are closest to a specified input based on a similarity metric (often the Euclidean distance) (Kapadia *et al.*, 2020).

In both classification and regression, the class of the input is determined by the majority class among the K neighbors, or a continuous value is predicted using their average (Jayalath, 2018; Kapadia *et al.*, 2020). The algorithm in classification determines the Euclidean distance between the input point x and all data points in the training set given a dataset with labeled instances (x_i, y_i), where x_i is the feature vector and y_i is the class label. After that, it chooses the k closest neighbors and, by majority vote, determines the class label. The prediction for regression tasks, in which the target variable is continuous, is derived from the mean of the target values of the k nearest neighbors. The mathematical ideas underlying the KNN algorithm are explained in basic terms by following these steps. Two hyperparameters that can be adjusted depending on the unique features of the dataset are the distance metric selection and the value of k (Jayalath, 2018; Kapadia *et al.*, 2020). The distance of the input points x and all data points x_i in the training dataset.

$$\mathrm{d}(x, x_i) = \sqrt{\sum_{j=1}^{n} (x_j - x_{ij})^2}$$

As the popularity of one-day international matches grows, it's critical to comprehend the potential predictors that have an impact on the final score. Among the more popular factors considered in the cricket literature are the home-field advantage, coin toss outcome, bat first or second, and day vs. night game format. The purpose of the article of Jayalath, 2018 is to concentrate on a thorough investigation of quantifying the relevance of those significant predictors using the widely used logistic regression and graphical "classification and regression tree" techniques (Jayalath, 2018). One of the more well-known cricket competitions in the world is the Indian Premier League (IPL), whose revenue rises each year, whose viewership has sharply increased, and whose betting market is expanding rapidly (Kapadia *et al.*, 2020). Because cricket is such a dynamic game that varies from ball to ball, bookmakers and bettors are encouraged to wager on the outcome of the match. This study examines ML technologies to address the problem of predicting cricket match results based on historical IPL match data. Filter-based approaches such as Information Gain, Wrapper, ReliefF, and Correlation-based Feature Selection have been used to find the dataset influential features. What's more, ML methods include Decision trees, KNN, Random Forest, and Naive Bayes (Kapadia *et al.*, 2020).

As analytical methods and tools develop, more and more researchers are using them to study a wide range of sports. Soccer, also known as association football, is thought to have about 4 billion fans, making it by far the most popular sport worldwide (Eryarsoy & Delen, 2019). Eryarsoy & Delen, 2019 objective is to create a model to forecast soccer (or association Football) game results (win-loss-draw) and identify the variables that affect game results. They tested several classifiers on 10 years of complete game-level data from the Turkish Super League, spanning the years 2007 to 2017, and identified the best effective techniques for outcome forecasting (Eryarsoy & Delen, 2019).

The area of sports analytics is becoming more and more widely used all around the world (Yiğit *et al.*, 2020). The value of football players is one of the unsolved issues in this area. To assist football clubs transfer decisions, this project aims to develop a model for evaluating the value of football players. The fundamental player characteristics that are presented in the video game football manager will serve as the foundation for the majority of the recommended models. To do this, various value assessment models are run using advanced supervised learning methods such as random forests, extreme gradient boosting, and ridge and lasso regressions on the individual statistics of 5316 players who are currently playing in 11 major leagues in Europe and South America (Yiğit *et al.*, 2020).

Summarizes the results of a study conducted to determine who will win an ODI cricket match following the first inning (Wickramasinghe, 2020). They create this prediction utilizing the Naive Bayes approach using the information gathered with 15 features, which are made up of characteristics linked to batting, team composition, bowling, and other factors. Their goal is to build an initial model, and then use feature selection methods including univariate, recursive elimination, and PCA to increase the accuracy of predicting the winner. They also look at how the right ratios between training sample size and testing sample size affect prediction accuracy. Based on the Bayes theorem, the Naïve Bayes algorithm is a supervised learning method that is utilized to solve classification problems.

An interesting study of using the Bayesian theorem in sports analytics is (Karlis & Ntzoufras, 2009). In this work, they take a new approach to how many goals each team scores. They focus on the difference between the number of goals, or the margin of victory, rather than explicitly modelling the overall number of goals. As opposed to just using the scores, modelling the inequalities has the potential to yield large gains. They first eliminate the correlation imposed by the competition between the competing teams, and then they reject the supposition that each team goal totals have a marginal Poisson distribution. The use of covariates in the Bayesian approach to Skellam's distribution is discussed (Karlis & Ntzoufras, 2009).

10.4.4 DEEP LEARNING

Due to its superior scalability and generalizability to more established ML and SL analysis techniques, DL has experienced tremendous growth (Bengio *et al.*, 2013). Researchers have successfully demonstrated that DL algorithms outperform traditional ML algorithms, particularly when it comes to learning complex feature representations and performance (Bengio *et al.*, 2013; Ghosh *et al.*, 2022). Applying particular shift-invariant transformation functions and nonlinear activation functions, the raw features are automatically learned in DL, helping to obtain better feature representation than the conventional learning algorithms and performance (Bengio *et al.*, 2013; Ghosh *et al.*, 2022, 2023).

DL is a subfield of ML that focuses on the use of neural networks, particularly deep neural networks, to model and solve complex tasks. Neural networks are composed of layers of interconnected nodes, and DL leverages architectures with many layers to learn hierarchical representations. A neural network is a computational model inspired by the structure and function of the human brain, consisting of layers of interconnected nodes (neurons). Also, it has an activation function, non-linear functions applied to the output of each node, allowing neural networks to model complex relationships (Deng, Yu, *et al.*, 2014). Furthermore, sports analytics uses supervised learning to analyze video (Johnson *et al.*, 2018). DL models, including convolutional neural networks (CNN), can be trained to automatically recognize and track these occurrences in real time by manually categorizing significant events in video data, such as goals, fouls, or assists (Vinyes Mora & Knottenbelt, 2017). This not only expedites the analytical process but also offers insightful information about player behaviour and strategy (Johnson *et al.*, 2018).

A CNN is a specialized neural network architecture designed for processing structured grid data, such as images. It utilizes convolutional layers to automatically and adaptively learn spatial hierarchies of features from the input data (Vinyes Mora & Knottenbelt, 2017). Researchers are gravitating towards RL and DL (CNN, recurrent neural network, gated recurrent unit), and other methods due to the aforementioned difficulties and developments in contemporary ML algorithms (Bengio *et al.*, 2013; Ghosh *et al.*, 2023). Additionally, researchers are creating frameworks that are more durable, generic, and scalable and that are simple to deploy and train in real-time applications (Bengio *et al.*, 2013; Ghosh *et al.*, 2023). They will present current developments in the sports area in DL and RL techniques (Ghosh *et al.*, 2023).

By applying specific nonlinear activation functions and shift-invariant transformation functions, the raw features are automatically learned in DL, guaranteeing improved feature representation in comparison to traditional learning algorithms (Bengio *et al.*, 2013; Ghosh *et al.*, 2023). Additionally, the researchers' topic expertise affects how accurate the handcrafted characteristics are (Ghosh *et al.*, 2023). As a result, the methods for feature extraction and selection are essential for building a strong data-driven ML model, particularly for applications in practical settings (Bengio *et al.*, 2013). DL made a significant contribution to the domain of athletic analytics.

One of the DL strategies that is still being developed is RL (Ghosh *et al.*, 2022). It depends on the direction the agent goes in an interactive environment when taking an action related to a state in order to get rewarded. Maximizing the payoff is the agent driving force and objective (Bengio *et al.*, 2013; Ghosh *et al.*, 2022). Exploration and exploitation, which both include enhancing or refining the attained reward score, are two methods for maximizing reward (Ghosh *et al.*, 2022). The exploration is based on a global search and involves examining the sample space (Ghosh *et al.*, 2023). RL can be mathematically represented as the Markov decision process (MDP tuple = (S, A, Si, R,)) where an agent conducts an action (A) in an interactive environment by starting from the current state (S) and moving to the successor state. Si, R represents the action-related reward function, whereas is the discount factor [0, 1] (Ghosh *et al.*, 2023).

Through interactions with its surroundings, an agent that uses RL to learn to make judgments. The agent acts, gets feedback in the form of incentives or penalties, and gradually gains the ability to make better decisions (Ghosh *et al.*, 2023). The enormous popularity of RL techniques has opened up new areas of study for researchers (Wang *et al.*, 2018). To analyze double teams in NBA games, Wang *et al.*, 2018 created a framework. Additionally, they developed a deep RL model that lowers the reward/score for the offense. Comparable to this, Luo, 2020 suggested a framework for creating a novel ranking system by the integration of inverse RL (apprenticeship learning) with Q-function learning (Luo, 2020).

Furthermore, they applied single-agent inverse RL for the multi-agent ice hockey Markov game using an alternating learning framework (Luo, 2020). They also employed a transfer learning strategy to move knowledge between the various reward functions for the same work (Luo, 2020). Similarly to this, Liu & Schulte, 2018 employed deep reinforcement to acquire an action-value Q function, which they subsequently applied to assess player behaviour in a variety of gaming settings. They employed dynamic long short-term memory architecture to incorporate the context signals and game history. They created the game impact meter, an innovative method for measuring how well players performed overall (Liu & Schulte, 2018).

Zhao *et al.*, 2019 suggested hierarchical learning with a multiagent reinforcement framework. A skill set and playing style like that of people in sports video games are the goals of the study (Zhao *et al.*, 2019). Two subproblems of hierarchical learning have been identified: high-level and low-level problems (Zhao *et al.*, 2019). Agents must perform similarly to humans in low-level problems, which is accomplished through imitation learning. The agents must adhere to a game plan, just like in the high-level problem, and RL makes this possible. Similar ideas were put out by (Jia *et al.*, 2020) in their study, where they suggested a deep multiagent RL framework and integrated curricula training system for an intense sports atmosphere centred around basketball. The fundamental objective of the project is to handle problems in asynchronous real-time, enabling training of both single and multiple agents (Ghosh *et al.*, 2023).

10.5 DISCUSSION

Sports analytics has emerged as a transformative force in the world of sports, revolutionizing the way teams, athletes, and organizations approach training, performance analysis, game strategy, fan engagement, and business operations (Barris & Button, 2008; Bengio *et al.*, 2013; Ghosh *et al.*, 2023; Maulud & Abdulazeez, 2020; Singh, 2020). This discussion explores various topics related to sports analytics, including its history, applications in different sports, advancements in technology, ethical considerations, challenges, and the future of sports analytics (Singh, 2020).

The history of sports analytics is lengthy and dates back several decades (Singh, 2020). Initially, it was largely confined to basic statistics, such as scoring averages and win-loss records (Singh, 2020). However, with the advent of computers and data collection technologies, sports analytics evolved into a more sophisticated discipline (Ghosh *et al.*, 2023). Billy Beane's use of data-driven player evaluation in baseball, as depicted in Michael Lewis's book "Moneyball", is often considered a turning point for sports analytics (Singh, 2020). This approach demonstrated the potential of using data to gain a competitive edge in player scouting and team management (Singh, 2020).

Sports analytics has found applications in a wide range of sports, from individual sports like tennis and golf to classic team sports like American football, basketball and soccer (Barris & Button, 2008; Bengio *et al.*, 2013; Ghosh *et al.*, 2023; Maulud & Abdulazeez, 2020; Singh, 2020). In team sports, analytics is used for player evaluation, tactical analysis, and injury prevention. For instance, soccer teams use tracking data to assess players positioning, sprint distances, and ball possession to optimize game strategies (Bengio *et al.*, 2013; Ghosh *et al.*, 2023). In individual sports, analytics helps athletes fine-tune their techniques and develop personalized training programs (Ghosh *et al.*, 2023).

Advancements in technology have been instrumental in propelling sports analytics to new heights (Ghosh *et al.*, 2023). The proliferation of wearable devices, such as GPS trackers and biometric sensors, has enabled the collection of real-time player data during practices and games. High-speed cameras and motion capture technology have made it possible to analyze athletes movements in detail, providing insights into biomechanics and injury prevention. Additionally, video analytics and computer vision techniques have streamlined game analysis, allowing coaches to identify patterns and weaknesses in opponents' strategies (Ghosh *et al.*, 2023).

It is noticed that several approaches are used to address real-world issues with business interest in sports. Among the many business applications in the several functional areas of management are strategy, operations, marketing, finance, leadership, and strategy. Discussions include squad composition, online marketing, match planning, bookmaker odds, social media analytics, fan base marketing, leadership choices, and other business-related subjects (Barris & Button, 2008). Social media marketing, sports injury prevention and game bidding are a few of these areas of rising interest. Data analytics is also used to assess sports injuries and develop injury prevention measures (Ghosh *et al.*, 2023; Maulud & Abdulazeez, 2020). These programs analyze secondary data on sports injuries to forecast injuries (Ghosh *et al.*, 2023; Maulud & Abdulazeez, 2020). This kind of research suggests ways to reduce sports injuries and consequently safeguard athletes health (Barris & Button, 2008; Bengio *et al.*, 2013; Ghosh *et al.*, 2023; Maulud & Abdulazeez, 2020; Singh, 2020).

According to the articles examined, machine-learning-based data analytics are becoming increasingly popular in sports marketing. To assist with fan-base marketing, sponsorship, promotions, social media, and other strategic, tactical, and operational issues in sports marketing, researchers have developed and deployed analytical methodologies (Ghosh *et al.*, 2023; Maulud & Abdulazeez, 2020). Since these studies have become more common in recent years, associations, clubs, managers and coaches are increasingly using data analytics in sports. According to some research, improved sport management and data analytics conceptualization will lead to improved applications (Ghosh *et al.*, 2023; Maulud & Abdulazeez, 2020). It is also found that, although the majority of data utilized in research to date comes from secondary datasets (which are proprietary), very few articles are based on publicly available data (Barris & Button, 2008; Bengio *et al.*, 2013; Ghosh *et al.*, 2023; Maulud & Abdulazeez, 2020; Singh, 2020).

The experiments also show that by continuously advancing data-gathering technologies, it is possible to properly utilize the enormous amount of data being acquired, both on and off the field (Bengio *et al.*, 2013; Singh, 2020). Different approaches are being successfully used to improve the interpretation and analysis of the data that is now available (Barris & Button, 2008; Maulud & Abdulazeez, 2020). These strategies are always being studied and improved upon in the literature. These uses, along with the reasons behind each, are provided in several research (Ghosh *et al.*, 2023).

The papers evaluated, technology and data analytics are something that organizations, teams, managers and coaches want to use in sports (Barris & Button, 2008; Bengio *et al.*, 2013; Ghosh *et al.*, 2023; Maulud & Abdulazeez, 2020; Singh, 2020). In conclusion, they discover that the research and practitioners both continue to emphasize the role that data analytics will play in sports in the future. According to some research, improved sports management and data analytics conceptualization will lead to improved applications. In addition, they discover that while sports analytics has received considerable attention from researchers, genuine sports analytics research is still in its infancy and urgently needs more study. Additionally, they discover that few articles are based on publicly accessible data, even though the vast bulk of data used in research to date comes from proprietary secondary sources.

With further developments within the fields of technology and data science, sports analytics is poised for intriguing future possibilities. Real-time data analysis to provide in-game insights instantly, improved AI and ML for more precise predictions and game strategies and expanding wearable technology integration to gather thorough player performance data are a few important next steps. The creation of increasingly complex models will also be prioritized to comprehend the complex interactions between players, teams, and game dynamics. Sports analytics will increasingly rely on big data techniques and cloud computing as data sources continue to expand to handle and interpret enormous amounts of information, resulting in more complex and data-driven decision-making processes in sports.

Sports analytics has used a variety of analytical approaches and techniques. Sports analytics research is happening more frequently, yet there are still some unresolved challenges, including (Ghosh *et al.*, 2023):

Identifying promising talent is essential in sports analytics. Predictive insights from extensive data help forecast athlete performances, aiding in the search for rising stars. Open Sports Big Data supplies sample data for analysis and decision-making, with robust privacy protections in place to safeguard athletes' personal information. The Unified Sports Big Data Platform merges these elements, creating an innovative environment where talent and technology collaborate to shape the future of sports through strategic planning and privacy commitment.

10.6 CONCLUSION

In this paper a review of the existing, to the best of our knowledge, articles on the subject of sports analytics is given. It covers how to produce, store, and use data at the outset. Following that, a study of sports analytics techniques was conducted. The publications were divided into two distinct ways in the paragraph that followed. The first method is contingent upon the sport covered in the paper. The second is based on the field, which concerns the publication across all sports.

We find that numerous analytical techniques and protocols have been applied in various sports-related scenarios. The increasing number of these types of studies in recent years suggests that sports analytics research is becoming more and more common. A variety of analytical approaches and tactics that have aided in the development of data analytics in the field of sports business management have been examined and analyzed in the literature. For this study, a thorough analysis of the literature on sports analytics from journals, conferences, and other sources was conducted. The analysis promotes scholarship and theoretical understanding by highlighting significant research on the subject.

Several helpful insights are produced by the examination of studies. The review indicates that several authors have offered illustrations of the various real-world contexts in which data analytics can be applied, such as fan base marketing, consumer attitudes, player bidding, sports injuries, player performance, promotions, and game bidding. Furthermore, a range of analytical methods have been used. These methods include descriptive analysis, optimization, multivariate regression, and even ML approaches (such as logistic regression, support vector machines, random forests, etc.).

REFERENCES

Albert, J., Bennett, J., & Cochran, J. J. (2005). *Anthology of statistics in sports*. SIAM.

Almujahed, S., Ongor, N., Tigmo, J., & Sagoo, N. (2013). Sports analytics: Designing a volleyball game analysis decision-support tool using big data. *2013 IEEE Systems and Information Engineering Design Symposium*, 19–24.

Aloufi, S., Alzamzami, F., Hoda, M., & El Saddik, A. (2018). Soccer fans sentiment through the eye of big data: The uefa champions league as a case study. *2018 IEEE Conference on Multimedia Information Processing and Retrieval (MIPR)*, 244–250.

Apostolou, K., & Tjortjis, C. (2019). Sports analytics algorithms for performance prediction. *2019 10th International Conference on Information, Intelligence, Systems and Applications (IISA)*, 1–4.

Arbues-Sanguesa, A., Martin, A., Fernandez, J., Ballester, C., & Haro, G. (2020). Using player's body-orientation to model pass feasibility in soccer. *Proceedings of the IEEE/CVF Conference on Computer Vision and Pattern Recognition (CVPR) Workshops*.

Arslan, H. A., Easley, R. F., Wang, R., & Yılmaz, Ö. (2022). Data-driven sports ticket pricing for multiple sales channels with heterogeneous customers. *Manufacturing & Service Operations Management*, *24*(2), 1241–1260.

Ashford, M., Abraham, A., & Poolton, J. (2021). Understanding a player's decision-making process in team sports: A systematic review of empirical evidence. *Sports*, *9*(5), 65.

Bacic, B., Meng, Q., & Chan, K. Y. (2017). Privacy preservation for esports: A case study towards augmented video golf coaching system. *2017 10th International Conference on Developments in eSystems Engineering (DeSE)*, 169–174.

Bai, Z., & Bai, X. (2021). Sports big data: Management, analysis, applications, and challenges. *Hindawi Complexity Volume 2021, Article ID 6676297*, 11 pages.

Barris, S., & Button, C. (2008). A review of vision-based motion analysis in sport. *Sports Medicine*, *38*, 1025–1043.

Bengio, Y., Courville, A., & Vincent, P. (2013). Representation learning: A review and new perspectives. *IEEE transactions on pattern analysis and machine intelligence*, *35*(8), 1798–1828.

Berrar, D. (2018). Bayes' theorem and naive bayes classifier. *Encyclopedia of bioinformatics and computational biology: ABC of bioinformatics*, *403*, 412.

Bharathan, S., Sundarraj, R. P., Abhijeet, S., & Ramakrishnan, S. (2015). A self-adapting intelligent optimized analytical model for team selection using player performance utility in cricket. *9th MIT Sloan Sports Analytics Conference, MIT, Boston*, 1–11.

Bigsby, K. G., & Ohlmann, J. W. (2017). Ranking and prediction of collegiate wrestling. *Journal of Sports Analytics*, *3*(1), 1–19.

Bishop, C. M. (2006). *Pattern recognition and machine learning*. Springer.

Brown, N. A., & Billings, A. C. (2013). Sports fans as crisis communicators on social media websites. *Public relations review*, *39*(1), 74–81.

Brunauer, R., Kremser, W., & Stöggl, T. (2020). From sensor data to coaching in alpine skiing–a software design to facilitate immediate feedback in sports. *Proceedings of the 12th International Symposium on Computer Science in Sport (IACSS 2019)*, 86–95.

Buzzacchi, L., Szymanski, S., & Valletti, T. M. (2003). Equality of opportunity and equality of outcome: Open leagues, closed leagues and competitive balance. *Journal of Industry, Competition and Trade*, *3*, 167–186.
Cao, C. (2012). Sports data mining technology used in basketball outcome prediction.
Cartigny, E., Fletcher, D., Coupland, C., & Bandelow, S. (2021). Typologies of dual career in sport: A cluster analysis of identity and self-efficacy. *Journal of Sports Sciences*, *39*(5), 583–590.
Cervone, D., D'Amour, A., Bornn, L., & Goldsberry, K. (2014). Pointwise: Predicting points and valuing decisions in real time with nba optical tracking data. *Proceedings of the 8th MIT Sloan Sports Analytics Conference, Boston, MA, USA*, *28*, 3.
Chambers, G. S., Venkatesh, S., & West, G. A. (2004). Automatic labeling of sports video using umpire gesture recognition. *Structural, Syntactic, and Statistical Pattern Recognition: Joint IAPR International Workshops, SSPR 2004 and SPR 2004, Lisbon, Portugal, August 18-20, 2004. Proceedings*, 859–867.
Chazan-Pantzalis, V. (2020). Sports analytics algorithms for performance prediction.
Chellatamilan, T., Ravichandran, M. M., & Kamalakkannan, K. (2015). Modern machine learning approach for volleyball winning outcome prediction. *Global Journal of Multidisciplinary Studies*, *4*(12), 63–71.
Chen, W.-J., Jhou, M.-J., Lee, T.-S., & Lu, C.-J. (2021). Hybrid basketball game outcome prediction model by integrating data mining methods for the national basketball association. *Entropy*, *23*(4), 477.
Chen, Y.-C. (2017). A tutorial on kernel density estimation and recent advances. *Biostatistics & Epidemiology*, *1*(1), 161–187.
Chmait, N., & Westerbeek, H. (2021). Artificial intelligence and machine learning in sport research: An introduction for non-data scientists. *Frontiers in Sports and Active Living*, *3*, 363.
Cintia, P., Rinzivillo, S., & Pappalardo, L. (2015). A network-based approach to evaluate the performance of football teams. *Machine learning and data mining for sports analytics workshop, Porto, Portugal.*
Combs, P. R., Ford, C. B., Nocera, M., Campbell, K. R., Marshall, S. W., Mihalik, J., Register-Mihalik, J., & Guskiewicz, K. M. (2019). Biomechanic data-driven behavior modification to reduce concussion risk in high school football athletes. *Orthopaedic Journal of Sports Medicine*, *7*(3_suppl), 2325967119S00105.
Coombs, W. T. (2007). *Ongoing crisis communication: Planning, managing, and responding*. Sage.
Cotterill, S. (2010). Pre-performance routines in sport: Current understanding and future directions. *International review of sport and exercise psychology*, *3*(2), 132–153.
Davenport, T. H. (2014). What businesses can learn from sports analytics. *MIT Sloan Management Review*.
Delibašić, B., Radovanović, S., Jovanović, M., Obradović, Z., & Suknović, M. (2018). Ski injury predictive analytics from massive ski lift transportation data. *Proceedings of the Institution of Mechanical Engineers, Part P: Journal of sports engineering and technology*, *232*(3), 208–217.
Deng, L., Yu, D., *et al.* (2014). Deep learning: Methods and applications. *Foundations and trends® in signal processing*, *7*(3–4), 197–387.
Dolatshah, M. (2018). Cleaning crowdsourced labels using oracles for statistical classification.
Drappi, C., & Co Ting Keh, L. (2019). Predicting golf scores at the shot level. *Journal of Sports Analytics*, *5*(2), 65–73.
Ehrenberg, H. R., Shin, J., Ratner, A. J., Fries, J. A., & Ré, C. (2016). Data programming with ddlite: Putting humans in a different part of the loop. *Proceedings of the Workshop on Human-In-the-Loop Data Analytics*, 1–6.
Elmore, R., & Urbaczewski, A. (2021). Loss aversion in professional golf. *Journal of Sports Economics*, *22*(2), 202–217.

Eryarsoy, E., & Delen, D. (2019). Predicting the outcome of a football game: A comparative analysis of single and ensemble analytics methods.

Escamilla-Fajardo, P., Núñez-Pomar, J. M., Calabuig-Moreno, F., & Gómez-Tafalla, A. M. (2020). Effects of the covid-19 pandemic on sports entrepreneurship. *Sustainability*, *12*(20), 8493.

Fleisig, G. S., Diffendaffer, A. Z., Ivey, B., & Aune, K. T. (2018). Do baseball pitchers improve mechanics after biomechanical evaluations? *Sports Biomechanics*, *17*(3), 314–321.

Fry, M. J., & Ohlmann, J. W. (2012). Introduction to the special issue on analytics in sports, part i: General sports applications.

Ghosh, I., Ramamurthy, S. R., Chakma, A., & Roy, N. (2022). Decoach: Deep learning-based coaching for badminton player assessment. *Pervasive and Mobile Computing*, *83*, 101608.

Ghosh, I., Ramasamy Ramamurthy, S., Chakma, A., & Roy, N. (2023). Sports analytics review: Artificial intelligence applications, emerging technologies, and algorithmic perspective. *Wiley Interdisciplinary Reviews: Data Mining and Knowledge Discovery*, e1496.

Giasemidis, G. (2020). Descriptive and predictive analysis of euroleague basketball games and the wisdom of basketball crowds. *arXiv preprint arXiv:2002.08465*.

Giles, B., Kovalchik, S., & Reid, M. (2020). A machine learning approach for automatic detection and classification of changes of direction from player tracking data in professional tennis. *Journal of sports sciences*, *38*(1), 106–113.

Goldman, M., & Rao, J. M. (2012). Effort vs. concentration: The asymmetric impact of pressure on nba performance. *Proceedings of the MIT Sloan sports analytics conference*, 1–10.

Gómez-Carmona, C. D., Mancha-Triguero, D., Pino-Ortega, J., & Ibáñez, S. J. (2021). Exploring physical fitness profile of male and female semiprofessional basketball players through principal component analysis—a case study. *Journal of Functional Morphology and Kinesiology*, *6*(3), 67.

Gomez-Lopez, M., Merino-Barrero, J. A., Manzano-Sánchez, D., & Valero-Valenzuela, A. (2019). A cluster analysis of high-performance handball players' perceived motivational climate: Implications on motivation, implicit beliefs of ability and intention to be physically active. *International Journal of Sports Science & Coaching*, *14*(4), 541–551.

Gong, X., & Wang, F. (2021). Classification of tennis video types based on machine learning technology. *Wireless Communications and Mobile Computing*, *2021*, 1–11.

Gudmundsson, J., & Horton, M. (2017). Spatio-temporal analysis of team sports. *ACM Computing Surveys (CSUR)*, *50*(2), 1–34.

Hamilton, M., Hoang, P., Layne, L., Murray, J., Padget, D., Stafford, C., & Tran, H. (2014). Applying machine learning techniques to baseball pitch prediction. *ICPRAM*, 520–527.

Harriss, D. J., & Atkinson, G. (2009). International journal of sports medicine–ethical standards in sport and exercise science research. *International Journal of Sports Medicine*, *30*(10), 701–702.

Hoege, J., Lansing, M., Nelson, S., Ungerleider, D., Iyer, R., Rhodes, C., Metzger, B., Worcester, P., Chandra, A., Leonard, J., *et al.* (2020). An interdisciplinary approach to sports analytics in a university setting. *2020 Systems and Information Engineering Design Symposium (SIEDS)*, 1–6.

Hong, C. C., & Calder, J. (2023). Ability to return to sports after early lateral ligament repair of the ankle in 147 elite athletes. *Knee Surgery, Sports Traumatology, Arthroscopy*, *31*(10), 4519–4525.

Hope, T. M. H. (2020). *Linear regression*. In *Machine learning* (67–81). Elsevier.

Hopkins, W., Marshall, S., Batterham, A., & Hanin, J. (2009). Progressive statistics for studies in sports medicine and exercise science. *Medicine+ Science in Sports+ Exercise*, *41*(1), 3.

Horvat, T., Havas, L., Srpak, D., & Medved, V. (2019). Data-driven basketball web application for support in making decisions. *icSPORTS*, 239–244.

Horvat, T., & Job, J. (2020). The use of machine learning in sport outcome prediction: A review. *Wiley Interdisciplinary Reviews: Data Mining and Knowledge Discovery*, *10*(5), e1380.

Hossain, H. S., Khan, M. A. A. H., & Roy, N. (2017). Soccermate: A personal soccer attribute profiler using wearables. *2017 IEEE International Conference on Pervasive Computing and Communications Workshops (PerCom Workshops)*, 164–169.

Hostrup, M., & Bangsbo, J. (2023). Performance adaptations to intensified training in top-level football. *Sports Medicine*, *53*(3), 577–594.

Hsu, C.-C., Chen, H.-T., Chou, C.-L., Ho, C.-P., & Lee, S.-Y. (2014). Trajectory based jump pattern recognition in broadcast volleyball videos. *Proceedings of the 22nd ACM international conference on Multimedia*, 1117–1120.

Irwin P. Beadle & Co., C. F., 141 William St. (1860). Beadle's dime base-ball player: A compendium of the game, comprising elementary instructions of this american game of ball, together with the revised rules and regulations for 1860, rules for the formation of clubs, names of the officers and delegates to the general convention, & c.. no. 1. *Beadle's dime base-ball player*, 40.

Jayal, A., McRobert, A., Oatley, G., & O'Donoghue, P. (2018). *Sports analytics: Analysis, visualisation and decision making in sports performance*. Routledge.

Jayalath, K. P. (2018). A machine learning approach to analyze odi cricket predictors. *Journal of Sports Analytics*, *4*(1), 73–84.

Jayanth, S. B., Anthony, A., Abhilasha, G., Shaik, N., & Srinivasa, G. (2018). A team recommendation system and outcome prediction for the game of cricket. *Journal of Sports Analytics*, *4*(4), 263–273.

Jensen, J. A., Wakefield, L., Cobbs, J. B., & Turner, B. A. (2016). Forecasting sponsorship costs: Marketing intelligence in the athletic apparel industry. *Marketing Intelligence & Planning*, *34*(2), 281–298.

Ji, R. (2020). Research on basketball shooting action based on image feature extraction and machine learning. *IEEE Access*, *8*, 138743–138751.

Jia, H., Hu, Y., Chen, Y., Ren, C., Lv, T., Fan, C., & Zhang, C. (2020). Fever basketball: A complex, flexible, and asynchronized sports game environment for multi-agent reinforcement learning. *arXiv preprint arXiv:2012.03204*.

Jlassi, M. M., Douik, A., & Messaoud, H. (2010). Objects detection by singular value decomposition technique in hybrid color space: Application to football images. *International Journal of Computers Communications & Control*, *5*(2), 193–204.

Joash Fernandes, C., Yakubov, R., Li, Y., Prasad, A. K., & Chan, T. C. (2020). Predicting plays in the national football league. *Journal of Sports Analytics*, *6*(1), 35–43.

Johnson, W. R., Alderson, J., Lloyd, D., & Mian, A. (2018). Predicting athlete ground reaction forces and moments from spatio-temporal driven cnn models. *IEEE Transactions on Biomedical Engineering*, *66*(3), 689–694.

Kapadia, K., Abdel-Jaber, H., Thabtah, F., & Hadi, W. (2020). Sport analytics for cricket game results using machine learning: An experimental study. *Applied Computing and Informatics*, *18*(3/4), 256–266.

Karg, A., Tamaddoni, A., McDonald, H., & Ewing, M. (2021). Predicting season ticket holder retention using rich behavioral data. *Journal of sport management*, *35*(5), 426–439.

Karlis, D., & Ntzoufras, I. (2009). Bayesian modelling of football outcomes: Using the skellam's distribution for the goal difference. *IMA Journal of Management Mathematics*, *20*(2), 133–145.

Kasiri, S., Fookes, C., Sridharan, S., & Morgan, S. (2017). Fine-grained action recognition of boxing punches from depth imagery. *Computer Vision and Image Understanding*, *159*, 143–153.

Kasiri-Bidhendi, S., Fookes, C., Morgan, S., Martin, D. T., & Sridharan, S. (2015). Combat sports analytics: Boxing punch classification using overhead depthimagery. *2015 IEEE International Conference on Image Processing (ICIP)*, 4545–4549.

Kitani, K. M., Okabe, T., Sato, Y., & Sugimoto, A. (2011). Fast unsupervised ego-action learning for first-person sports videos. *CVPR 2011*, 3241–3248.

Kniffin, K. M., Howley, T., & Bardreau, C. (2017). Putting muscle into sports analytics: Strength, conditioning, and ice hockey performance. *The Journal of Strength & Conditioning Research*, *31*(12), 3253–3259.

Knobbe, A., Orie, J., Hofman, N., Van der Burgh, B., & Cachucho, R. (2017). Sports analytics for professional speed skating. *Data Mining and Knowledge Discovery*, *31*(6), 1872–1902.

Kumar, G. (2013). Machine learning for soccer analytics. *University of Leuven*.

Lacey, R., & Kennett-Hensel, P. (2016). How expectations and perceptions of corporate social responsibility impact nba fan relationships. *Sport Marketing Quarterly*, *25*(1).

Leman, S. C., & House, L. (2012). Improving mr. miyagi's coaching style: Teaching data analytics with interactive data visualizations. *Chance*, *25*(2), 4–12.

Liu, C. H., Lin, Q., & Wen, S. (2018). Blockchain-enabled data collection and sharing for industrial iot with deep reinforcement learning. *IEEE Transactions on Industrial Informatics*, *15*(6), 3516–3526.

Liu, G., & Schulte, O. (2018). Deep reinforcement learning in ice hockey for context-aware player evaluation. *arXiv preprint arXiv:1805.11088*.

Liu, H. (2019a). Opportunities, challenges and countermeasures for the development of china's sports industry in the era of big data, *Journal of Physics Conference, vol. 1237, Article ID 022012,*

Liu, H. (2019b). Retracted: Opportunities, challenges and countermeasures for the development of china's sports industry in the era of big data. *Journal of Physics: Conference Series*, *1237*(2), 022012.

Liu, Y. (2020). Teaching effect and improvement model of college basketball sports based on big data analysis. *Journal of Physics: Conference Series*, *1533*(4), 042056.

Luo, Y. (2020). Inverse reinforcement learning for team sports: Valuing actions and players.

Macdonald, B. (2012). An expected goals model for evaluating nhl teams and players. *Proceedings of the 2012 MIT Sloan Sports Analytics Conference*.

Majumdar, A., Bakirov, R., Hodges, D., Scott, S., & Rees, T. (2022). Machine learning for understanding and predicting injuries in football. *Sports Medicine-Open*, *8*(1), 1–10.

Manasis, V., & Ntzoufras, I. (2014). Between-seasons competitive balance in european football: Review of existing and development of specially designed indices. *Journal of Quantitative Analysis in Sports*, *10*(2), 139–152.

Manzenreiter, W., & Horne, J. (2013). *Global governance in world sport and the 2002 world cup korea/japan*. In *Japan, korea and the 2002 world cup* (1–25). Routledge.

Maszczyk, A., Gołaś, A., Pietraszewski, P., Roczniok, R., Zajac, A., & Stanula, A. (2014). Application of neural and regression models in sports results prediction. *Procedia-Social and Behavioral Sciences*, *117*, 482–487.

Maulud, D., & Abdulazeez, A. M. (2020). A review on linear regression comprehensive in machine learning. *Journal of Applied Science and Technology Trends*, *1*(4), 140–147.

Meehan I. I. I., W. P., d'Hemecourt, P., Collins, C. L., & Comstock, R. D. (2011). Assessment and management of sport-related concussions in united states high schools. *The American journal of sports medicine*, *39*(11), 2304–2310.

Miller, T. W. (2015). *Sports analytics and data science: Winning the game with methods and models*. FT press.

Mishra, S. P., Sarkar, U., Taraphder, S., Datta, S., Swain, D., Saikhom, R., Panda, S., & Laishram, M. (2017). Multivariate statistical data analysis-principal component analysis (pca). *International Journal of Livestock Research*, *7*(5), 60–78.

Mondello, M., & Kamke, C. (2014). The introduction and application of sports analytics in professional sport organizations. *Journal of Applied Sport Management*, *6*(2), 11.

Morgulev, E., Azar, O. H., & Lidor, R. (2018). Sports analytics and the big-data era. *International Journal of Data Science and Analytics*, *5*, 213–222.

Mueller, S. Q. (2020). Pre-and within-season attendance forecasting in major league baseball: A random forest approach. *Applied Economics*, *52*(41), 4512–4528.

Mun, K., Cha, B., Lee, J., Kim, J., & Jo, H. (2023). Competenet: Siamese networks for predicting win-loss outcomes in baseball games. *2023 IEEE International Conference on Big Data and Smart Computing (BigComp)*, 1–8.

Muraoka, S., DeLeo, A. M., Yang, Z., Tatebe, H., Yukawa-Takamatsu, K., Ikezu, S., Tokuda, T., Issadore, D., Stern, R. A., & Ikezu, T. (2021). Proteomic profiling of extracellular vesicles separated from plasma of former national football league players at risk for chronic traumatic encephalopathy. *Aging and disease*, *12*(6), 1363.

Myer, G. D., Ford, K. R., Brent, J. L., & Hewett, T. E. (2012). An integrated approach to change the outcome part i: Neuromuscular screening methods to identify high acl injury risk athletes. *Journal of strength and conditioning research/National Strength & Conditioning Association*, *26*(8), 2265.

Naraine, M. L., Bakhsh, J. T., & Wanless, L. (2022). The impact of sponsorship on social media engagement: A longitudinal examination of professional sport teams. *Sport Marketing Quarterly*, *31*(3), 239–252.

Nieto, R. M., & Sánchez, J. M. M. (2013). An automatic system for sports analytics in multi-camera tennis videos. *2013 10th IEEE International Conference on Advanced Video and Signal Based Surveillance*, 438–442.

Núñez-Pomar, J. M., Escamilla-Fajardo, P., & Prado-Gascó, V. (2020). Relationship between entrepreneurial orientation and social performance in spanish sports clubs. the effect of the type of funding and the level of competition. *International Entrepreneurship and Management Journal*, *16*(3), 981–999.

O'Donoghue, P. (2013). *Statistics for sport and exercise studies: An introduction*. Routledge.

Okagbue, H. I., Erondu, E. C., Atayero, A. A., Oguntunde, P. E., Opanuga, A. A., Olawande, T. I., Ijezie, O. A., & Eze, G. A. (2018). Statistical analysis of frequencies of opponents eliminations in royal rumble wrestling matches, 1988–2018. *Data in brief*, *19*, 1458–1465.

Pan, L. (2019). A big data-based data mining tool for physical education and technical and tactical analysis. *International Journal of Emerging Technologies in Learning (Online)*, *14*(22), 220.

Pantzalis, V. C., & Tjortjis, C. (2020). Sports analytics for football league table and player performance prediction. *2020 11th International Conference on Information, Intelligence, Systems and Applications (IISA*, 1–8.

Polk, T., Jäckle, D., Häußler, J., & Yang, J. (2019). Courttime: Generating actionable insights into tennis matches using visual analytics. *IEEE Transactions on Visualization and Computer Graphics*, *26*(1), 397–406.

Premalatha, M., Lakshmi, C. V., & Eu, W. A. (2013). Svm trade-off between maximize the margin and minimize the variables used for regression. *International Journal of Pure and Applied Mathematics*, *87*(6), 741–750.

Pyne, D. B., Verhagen, E. A., & Mountjoy, M. (2014). Nutrition, illness, and injury in aquatic sports. *International journal of sport nutrition and exercise metabolism*, *24*(4), 460–469.

Ratner, A., Bach, S. H., Ehrenberg, H., Fries, J., Wu, S., & Ré, C. (2017). Snorkel: Rapid training data creation with weak supervision. *Proceedings of the VLDB Endowment. International Conference on Very Large Data Bases*, *11*, 269.

Rekatsinas, T., Chu, X., Ilyas, I. F., & Ré, C. (2017). Holoclean: Holistic data repairs with probabilistic inference. *arXiv preprint arXiv:1702.00820*.

Rentzmann, S., & Wuthrich, M. V. (2019). Unsupervised learning: What is a sports car? *Available at SSRN 3439358*.

Reyers, M., & Swartz, T. B. (2023). Quarterback evaluation in the national football league using tracking data. *AStA Advances in Statistical Analysis*, *107*(1-2), 327–342.

Ribeiro, A. S., Lima, F., Kraus, S., & Calabuig, F. (2022). Tournaments within football teams: Players' performance and wages. *Economic research-Ekonomska istraživanja*, *35*(1), 4884–4901.

Rodenberg, R., & Feustel, E. D. (2014). Forensic sports analytics: Detecting and predicting match-fixing in tennis. *The Journal of Prediction Markets*, *8*(1), 77–95.

Rodenberg, R. M., Sackmann, J., & Groer, C. (2016). Tennis integrity: A sports law analytics review. *The International Sports Law Journal*, *16*, 67–81.

Rodgers, W., Murray, J. M., Stefanidis, A., Degbey, W. Y., & Tarba, S. Y. (2023). An artificial intelligence algorithmic approach to ethical decision-making in human resource management processes. *Human Resource Management Review*, *33*(1), 100925.

Roh, Y., Heo, G., & Whang, S. E. (2019). A survey on data collection for machine learning: A big data-ai integration perspective. *IEEE Transactions on Knowledge and Data Engineering*, *33*(4), 1328–1347.

Russell, J. L., McLean, B. D., Impellizzeri, F. M., Strack, D. S., & Coutts, A. J. (2021). Measuring physical demands in basketball: An explorative systematic review of practices. *Sports Medicine*, *51*, 81–112.

Sanakoyeu, A., Bautista, M. A., & Ommer, B. (2018). Deep unsupervised learning of visual similarities. *Pattern Recognition*, *78*, 331–343.

Sarlis, V., & Tjortjis, C. (2020). Sports analytics evaluation of basketball players and team performance. *Information Systems*, *93*, 101562.

Schumaker, R. P., Solieman, O. K., Chen, H., *et al.* (2010). *Sports data mining* (*26*). Springer.

Sharma, D., & Sharma, V. (2023). Machine learning-based model for injury-free sports training. *International journal of convergence in healthcare*, *3*(1), 1–5.

Sheng, V. S., Provost, F., & Ipeirotis, P. G. (2008). Get another label? improving data quality and data mining using multiple, noisy labelers. *Proceedings of the 14th ACM SIGKDD international conference on Knowledge discovery and data mining*, 614–622.

Shlens, J. (2014). A tutorial on principal component analysis. *arXiv preprint arXiv:1404.1100*.

Siddiqa, A., Hashem, I. A. T., Yaqoob, I., Marjani, M., Shamshirband, S., Gani, A., & Nasaruddin, F. (2016). A survey of big data management: Taxonomy and state-of-the-art. *Journal of Network and Computer Applications*, *71*, 151–166.

Singh, N. (2020). Sport analytics: A review. *learning*, *9*, 11.

Smolkin, O. (2019). *Development of analytical system for sport fencing* [Doctoral dissertation, Ukrainian Catholic University].

Socolow, B., & Jolly, I. (2017). Game-changing wearable devices that collect athlete data raise data ownership issues. *World Sports Advocate*, *15*(7), 15–17.

Solntsev, I. (2014). Application of income approach for valuation of football club. *Social networks and the Economics of sports*, 9–40.

Sun, X., Davis, J., Schulte, O., & Liu, G. (2020). Cracking the black box: Distilling deep sports analytics. *Proceedings of the 26th acm sigkdd international conference on knowledge discovery & data mining*, 3154–3162.

Suthaharan, S. (2016). Machine learning models and algorithms for big data classification. *Integr. Ser. Inf. Syst, 36*, 1–12.

Synrem, W. (2019). Influence of technology on sports: A policy based perspective. *NUALS LJ, 13*, 51.

Terrell, G. R., & Scott, D. W. (1992). Variable kernel density estimation. *The Annals of Statistics*, 1236–1265.

Teune, B., Spencer, B., Sweeting, A. J., Woods, C., Inness, M., & Robertson, S. (2021). Application of a continuous pressure metric for australian football. *Journal of Sports Sciences, 39*(13), 1548–1554.

Thornton, J. S., Frémont, P., Khan, K., Poirier, P., Fowles, J., Wells, G. D., & Frankovich, R. J. (2016). Physical activity prescription: A critical opportunity to address a modifiable risk factor for the prevention and management of chronic disease: A position statement by the canadian academy of sport and exercise medicine. *British journal of sports medicine, 50*(18), 1109–1114.

Tian, C., De Silva, V., Caine, M., & Swanson, S. (2019). Use of machine learning to automate the identification of basketball strategies using whole team player tracking data. *Applied Sciences, 10*(1), 24.

Tracy, R., Xia, H., Rasla, A., Wang, Y.-F., & Singh, A. (2023). Graph encoding and neural network approaches for volleyball analytics: From game outcome to individual play predictions. *arXiv preprint arXiv:2308.11142*.

Tuyls, K., Omidshafiei, S., Muller, P., Wang, Z., Connor, J., Hennes, D., Graham, I., Spearman, W., Waskett, T., Steel, D., *et al.* (2021). Game plan: What ai can do for football, and what football can do for ai. *Journal of Artificial Intelligence Research, 71*, 41–88.

Van Haaren, J., Ben Shitrit, H., Davis, J., & Fua, P. (2016). Analyzing volleyball match data from the 2014 world championships using machine learning techniques. *Proceedings of the 22nd ACM SIGKDD International Conference on Knowledge Discovery and Data Mining*, 627–634.

Vats, K., McNally, W., Walters, P., Clausi, D. A., & Zelek, J. S. (2022). Ice hockey player identification via transformers and weakly supervised learning. *Proceedings of the IEEE/CVF Conference on Computer Vision and Pattern Recognition*, 3451–3460.

Veal, A. J., & Darcy, S. (2014). *Research methods in sport studies and sport management: A practical guide*. Routledge.

Vinyes Mora, S., & Knottenbelt, W. J. (2017). Deep learning for domain-specific action recognition in tennis. *Proceedings of the IEEE conference on computer vision and pattern recognition workshops*, 114–122.

Vleugels, R., Van Herbruggen, B., Fontaine, J., & De Poorter, E. (2021). Ultra-wideband indoor positioning and imu-based activity recognition for ice hockey analytics. *Sensors, 21*(14), 4650.

Wang, J., Fox, I., Skaza, J., Linck, N., Singh, S., & Wiens, J. (2018). The advantage of doubling: A deep reinforcement learning approach to studying the double team in the nba. *arXiv preprint arXiv:1803.02940*.

Wang, T., Ke, H., Zheng, X., Wang, K., Sangaiah, A. K., & Liu, A. (2019). Big data cleaning based on mobile edge computing in industrial sensor-cloud. *IEEE Transactions on Industrial Informatics, 16*(2), 1321–1329.

Wäsche, H., Dickson, G., Woll, A., & Brandes, U. (2017). Social network analysis in sport research: An emerging paradigm. *European Journal for Sport and Society, 14*(2), 138–165.

Wickramasinghe, I. (2020). Naive bayes approach to predict the winner of an odi cricket game. *Journal of Sports Analytics, 6*(2), 75–84.

Worsey, M. T. O., Espinosa, H. G., Shepherd, J. B., & Thiel, D. V. (2020). An evaluation of wearable inertial sensor configuration and supervised machine learning models for automatic punch classification in boxing. *IoT, 1*(2), 360–381.

Xu, C., Wang, J., Wan, K., Li, Y., & Duan, L. (2006). Live sports event detection based on broadcast video and web-casting text. *Proceedings of the 14th ACM international conference on Multimedia*, 221–230.

Yiğit, A. T., Samak, B., & Kaya, T. (2020). Football player value assessment using machine learning techniques. *Intelligent and Fuzzy Techniques in Big Data Analytics and Decision Making: Proceedings of the INFUS 2019 Conference, Istanbul, Turkey, July 23-25, 2019*, 289–297.

Yu, H., Yang, J., & Han, J. (2003). Classifying large data sets using svms with hierarchical clusters. *Proceedings of the ninth ACM SIGKDD international conference on Knowledge discovery and data mining*, 306–315.

Yu, X., & Farin, D. (2005). Current and emerging topics in sports video processing. *2005 IEEE International Conference on Multimedia and Expo*, 526–529.

Zambom, A. Z., & Ronaldo, D. (2013). A review of kernel density estimation with applications to econometrics. *International Econometric Review*, *5*(1), 20–42.

Zhang, C. (2015). *Deepdive: A data management system for automatic knowledge base construction* [Doctoral dissertation, The University of Wisconsin-Madison].

Zhao, Y., Borovikov, I., Rupert, J., Somers, C., & Beirami, A. (2019). On multi-agent learning in team sports games. *arXiv preprint arXiv:1906.10124*.

11 A New Approach to Basketball Game Predictions by Simulating Games Based on the Predicted Numbers and Types of Plays

Emiel Platjouw
Signpost Belgium

Abdolnasser Sadeghkhani
North Carolina Agricultural and Technical State University
E-Mail: asadeghkhani@ncat.edu. ORCID: 0000-0002-3074-0443.

Marica Manisera
University of Brescia
E-Mail: marica.manisera@unibs.it. ORCID: 0000-0002-2982-0243.

Paola Zuccolotto
University of Brescia
E-Mail: paola.zuccolotto@unibs.it. ORCID: 0000-0003-4399-7018.

Christophe Ley
University of Luxembourg
E-Mail: christophe.ley@uni.lu. ORCID: 0000-0002-2290-8437.

11.1 INTRODUCTION

The rise in importance of statistics and mathematical modelling in sports within the last decades, combined with the ever–increasing availability of data, means a lot of effort has been put into predicting the outcome of games, both in terms of the winning team and final score Stekler *et al.*, 2010. In the context of basketball, various techniques have been proposed. Early contributions date back to the seventies and propose a regression-based method only requiring past scores and opponents Stefani, 1977. More modern approaches include k–nearest neighbours Horvat *et al.*, 2018, naive Bayes classifier Miljković *et al.*, 2010 Cao, 2012 Zimmermann *et al.*, 2013, hierarchical Bayesian models Page *et al.*, 2007, probit regression Caudill, 2003a, decision trees and random forests Zimmermann *et al.*, 2013, support vector machines Beckler *et al.*, 2013, neural networks

DOI: 10.1201/9781032664330-11

Loeffelholz *et al.*, 2009a Loeffelholz *et al.*, 2009b Zimmermann *et al.*, 2013 Beckler *et al.*, 2013, and linear and logistic regression Loeffelholz *et al.*, 2009b. Manner, 2016b extends the traditional models to allow for heteroscedasticity and introduces a dynamic state space model to take account of team strengths. These models are usually based on the traditional box score summary statistics that are commonly recorded for each game, and aim to classify each opposing team as either winner or loser. The Four Factors, more advanced statistics obtained from box scores, are also used in this framework (see for example Teramoto & Cross, 2010,Baghal, 2012).

Play-by-play data contain more detailed descriptions of the events happening in a basketball game compared to traditional box score statistics. Multiple studies have already exploited this additional information in forecasting the results of basketball games by simulating the games using a Markov chain model Vračar *et al.*, 2016 Štrumbelj & Vračar, 2012, an alternating renewal–reward process model Peuter, 2013, continuous–time anti–persistent random walk model Gabel & Redner, 2012, or a Poisson regression Necati *et al.*, 2013. Lopez & Matthews, 2015 use simulations to quantify the degree to which luck played a role in the success of a model by generating tournament outcomes under different sets of true underlying game probabilities and Ruiz & Perez-Cruz, 2015 simulate bets on six online betting houses to check whether their Poisson-based model has the potential to provide greater benefits than those based on the implicit probabilities of the betting houses. In this paper, we will also simulate basketball games for the purposes of prediction.

Besides already mentioned papers, it is interesting to examine the proposals made in the context of predicting whole tournament outcomes. Compared to single game prediction, the difficulties are clearly inflated, so the proposed solutions for tournaments can greatly help to develop better prediction models also for single games. In basketball this challenge typically occurs with the NCAA tournament, the prediction of which is a competition for millions of people each year. In this context, Lopez & Matthews, 2015 show that a combination of simple statistical methods (logistic regression) with informative data (point spreads set by Las Vegas sportsbooks with possession based team efficiency metrics) can meet or exceed the accuracy of more complex algorithms. Gupta, 2015 proposes a dual-proportion probability model built around the constraint of two teams with a combined 100% probability of winning a given game. Yuan *et al.*, 2015 present the results of a team of modellers who mix several different methods. Other studies in the context of the NCAA tournament can be found in Schwertman *et al.*, 1991, Harville, 2003, Caudill, 2003b, Carlin, 2005, Stekler & Klein, 2012, and many others.

In this paper, we take a new angle on the problem of predicting match outcomes. We aim at predicting the final score, from which one of course directly derives the winner of the game. To this end, we need to analyze the dynamics of a basketball game. According to Dean Oliver's seminal book "Basketball on Paper" Oliver, 2004, possessions form the basis of a lot of advanced basketball analytics, directly obtained from play–by–play data, as formalized in the milestone work Kubatko *et al.*, 2007. The concept allows a team to score an unlimited amount of points in a single possession if it keeps on getting fouled after a scored field goal, misses the free throw, and gets the offensive rebound. Additionally, the team can play a whole quarter in a single possession if it also gets the offensive rebound after each failed shot. However, such a situation is extremely rare, yet some interesting sequences have been recorded. For instance, the Miami Heat recorded a 90 second possession against the Brooklyn Nets in the Eastern Conference Finals in 2014, with four missed field goals and four offensive rebounds, eventually scoring a lay–up.

From a simulation viewpoint, a concept that better captures the dynamics of a basketball game is the concept of plays, which is essentially the same as a possession, except that a new play starts after an offensive rebound. The length of a play is therefore bounded by the length of the shot–clock, and the number of points scored during one play ranges from zero to four. By predicting the different number of plays of each opposing team we are able to predict the scores and hence the winner. By predicting the different types of plays in different leagues, such as the NBA and the EuroLeague, we also can identify the most influential factors of each team and use them for compression purposes.

Season	Total Games	Teams	Regular Games
2000-2001	1189	29	82
2001-2002	1189	29	82
2002-2003	1189	29	82
2003-2004	1189	29	82
2004-2005	1230	30	82
2005-2006	1230	30	82
2006-2007	1230	30	82
2007-2008	1230	30	82
2008-2009	1230	30	82
2009-2010	1230	30	82
2010-2011	1230	30	82
2011-2012	990	30	66
2012-2013	1229	30	82
2013-2014	1230	30	82
2014-2015	1230	30	82
2015-2016	1230	30	82
2016-2017	1230	30	82
2017-2018	1230	30	82
2018-2019	1230	30	82

Season	Total Games	Teams	Regular Games
2000-2001	120	24	10
2001-2002	224	32	14
2002-2003	168	24	14
2003-2004	168	24	14
2004-2005	168	24	14
2005-2006	168	24	14
2006-2007	168	24	14
2007-2008	168	24	14
2008-2009	120	24	10
2009-2010	120	24	10
2010-2011	120	24	10
2011-2012	120	24	10
2012-2013	120	24	10
2013-2014	120	24	10
2014-2015	120	24	10
2015-2016	120	24	10
2016-2017	240	16	30
2017-2018	240	16	30
2018-2019	240	16	30

Figure 11.1 Number of teams, regular games and resulting total number of games played in each season of the NBA (left) and EuroLeague (right).

Since the frequency of scoring follows a Poisson–like process Gabel & Redner, 2012 Necati *et al.*, 2013, we use both a Poisson regression and a negative binomial regression to predict the different number of plays. Afterwards a compound Poisson process is used to simulate each game to find the most likely scoring outcome and predict the winner.

11.2 METHODS

11.2.1 DATASET

We consider games from the NBA for the seasons 2001-2002 till 2019-2020 and the EuroLeague for the seasons 2008-2009 till 2019-2020 (because no play-by-play data was available before). The data used for this research is available at https://www.basketball-reference.com and www.euroleague.net. For obtaining the data, we used the `rvest` package Wickham, 2022 in R, except for the EuroLeague for which play–by–play data was obtained using the `Eurolig` package Sergio, 2019. Various data cleaning and checking operations were conducted. Figure 11.1 provides details on the number of teams, regular games and resulting total number of games in both leagues.

For the sake of a better understanding of the data, we now provide formal definitions of a possession and a play and give further insight into the datasets we are working with.

11.2.1.1 Play-by-play data

A *possession* starts when a team gains control of the ball and ends when that team gives up the control. The latter happens (i) after the team scores field goals or free throws, thereby leading to the other team taking the ball out of bounds, (ii) by defensive rebounds from the opponent, or (iii) by turnovers (for more information, see Kubatko *et al.*, 2007).

On the other hand, a *play* starts when (i) a team gains control of the basketball after the other team scores and the ball goes out of bounds, (ii) after defensive rebounds, (iii) turnovers, and (iv) offensive rebounds.

In each game one can extract the i–point plays ($i$pp) information from the raw play–by–play data for each team for $i = 0, 1, \dots, 4$, where i–point plays refers to the number of plays resulting in i point(s) for the corresponding team. Such a 0pp play can result from either a turnover, a missed field

Table 11.1
The game between the Atlanta Hawks (ATL, home team) and the Charlotte Hornets (CHA, opponent team) played in season 2000 – 2001. The Atlanta Hawks scored 30 field goals (FG) and the Charlotte Hornets scored 35 field goals.

Team	Season	Team_FG	Opp_FG
ATL	2001	30	35
CHA	2001	35	30

goal, missed free throw after a technical foul, or missing all the free throws after a foul resulting in two or three free throws. Similarly, a 1pp play occurs from a scored free throw after a technical foul, or a single scored free throw after a foul resulting in two or three free throws. A 2pp play results from a scored two–point field goal, a missed free throw after a fouled and scored two–point field goal, or from scoring two free throws after a foul resulting in two or three free throws. A 3pp play can be obtained through either a scored three-point shot, a scored free throw after a fouled and scored two–point field goal, a missed free throw after a fouled and scored three–point shot, or by scoring three free throws after a fouled three–point shot. Finally, a 4pp play can result from a scored free throw after a fouled and scored three–point shot.

Moreover, the total plays (TP) is the total number of plays and can also be obtained as the sum of the five kinds of plays described above, and a team possession (TMP) is the number of times a team possesses the ball and corresponds to possession as described above.

11.2.1.2 Final Dataset

Each game is split into the home and the visiting team. Table 11.1 illustrates a simplified version of the statistics for a game between the Atlanta Hawks (ATL, home team) and the Charlotte Hornets (CHA, opponent or visiting team). We refer the reader to the Appendix for a full list of all variables/statistics used.

11.2.2 REGRESSION ANALYSIS

11.2.2.1 Poisson and Negative Binomial Regression

When trying to predict the $i$pp, $i = 0, \ldots, 4$, along with the TP during a basketball game, we are analysing the number of occurrences of a certain event over a specific period of time, that is, we are dealing with count data. Two common methods to predict count data are Poisson regression (PR) and negative binomial regression (NBR) (see, e.g., Cameron & Trivedi, 2013).

As a brief reminder, the Poisson distribution $\mathrm{Po}(\mu)$, with parameter $\mu > 0$ equaling both the mean and variance, has the probability mass function (pmf)

$$P(Y = y \mid \mu) = \frac{\mu^y}{y!} e^{-\mu}, \; y \in \mathbb{N}. \tag{11.1}$$

The negative binomial (NB) distribution is an interesting alternative to the Poisson distribution because it allows for over–dispersion (i.e., the variance exceeds the mean). Denoted by $\mathrm{NB}(\mu, \theta)$, it has two parameters, the mean parameter $\mu > 0$ and the over–dispersion parameter $\theta > 0$, and its pmf is given by

$$P(Y = y \mid \mu, \theta) = \frac{\Gamma(\theta + y)}{\Gamma(\theta) y!} \left(\frac{\mu}{\theta + \mu}\right)^y \left(\frac{\theta}{\theta + \mu}\right)^\theta, \; y \in \mathbb{N}, \tag{11.2}$$

where $\Gamma(\cdot)$ is the gamma function. The pmf in (11.2) converges to the Poisson distribution with the parameter μ in (11.1) as $\theta \to \infty$. The expectation and variance of the NB are given by μ and $\mu + \mu^2/\theta$, respectively; for more information, see, e.g., Hilbe, 2014.

Let $(y_i, \mathrm{x_i})$, for $i = 1, \ldots, n$, be independent observations from either model (11.1) or (11.2), and $\mathrm{x_i} = (x_{i1}, \ldots, x_{ip})' \in \mathbb{R}^p$ is a corresponding set of p covariates. In PR and NBR, the covariates are introduced via the link function $\log \mu_i = \log(\mathrm{x_i}'\beta)$ for $i = 1, \ldots, n$, with parameters $\beta = (\beta_1, \ldots, \beta_p)' \in \mathbb{R}^p$, and we write $y_i \mid \mathrm{x_i} \sim \mathrm{Po}(\mu_i)$ or $y_i \mid \mathrm{x_i} \sim \mathrm{NB}(\mu_i, \theta)$, depending on the chosen model. The parameters (β for Poisson, β and θ for NB) are estimated by penalized Poisson regression (PPR) or penalized NB regression (PNBR). The reason for this choice is that the number of covariates considered in the prediction model is high; so high that in multiple models for the EuroLeague, for instance, the number of predictor variables exceeds the number of observations. In addition, multiple predictor variables are strongly correlated with each other, which causes multicollinearity problems. A solution to these problems is using elastic net regularization by adding a L_1- and a L_2 penalty term to the log-likelihood function. The objective functions in PPR (see, e.g., Friedman *et al.*, 2010; Mwikali *et al.*, 2019) and PNBR (see, e.g., Lehman & Archer, 2019; Xie & Xiao, 2020) are equivalently defined as $\ell(\xi) + \lambda P_\alpha(\xi)$, where ξ stands for the parameters of the model, $\ell(\xi)$ represents the logarithm of the likelihood function of the model with parameters ξ, $\lambda \geq 0$ is a tuning parameter, and $P_\alpha(\cdot)$ for $0 \leq \alpha \leq 1$ is a compromise between L_1 and L_2 penalties Friedman *et al.*, 2010, known as the elastic net penalty, and is given by

$$P_\alpha(\xi) = \alpha\,||\xi||_1 + (1-\alpha)\,||\xi||_2^2, \tag{11.3}$$

where $||\xi||_r = (\sum_{i=1}^{p} |\xi_i|^r)^{\frac{1}{r}}$ is the L_r–norm for $r \in [1, \infty)$. Note that P_α for $\alpha = 0$ and 1, respectively, yields Lasso and ridge penalties. The estimator $\hat{\xi}$ corresponding to the elastic net penalized maximum log–likelihood method is thus obtained using

$$\arg\max_{\xi \in \mathbb{R}^p} \left(\ell(\xi) - \lambda(\alpha\,||\xi||_1 + (1-\alpha)\,||\xi||_2^2)\right),$$

where the tuning parameters λ and α are determined by means of cross–validation Fan & Tang, 2012.

11.2.2.2 Evaluation Measures

To train and evaluate the PPR and PNBR models, we use deviance. The deviance is a goodness-of-fit measure for generalized linear models McCullagh & Nelder, 1989, and generalizes the residual sum of squares in the linear regression model. It is proportional to twice the difference between the maximum achievable log-likelihood $l(\mu)$, associated with the saturated model, that is a perfect model for the sample at hand, and the log–likelihood $l(\hat{\mu})$ of the fitted model Liur, 2019 Cameron & Windmeijer, 1996 Agresti, 2003

$$D(y;\hat{\mu}) = 2\left[l(\mu) - l(\hat{\mu})\right]. \tag{11.4}$$

By substituting the log-likelihood of the fitted models PR and NBR in (11.4), we obtain the deviance D for the PPR and the PNBR.

In order to situate the performance of our best models with respect to the literature, we shall also consider the accuracy as evaluation measure (but not at the level of training). The accuracy is simply defined as the proportion of correctly predicted match outcomes.

11.2.3 THE DIFFERENT REGRESSION MODELS

11.2.3.1 General Model

First, a model was trained to predict each number of plays using data from the all seasons except the last two for each respective league. The last two seasons (2018-2019, 2019-2020) serve as test data.

The training data were then first split into a training and validation set in an 80/20 ratio. A tuning grid was created to train for the best value for α in the elastic net penalty. Values ranging from zero to one in steps of 0.1 were evaluated. For each value of α, the tuning parameter λ was trained using five-fold cross-validation on the training set. The model with the optimal value for λ, in other words the model with the lowest deviance, was then selected to be the optimal model for that value of α. The value of α whose optimal model had the lowest deviance on the validation set was then finally chosen. The chosen model got subsequently used to predict the different kinds of plays for all the games in the last two seasons, i.e., the test set.

11.2.3.2 Team Models

The same protocol as described in Subsection 11.2.3.1 was applied to create a model for each separate team ("Team Models 1"). Tuning the parameter α in (11.3) generally returned an optimal value of $\alpha = 0$ or $\alpha = 0.1$. A value of $\alpha = 0$ means the elastic net penalty is effectively the ridge penalty and no variable selection is performed.

As we deemed variable selection important for identifying the most important variables, we re-trained the models for each team by this time setting α to 0.1 for each team, resulting in "Team Models 2". This also had the benefit of saving considerable amounts of computing time.

All predicted values for EuroLeague teams with no 4pp in their EuroLeague history, or without enough data to train a model to predict their 4pp, were set to zero for the Compound Poisson simulation described in Subsection 11.2.4. Some teams playing in the EuroLeague test seasons did not play any EuroLeague games in one of the previous seasons used for training, hence no team model could be constructed to predict their plays. The average of each play over their previous games played in the season was taken to replace the model prediction for these teams.

11.2.3.3 No Prediction Models

In order to quantify the value of the different types of plays to use in the simulations, a further model was set up. In this model, the different types of plays used in the simulations are not predicted using prediction models but are simply the average number of TP and $i$pp ($i = 0, \ldots, 4$) of the team so far that season.

11.2.4 FINAL STEP: GAME SIMULATIONS

Finally, each game is simulated 10,000 times using a Compound Poisson process $Y = \sum_{i=1}^{T} X_i$, where T is the number of events following a discrete distribution, and X_i is the magnitude of the outcome of the i_{th} event Zhang, 2012. In our concrete situation, Y represents the points scored by a team during the game, while T is the total number of plays by that team during the game and T follows a Poisson distribution with mean equal to the predicted total plays (TP). Finally, X_i is the number of points scored per play (0, 1, 2, 3 or 4), with the outcome of each play randomly sampled from a multinomial distribution with $k = 5$, one for each possible type of play ($i$pp), and the multinomial probabilities are given by the ratios of the predicted amount for each type divided by the predicted total number of plays (Nota bene: a draw from the multinomial is only valid when the sum of predicted $i$pp's equals the predicted TP, if not then we need to resample).

Figure 11.2 illustrates an example of how a single simulation looks like for an NBA team. In this simulation, the team has 95 plays (95 points are plotted) over the course of a 48 minutes game and scores 88 points. The sequence of plays follows a Poisson process, as reflected by the distribution of the points along the x–axis. The value of each play is shown by the colour and shape of each individual point. The score of the team at a certain time in the game is represented by the cumulative value of the points achieved up to that moment and is shown on the y–axis.

Each of these 10,000 simulations (related to a given game) results in the simulated amount of points scored by the team at the end of a given game. The same is done for the opponent and, based

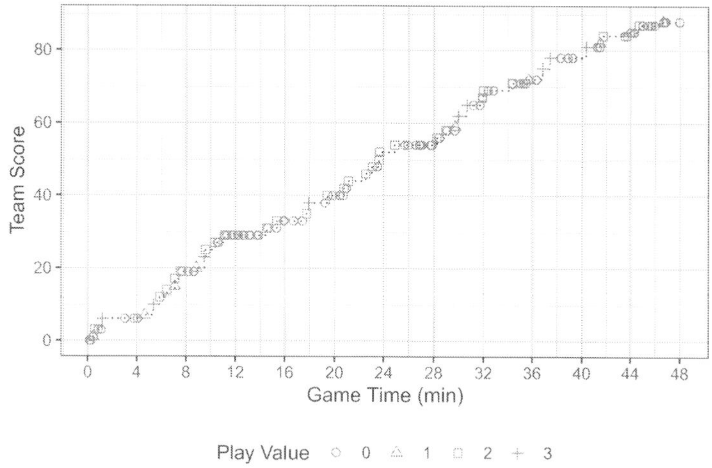

Figure 11.2 A single simulation for an NBA team. The team scored 88 points over 48 minutes in 95 plays. The colour and shape of each point represent the outcome of the play (orange circle = 0pp, green triangle = 1pp, blue square = 2pp, purple plus = 3pp).

on the score difference, the winner of a single simulation is identified. The predicted winner of the game is then the team that won the most simulations.

11.3 MAIN RESULTS

An initial analysis (whose details can be provided by the authors upon request) showed that, as expected, the TP, 0pp and 2pp are strongly correlated with the number of overtimes. This is not surprising, because adding minutes to the game will cause more plays to be made (TP), and 0pp and 2pp are the most common plays in a game. The number of overtimes a game will have is, however, not known beforehand and very difficult to predict.

From our training and test sets, we noticed that the obtained prediction models for the different types of plays contained little to no overfitting, as we found small to non-existing differences in the deviance between training and test set. There were even less differences for the NBA games than for the EuroLeague games. We do no provide values for the deviances here as the numbers have no clear meaning, but overall the predictions for the different types of plays were rather poor (which, as we shall see later, does not pose a problem to the prediction of game outcomes, our main interest). We shall, in the next section, identify and compare the trends and factors that influence the different types of plays in both leagues. This helps validating the models and identifying differences between the two leagues. Those factors are identified as those that contribute significantly to more than 10 prediction models for the NBA, and 5 for the EuroLeague which has less teams. After that, we present the simulations and game prediction results.

11.3.1 TRENDS AND INFLUENTIAL VARIABLES

11.3.1.1 Total Plays (TP)

Figure 11.3 shows a slightly increasing trend in the total number of plays in the NBA, starting around the 2011-2012 season. No clear trend is visible for the EuroLeague. The pace in the NBA is always higher than in the EuroLeague, as seen by the higher average total number of plays, of about 0.2 plays per minute.

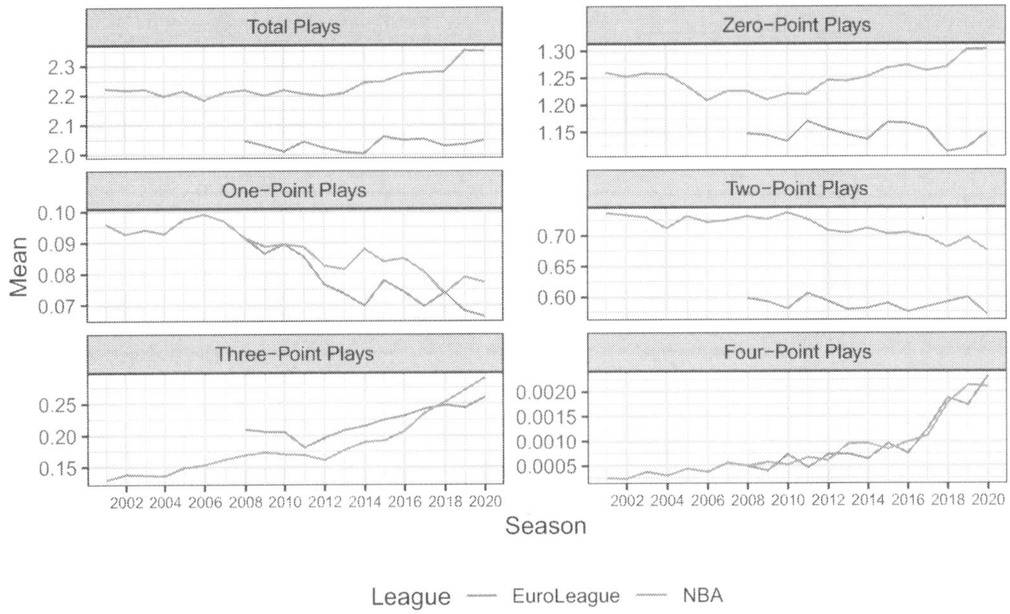

Figure 11.3 Mean number of the different types of plays per minute, per season for both EuroLeague (red) and NBA (blue).

In both leagues, playing against fast-paced, offensive opponents or teams known to have a faster play style such as the Golden State Warriors or the Phoenix Suns, resulted in a higher number of plays for the team. The opposite is also shown by the negative influence of slower, defensive opponents such as the Utah Jazz, San Antonio Spurs and Detroit Pistons on the team's number of plays. Furthermore, both home advantage and game type had an impact on the total number of plays in both leagues. If the team was playing at home, it tended to get more plays than when it was not. On the other hand, playing a playoff game resulted in less plays, a slower pace, compared to playing a regular season game.

In the NBA, playing against opponents with a high average number of turnovers, points, and blocks resulted in more plays for the team while the opposite was true for playing against teams with a high average number of offensive rebounds. In the EuroLeague, a higher average number of field-goal attempts, points, and steals of the opponent resulted in more plays for the team.

11.3.1.2 Zero-Point Plays (0pp)

More than 50% of the plays in a game result in zero points. The number of 0pp therefore follows a very similar trend as the TP (Figure 11.3). The difference in average 0pp per minute between both leagues is around 0.1 plays per minute.

As expected, many of the factors influencing the total number of plays influence the number of 0pp as well. Playing against fast–paced teams results in a higher number of 0pp while playing against slower–paced teams results in less 0pp. If the opponent averages high numbers in its defensive statistics, such as steals, defensive rebounds, and blocks, the team will have more 0pp. In the EuroLeague it seems that a high average number of blocks of the opponent seems to be much more important than a high average number of steals or defensive rebounds as found in the NBA. In the NBA, there is also a clear effect of strength difference between teams on the number of 0pp, where strong teams with a lot of wins have less 0pp on average, especially when playing against weak opponents with a lot of losses.

Unsurprisingly, there exists a strong negative influence of home advantage on the number of 0pp in both leagues. In other words, teams playing at home have less 0pp than teams playing away.

11.3.1.3 One-Point Plays (1pp)

The average 1pp per minute is lower in the EuroLeague than in the NBA in most seasons, and equal in three seasons but never higher (Figure 11.3). The difference, however, is very small, of about 0.018 1pp per minute. There exists a decreasing trend in 1pp in both leagues, with the trend starting in the 2005-2006 season.

It can be seen that playing against more aggressive opponents who commit a lot of personal fouls results in more 1pp for the team. Playing at home results in more 1pp in both leagues. Finally, if a team has a high free–throw percentage, this positively impacts on 1pp.

11.3.1.4 Two-Point Plays (2pp)

There is no upward trend in the number of 2pp over time as seen in the average TP and 0pp. It stays more or less equal in both leagues, with a small decline from the 2009-2010 season onwards in the NBA (Figure 11.3). Again, the average number of 2pp per minute is noticeably lower for the EuroLeague compared to the NBA, with more than 0.1 2pp per minute.

Similar to the number of TP and 0pp, playing against fast-paced teams will result in more 2pp in both leagues, as well as playing against opponents scoring a lot of points. The average number of steals of the opponent was the most common detrimental variable to the number of 2pp in the NBA, coincidentally also being one of the most common positive variables to predict the number of 0pp in the NBA, and also had a negative impact in the EuroLeague. This connection with the number of 0pp is also found in the effects of home advantage in both leagues and playing against weak teams in the NBA. Both had a negative effect on the number of 0pp and a positive effect on the number of 2pp.

Finally, game type was the second most negatively impactful variable in both leagues and has a similar influence as for the TP. Teams tend to play slower or have less plays in the playoffs compared to the regular season and also have less 2pp.

11.3.1.5 Three-Point Plays (3pp)

In contrast to the previous types of plays, the average 3pp per minute was higher in the EuroLeague than in the NBA until the 2017-2018 season, when the NBA overtook the EuroLeague. This difference, however, is relatively small, of about 0.04 3pp per minute (Figure 11.3). The NBA shows a very strong increase in the number of 3pp from the 2011-2012 season onward. The average number of 3pp per minute more than doubled between the 2000-2001 and 2019-2020 seasons. The EuroLeague also shows an increasing trend, although less pronounced than in the NBA, but with a noticeable drop in average 3pp per min in the 2010-2011 season.

Unlike for the other types of plays, variables positively influencing the number of 3pp of a team in a game were less related to the statistics of an opponent but related to the team's three–point statistics. Teams that started to attempt more three-point shots and less two-point shots had more 3pp. In the NBA, playing against more fast-paced teams and bad teams also facilitated the number of 3pp of the team.

In the NBA, playing against defensive teams, such as the San Antonio Spurs and the Detroit Pistons, will result in less 3pp, as will playing against teams with on average a high number of blocks and offensive rebounds.

Table 11.2
Prediction accuracies of the game outcomes through simulations using the predicted plays of the PPR and the PNBR models

Model	NBA PPR	NBA PNB	EuroLeague PPR	EuroLeague PNB
Team Models 1	0.643	0.636	0.689	0.652
Team Models 2	0.640	0.633	0.672	0.660
General Model	0.633	0.619	0.699	0.697
No Prediction Model	0.572	0.572	0.604	0.604

11.3.1.6 Four-Point Plays (4pp)

A 4pp is a rare event that only happens when a three-point shot is made and fouled, and the subsequent free throw scored. Intuitively, the number of 4pp is thus dependent on the number of three-point shots made and attempted by the team. A strong increasing trend in the average 4pp per minute in both leagues follows the increase in number of 3pp, and is approximately the same in both leagues (Figure 11.3).

11.3.2 GAME SIMULATIONS AND GAME PREDICTIONS

As we can see from Table 11.2, the prediction accuracies (computed considering the games in all the seasons in the data set) using the predicted plays from the PPR models are higher overall than those from the PNBR models. Accounting for overdispersion in the data thus does not improve the prediction accuracy, which explains why we focus in the sequel on the results using the PPR.

Year-by-year prediction accuracies are given in Tables 11.3 and 11.4 for the NBA and EuroLeague, respectively. Using the plays predicted by the Team Models 1 or 2 gives the best results for predicting outcomes in most of the training seasons 2000-2018 for the NBA, while the General Model yields the highest game prediction accuracies for the test seasons in the NBA and for all the seasons in the EuroLeague. The latter fact is corroborated by the findings of Table 11.2.

From Tables 11.3 and 11.4 we can see that the prediction accuracy is higher in the EuroLeague. We correctly predict 1937 of the 2772 games (69.9%, Table 11.2) in all seasons, 179 of the 252 games (71.0%, Table 11.4) in our test season 2018-2019, and 155 of the 243 games (63.8%, Table 11.4) in our test season 2019-2020. In the NBA 16281 of the 25330 games (64.3%, Table 11.2) are correctly predicted in all seasons, 810 of the 1296 games (62.5%, Table 11.3) in our test season 2018-2019, and 662 of the 1127 games (58.7%, Table 11.3) in our test season 2019-2020. We did achieve higher accuracies in some of our training seasons, with the highest in the NBA being 69.1% for the 2012-2013 season and 75.6% in the EuroLeague for the 2007-2008 season. Figure 11.4 shows two examples of simulations for individual games and the resulting probability distribution of the scores of each team.

11.4 DISCUSSION

11.4.1 TRENDS

In general, there has been a steady increase in pace, i.e., in the average number of plays per game, in the NBA over the last decade, and this pace was always higher than in the EuroLeague. This

Table 11.3

Prediction accuracies of the game outcomes through simulations using the predicted plays of the different PPR models for all the NBA seasons.

Model	2001	2002	2003	2004	2005	2006	2007	2008	2009	2010
Team Models 1	0.625	0.627	0.635	0.637	0.657	0.629	0.618	0.669	0.668	0.658
Team Models 2	0.620	0.628	0.644	0.639	0.650	0.643	0.617	0.655	0.666	0.663
General Model	0.613	0.611	0.647	0.635	0.650	0.613	0.615	0.659	0.669	0.642
No Prediction Model	0.579	0.596	0.569	0.528	0.575	0.544	0.526	0.595	0.569	0.562
Model	**2011**	**2012**	**2013**	**2014**	**2015**	**2016**	**2017**	**2018**	**2019**	**2020**
Team Models 1	0.658	0.670	0.691	0.647	0.662	0.688	0.632	0.638	0.602	0.533
Team Models 2	0.648	0.672	0.682	0.643	0.662	0.681	0.628	0.632	0.596	0.524
General Model	0.656	0.640	0.660	0.615	0.622	0.638	0.610	0.637	0.625	0.587
No Prediction Model	0.560	0.589	0.590	0.579	0.600	0.577	0.577	0.586	0.581	0.564

Table 11.4

Prediction accuracies of the game outcomes through simulations using the predicted plays of the different PPR models for all the EuroLeague seasons.

Model	2008	2009	2010	2011	2012	2013	2014	2015	2016	2017	2018	2019	2020
Team Models 1	0.681	0.705	0.750	0.655	0.729	0.718	0.714	0.748	0.685	0.669	0.687	0.623	0.617
Team Models 2	0.676	0.682	0.738	0.593	0.712	0.680	0.693	0.739	0.676	0.657	0.671	0.675	0.556
General Model	0.756	0.722	0.727	0.672	0.747	0.697	0.718	0.748	0.681	0.657	0.635	0.710	0.638
No Prediction Model	0.610	0.608	0.599	0.582	0.641	0.660	0.610	0.608	0.609	0.5140	0.528	0.663	0.597

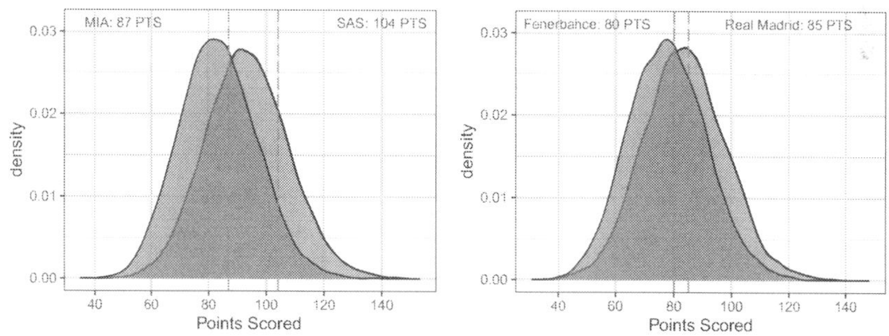

Figure 11.4 Simulation of the 2014 NBA Finals between the San Antonio Spurs (SAS, home) and the Miami Heat (MIA, visiting) (left panel). Simulation of the 2018 EuroLeague Finals between Real Madrid (home) and Ulker Fenerbahce (visiting) (right panel).

increase in pace was also found in Mandić *et al.*, 2019. This is due to the so-called three-point revolution the NBA undergoes, with teams realizing the potential of three-point shots and gradually taking more advantage of it. Three-point shooting effectiveness is generally lower than two-point shooting effectiveness, which would explain the rise in zero-point plays in the form of missed shots. Consequently, the increased attempts of three-point shots also explains the decrease in two-point field goal attempts as more plays that were previously two-point attempts are now replaced by three-point attempts.

This change also results in less fouls as fouls are more often committed within the three-point line. Less fouls means less free throw attempts resulting in a decrease in one-point plays. Another explanation for the decrease in one-point plays found in both leagues would be the increase in accuracy in free throw shooting. In recent years, the free throw accuracy rose to an average of 75% in each league. This means that a personal foul resulting in free throw attempts results in 1.5 points per possession on average, which is higher than the 1 point per possession for two-point attempts and 1.1 points per possession for three-point attempts (Mandić *et al.*, 2019). Preventing fouls that result in free throw attempts would thus be a sound tactical decision.

The increasing trend in the number of three-point shots is also present, although less dramatically, in the EuroLeague, as also found by Mandić *et al.*, 2019 and Štrumbelj *et al.*, 2013. In the season 2010-2011, however, we saw a major decline in the number of three-point plays compared to the previous seasons, combined with a bump of zero- and two-point plays. This coincides with the major rule changes in the EuroLeague, and several other European and most national basketball federations Štrumbelj *et al.*, 2013. These rule changes included, among others, moving the three-point arc 6.75 meters away from the basket (compared to 6.25 meters previously), adding a no-charge semicircle under the basket, and changes in how and when the 24-second shot-clock is reset. The adjustment period needed for players to adapt to this new distance would explain the decline in three-point plays.

Interesting to see is that the EuroLeague consistently had on average more three-point plays than the NBA, except in the last 3 seasons of our data set, while this was never the case in the other types of plays. They even shot and scored on average more three-point shots during the dip after the rule change in the 2010-2011 season. One reason could be that scoring three-point shots is easier in the EuroLeague compared to the NBA because the distance of the three-point arc from the basket has always been shorter in the EuroLeague compared to the 7.24 meters (6.71 meters at the corners) in the NBA. There was, however, no significant difference in shooting percentage between both leagues according to Mandić *et al.*, 2019. The strong increase in NBA three-point plays also seems to counteract that argument. Another argument brought up by Chung, 2019 is that the NBA is a copycat league and tends to be risk averse.

The importance of three-point shots has been known for years, if not decades, through basketball analytics. The EuroLeague could be considered a league that puts more emphasis on the tactical aspects of basketball, while the NBA places a lot of value on athleticism. The three-point field goal percentage and three-point field goals made have been a deciding factor for winning a game for many years in the EuroLeague Çene, 2018 Dogan & Ersoz, 2019. The EuroLeague teams can therefore be seen to have been quicker to adopt a new strategy, while the NBA teams appear to have waited in order to first see the evidence of the tactic in the winning teams before adopting it themselves Chung, 2019. The success of the record-breaking Golden State Warriors since the season 2014-2015, led by career three-point field goal percentage record holder Steve Kerr as rookie head coach and with three-point shooting machines Stephen Curry and Klay Thompson as players, was then likely the ultimate catalyst that changed the NBA into the three-point league that it is today.

11.4.2 INFLUENTIAL VARIABLES

Let us now analyze the different factors influencing the play predictions.

- Home advantage plays a big role in both leagues. It was the most important factor to reduce the number of zero-point plays and to improve the number of two-point plays. Home advantage also had a positive influence for many teams on the total plays, one-point plays, and three-point plays. These findings are in line with Harris & Roebber, 2019, who found that the key factor in home advantage is the style of play. Teams that take more two-point and free throw shots have larger home advantages in the NBA. They argue that the rise in three-point shooting in recent years could partially explain the gradual decline in home advantage. Similar results were found in the EuroLeague by Pojskic *et al.*, 2011. A difference with the NBA is that the influence of home advantage was reinforced in the EuroLeague by more consecutive road games. Being on the road a longer time for consecutive road games has an impact on the performance of teams. Travelling is tiring, a fact that is exacerbated by crossing time zones Swartz & Arce, 2014 Entine & Small, 2008. NBA franchises have bigger budgets than EuroLeague teams, with the lowest valued NBA franchise being worth 1.3 billion dollars Badenhausen *et al.*, 2020, while the highest valued EuroLeague team is only worth 41 million Euro Askounis, 2019. This difference in budget could mean a difference in travel comfort between both leagues, and hence explain the stronger home advantage in the EuroLeague.

- A difference between the two leagues are the factors influencing zero-point plays. In the NBA, the average number of steals of the opponent is the most important factor, followed closely by defensive rebounds and blocks, which are basically all the most important defensive statistics. The importance of steals is highlighted by their negative impact on the number of two-point plays. In the EuroLeague, the average number of blocks of the opponent is the most important factor, with defensive rebounds and steals being much further down the list.
- In general, the opponent team is a major factor. Playing against fast-paced opponents will result in more plays for a team, playing against good rebounding teams will result in less plays, facing good defensive teams will result in more zero-point plays and less two- and three-point plays, while matching up against aggressive teams committing a lot of personal fouls will result in more one-point plays. The exception, however, lies in the average number of three-point plays. A team that attempts more three-point shots and/or has a higher three-point shot rate in a season will have more three-point plays in their games that season.

11.4.3 GAME SIMULATIONS

Our simulations worked very well in predicting outcomes of both NBA and EuroLeague games. The prediction accuracy of our NBA training set, 64.3%, and of our NBA test season 2018-2019, 62.5%, is similar to or even higher than the achieved accuracies by several papers using various machine learning methods Miljković *et al.*, 2010 Cao, 2012 Loeffelholz *et al.*, 2009b Peuter, 2013 Zimmermann, 2016 Manner, 2016a Job & Horvat, 2019 Štrumbelj & Vračar, 2012. The betting market and the experts are still slightly more accurate than our simulation model (68.7%–69.4%) Loeffelholz *et al.*, 2009b Peuter, 2013 Manner, 2016a, although we did achieve similar accuracies for some seasons. Considering that we found a model generally applicable to multiple seasons for multiple teams compared to most papers that focus on a single season and/or a single team, our results are very promising. They can certainly be further improved by finding prediction models that better predict the different types of plays.

The prediction accuracy of EuroLeague games is generally higher in the literature, which was also the case for us with 69.9% accuracy for our training set, and 71.0% accuracy for our 2018-2019 test season. Additionally, we achieved a prediction accuracy of around 75% for multiple individual seasons. This is better than most papers attempting to predict EuroLeague game outcomes. Giasemidis, 2009 compared several different machine learning methods to predict the outcome of EuroLeague games and did not exceed an accuracy of 67%. Additionally, he demonstrated the "wisdom of the basketball crowd", where he asked members of a basketball forum to predict the winner of games, which was decided by majority vote. This resulted in an accuracy of 73%, higher than the machine learning models he used. Necati *et al.*, 2013 used a Poisson regression to predict the scores of EuroLeague games and obtained an accuracy of 73% in their test season 2011-2012, while we achieved an accuracy of 74.7% for that same season.

The accuracy of our second test season, season 2019-2020, was much lower than that of our training seasons or the test season 2018-2019 for both leagues. The 2019-2020 season was special due to the COVID-19 global pandemic, which changed so many factors. The EuroLeague cancelled their 2019-2020 season and played their last game on March 6. The NBA suspended their regular season on March 11. The season was resumed on July 30 but only 22 teams played eight additional regular season games to determine playoff seeding, after which 16 teams entered the playoffs. All these games took place in the so-called NBA Bubble, which was an isolation zone specifically created for these playoffs.

Even our simulations based on the teams' seasonal average number of plays did not perform that badly, with 57.2% overall accuracy for the NBA and 60.4% overall accuracy for the EuroLeague. This makes this approach so interesting because there is no need for advanced statistical modelling to get very fast and somewhat decent results. If you have the data, which is freely available on the internet in the form of basic play–by–play data, simulating a game takes only seconds, and its interpretation is very simple. The simulations are also easily visualizable with two curves representing the distribution of the amount of points scored by each opposing team. This can serve as an extra tool when trying to predict single games. Finally, the fact that our game prediction accuracy is clearly higher when using the plays predicted from our relatively poorly performing play prediction models is quite promising: improving these prediction models will likely increase the accuracy even more.

11.5 CONCLUSION

The prediction models attempting to predict the different types of plays did not give very accurate predictions. Much of the variation was left unexplained, with the main identified excluded variable for the total plays, zero-point plays and two-point plays being the number of overtimes. The analysis still provided some meaningful insights in the factors affecting the different kinds of plays and their differences between the NBA and the EuroLeague. Home advantage plays a big role in both leagues, especially for the zero-point plays and the two-point plays. The impact of consecutive road games reinforced the influence of home advantage in the EuroLeague while it did less so in the NBA. In general, the statistics of the opponent had more impact on the different types of plays than the statistics of the team itself, except for the three-point plays.

The simulations performed very well in predicting the game winners in both NBA and EuroLeague. The prediction accuracy for the NBA was comparable to, if not better than, most prediction models in the literature, but still performed slightly worse than the experts and the betting market. The prediction accuracy for the EuroLeague was generally higher than other prediction models found. The simulations based on the non-predicted play data also performed reasonably well. This shows the strength of that method, as it can give decent predictions without resorting to complex machine learning models. It is fast, easy to use, interpret and visualize. The increase in accuracy using the play data from relatively poorly performing prediction models shows that there is room for improvement. If the prediction models can be optimized, the prediction accuracy is expected to become even better than it already is. Compound Poisson simulation is thus a powerful and accessible way of predicting outcomes of basketball games and worth further research.

REFERENCES

Agresti, A. (2003). *An introduction to Categorical Data Analysis*. John Wiley & Sons.

Askounis, J. (2019, January). List of season budgets of Euroleague teams according to L'Equipe [Accessed on 2012-08-17].

Badenhausen, K., Ozanian, M., & Settimi, C. (2020, February). NBA team values 2020: Lakers and Warriors join Knicks in rarefied $4 billion club [Accessed on 2012-08-17].

Baghal, T. (2012). Are the" four factors" indicators of one factor? An application of structural equation modeling methodology to NBA data in prediction of winning percentage. *Journal of Quantitative Analysis in Sports*, *8*(1).

Beckler, M., Wang, H., & Papamichael, M. (2013). NBA *Oracle* (tech. rep.). Carnegie Mellon University.

Cameron, A., & Windmeijer, F. (1996). R-squared measures for count data regression models with applications to health-care utilization. *Journal of Business & Economic Statistics*, *14*(2), 209–20. https://EconPapers.repec.org/RePEc:bes:jnlbes:v:14:y:1996:i:2:p:209-20.

Cameron, A., & Trivedi, P. (2013). *Regression analysis of count data, second edition*. Cambridge University Press.

Cao, C. (2012). *Sports data mining technology used in basketball outcome prediction* [Master's thesis, Technological University Dublin].

Carlin, B. P. (2005). *Improved* NCAA *basketball tournament modeling via point spread and team strength information*. In *Anthology of statistics in sports* (149–153). SIAM.

Caudill, S. B. (2003a). Predicting discrete outcomes with the maximum score estimator: The case of the NCAA men's basketball tournament. *International Journal of Forecasting*, *19*(2), 313–317. https://EconPapers.repec.org/RePEc:eee:intfor:v:19:y:2003:i:2:p:313-317.

Caudill, S. B. (2003b). Predicting discrete outcomes with the maximum score estimator: The case of the ncaa men's basketball tournament. *International Journal of Forecasting*, *19*(2), 313–317.

Çene, E. (2018). What is the difference between a winning and a losing team: Insights from Euroleague basketball. *International Journal of Performance Analysis in Sport*, *18*(1), 55–68. https://doi.org/10.1080/24748668.2018.1446234.

Chung, J. (2019). Explaining the trends of NBA strategy through the lens of human risk tolerance. *International Journal of Scientific & Engineering Research*, *10*(1), 1012–1031. https://doi.org/10.1080/24748668.2018.1446234.

Dogan, I., & Ersoz, Y. (2019). The important game-related statistics for qualifying next rounds in Euroleague. *Montenegrin Journal of Sports Science and Medicine*, *8*(1), 43–50.

Entine, O., & Small, D. S. (2008). The role of rest in the NBA home-court advantage. *Journal of Quantitative Analysis in Sports*, *4*(2), 1–11. https://EconPapers.repec.org/RePEc:bpj:jqsprt:v:4:y:2008:i:2:n:6.

Fan, Y., & Tang, C. Y. (2012). Tuning parameter selection in high dimensional penalized likelihood. *Journal of the Royal Statistical Society: Series B (Statistical Methodology)*, *75*(3), 531–552. http://dx.doi.org/10.1111/rssb.12001.

Friedman, J., Hastie, T., & Tibshirani, R. (2010). Regularization paths for Generalized Linear Models via coordinate descent. *Journal of Statistical Software*, *33*(1), 1–22. https://www.jstatsoft.org/v33/i01/.

Gabel, A., & Redner, S. (2012). Random walk picture of basketball scoring. *Journal of Quantitative Analysis in Sports*, *8*(1). http://dx.doi.org/10.1515/1559-0410.1416.

Giasemidis, G. (2009). Descriptive and predictive analysis of Euroleague basketball games and the wisdom of basketball crowds.

Gupta, A. A. (2015). A new approach to bracket prediction in the NCAA men's basketball tournament based on a dual-proportion likelihood. *Journal of Quantitative Analysis in Sports*, *11*(1), 53–67.

Harris, A. R., & Roebber, P. J. (2019). NBA team home advantage: Identifying key factors using an artificial neural network. *PLOS ONE*, *14*(7), 1–9. https://doi.org/10.1371/journal.pone.0220630.

Harville, D. A. (2003). The selection or seeding of college basketball or football teams for postseason competition. *Journal of the American Statistical Association*, *98*(461), 17–27.
Hilbe, J. M. (2014). *Modeling count data*. Cambridge University Press.
Horvat, T., Job, J., & Medved, V. Prediction of Euroleague games based on supervised classification algorithm k-Nearest Neighbours. In: In *Proceedings of the 6th international congress on sport sciences research and technology support - volume 1: K-bios*. INSTICC. SciTePress, 2018, 203–207.
Job, J., & Horvat, T. (2019). Importance of the training dataset length in basketball game outcome prediction by using naive classification machine learning methods. *Elektrotehniški Vestnik*, *86*(4), 197–202.
Kubatko, J., Oliver, D., Pelton, K., & Rosenbaum, D. (2007). A starting point for analyzing basketball statistics. *Journal of Quantitative Analysis in Sports*, *3*(3), 1–1.
Lehman, R. R., & Archer, K. J. (2019). Penalized negative binomial models for modeling an overdispersed count outcome with a high-dimensional predictor space: Application predicting micronuclei frequency. *PLOS ONE*, *14*(1), 1–21. https://doi.org/10.1371/journal.pone.0209923.
Liur, E. (2019). *Deviance of Poisson regression* [Master's thesis, Mary MacKillop Institute for Health Research, ACU].
Loeffelholz, B., Bednar, E., & Bauer, K. W. (2009a). Predicting NBA games using neural networks. *Journal of Quantitative Analysis in Sports*, *5*(1).
Loeffelholz, B., Bednar, E., & Bauer, K. W. (2009b). Predicting nba games using neural networks. *Journal of Quantitative Analysis in Sports*, *5*(1).
Lopez, M. J., & Matthews, G. J. (2015). Building an NCAA men's basketball predictive model and quantifying its success. *Journal of Quantitative Analysis in Sports*, *11*(1), 5–12.
Mandić, R., Jakovljević, S., Erčulj, F., & Štrumbelj, E. (2019). Trends in NBA and Euroleague basketball: Analysis and comparison of statistical data from 2000 to 2017. *PLOS ONE*, *14*(10), 1–17. https://doi.org/10.1371/journal.pone.0223524.
Manner, H. (2016a). Modeling and forecasting the outcomes of NBA basketball games. *Journal of Quantitative Analysis in Sports*, *12*, 31–41.
Manner, H. (2016b). Modeling and forecasting the outcomes of nba basketball games. *Journal of Quantitative Analysis in Sports*, *12*(1), 31–41.
McCullagh, P., & Nelder, J. A. (1989). *Generalized linear models, second edition*. Chapman & Hall. https://books.google.com/books/about/Generalized_Linear_Models_Second_Edition.html?id=h9kFH2_FfBkC&hl=en.
Miljković, D., Gajić, L., Kovacevic, A., & Konjovic, Z. The use of data mining for basketball matches outcomes prediction. In: In IEEE *8th International Symposium on Intelligent Systems and Informatics, subotica, serbia*. 2010, 309–312.
Mwikali, J., Mwalili, S., & Wanjoya, A. (2019). Penalized Poisson regression model using elastic net and least absolute shrinkage and selection operator (lasso) penality. *International Journal of Data Science and Analysis*, *5*(5), 99–103. https://doi.org/10.11648/j.ijdsa.20190505.14.
Necati, E., Ermis, E., & Tasmektepligil, M. Y. (2013). Basketball Turkish Airlines Euro League 2011-12 season Poisson regression simulation modeling. *International Journal of Academic Research*, *5*(5), 151–158.
Oliver, D. (2004). *Basketball on paper: Rules and tools for performance analysis*. Brassey's, Incorporated. https://books.google.be/books?id=Xh2iSGCqJJYC.
Page, G. L., Fellingham, G. W., & Reese, C. S. (2007). Using box-scores to determine a position's contribution to winning basketball games. *Journal of Quantitative Analysis in Sports*, *3*(4).
Peuter, C. D. (2013). *Modeling basketball games as alternating renewal-reward processes and predicting match outcomes* [Master's thesis, Duke University, Durham NC].

Pojskic, H., Šeparović, V., & Užičanin, E. (2011). Modelling home advantage in basketball at different levels of competition. *Acta Kinesiologica*, *5*(1), 25–30.

Ruiz, F. J. R., & Perez-Cruz, F. (2015). A generative model for predicting outcomes in college basketball. *Journal of Quantitative Analysis in Sports*, *11*(1), 39–52.

Schwertman, N. C., McCready, T. A., & Howard, L. (1991). Probability models for the NCAA regional basketball tournaments. *The American Statistician*, *45*(1), 35–38.

Sergio, O. P. (2019). Eurolig.

Stefani, R. T. (1977). Football and basketball predictions using least squares. IEEE T*ransactions on* S*ystems,* M*an, and* C*ybernetics*, *7*(2), 117–21.

Stekler, H. O., & Klein, A. (2012). Predicting the outcomes of NCAA basketball championship games. *Journal of Quantitative Analysis in Sports*, *8*(1).

Stekler, H. O., Sendor, D., & Verlander, R. (2010). Issues in sports forecasting. *International Journal of Forecasting*, *26*(3), 606–621.

Štrumbelj, E., & Vračar, P. (2012). Simulating a basketball match with a homogeneous Markov model and forecasting the outcome. *International Journal of Forecasting*, *28*(2), 532–542. https://EconPapers.repec.org/RePEc:eee:intfor:v:28:y:2012:i:2:p:532-542.

Štrumbelj, E., Vračar, P., Robnik-Šikonja, M., Dežman, B., & Erčulj, F. (2013). A decade of Euroleague basketball: An analysis of trends and recent rule change effects. *Journal of Human Kinetics*, *38*(2), 183–189. https://pubmed.ncbi.nlm.nih.gov/24235993.

Swartz, T. B., & Arce, A. (2014). New insights involving the home team advantage. *International Journal of Sports Science & Coaching*, *9*(4), 681–692. https://doi.org/10.1260/1747-9541.9.4.681.

Teramoto, M., & Cross, C. L. (2010). Relative importance of performance factors in winning NBA games in regular season versus playoffs. *Journal of Quantitative Analysis in Sports*, *6*(3).

Vračar, P., Štrumbelj, E., & Kononenko, I. (2016). Modeling basketball play-by-play data. *Expert Systems with Applications*, *44*, 58–66. http://www.sciencedirect.com/science/article/pii/S095741741500617X.

Wickham, H. (2022). *Rvest: Easily harvest (scrape) web pages* [https://rvest.tidyverse.org/, https://github.com/tidyverse/rvest].

Xie, F., & Xiao, Z. (2020). Consistency of L1 penalized Negative Binomial regressions. *Statistics & Probability Letters*, *165*(C). http://www.sciencedirect.com/science/article/pii/S016771522030119X.

Yuan, L.-H., Liu, A., Yeh, A., Kaufman, A., Reece, A., Bull, P., Franks, A., Wang, S., Illushin, D., & Bornn, L. (2015). A mixture-of-modelers approach to forecasting NCAA tournament outcomes. *Journal of Quantitative Analysis in Sports*, *11*(1), 13–27.

Zhang, Y. W. (2012). *Cplm: Compound poisson linear models*.

Zimmermann, A. (2016). Basketball predictions in the NCAAB and NBA: Similarities and differences. *Statistical Analysis and Data Mining*, *9*(5), 350–364. https://doi.org/10.1002/sam.11319.

Zimmermann, A., Moorthy, S., & Shi, Z. Predicting NCAAB match outcomes using ml techniques - some results and lessons learned (J. Davis, J. V. Haaren, & A. Zimmermann, Eds.). In: *Proceedings of the 2nd* W*orkshop on* M*achine* L*earning and* D*ata* M*ining for sports analytics co-located with 2013* E*uropean conference on machine learning and principles and practice of knowledge discovery in databases (ECML PKDD 2013), prague, czech republic, september 27th, 2013* (J. Davis, J. V. Haaren, & A. Zimmermann, Eds.). Ed. by Davis, J., Haaren, J. V., & Zimmermann, A. *1969*. CEUR Workshop Proceedings. CEUR-WS.org, 2013, 69–78. http://ceur-ws.org/Vol-1969/paper-09.pdf.

APPENDIX

Variables used in the models

game_id	Unique ID of the game
Season	Year in which the game,Äôs season finishes
GameType	Regular season game or playoff game
Team	Abbreviation of the team
Team_FG	Average number of field goals made by the team this season
Team_FGA	Average number of field goals attempted by the team this season
Team_FG_perc	Average percentage of field goals made by the team this season
Team_2P	Average number of two-point field goals made by the team this season
Team_2PA	Average number of two-point field goals attempted by the team this season
Team_2P_perc	Average percentage of two-point field goals made by the team this season
Team_3P	Average number of three-point field goals made by the team this season
Team_3PA	Average number of three-point field goals attempted by the team this season
Team_3P_perc	Average percentage of three-point field goals made by the team this season
Team_FT	Average number of free throws made by the team this season
Team_FTA	Average number of free throws attempted by the team this season
Team_FT_perc	Average percentage of free throws made by the team this season
Team_ORB	Average number of offensive rebounds by the team this season
Team_DRB	Average number of defensive rebounds by the team this season
Team_TRB	Average number of total rebounds by the team this season
Team_AST	Average number of assists by the team this season
Team_STL	Average number of steals by the team this season
Team_BLK	Average number of blocks by the team this season
Team_TOV	Average number of turnovers by the team this season
Team_PF	Average number of personal fouls by the team this season
Team_PTS	Average number of points made by the team this season
Team_TS_perc	Average true shooting percentage of the team this season
Team_eFG_perc	Average effective field goal percentage of the team this season
Team_3Par	Average 3-point attempt rate of the team this season
Team_FTr	Average free throw attempt rate of the team this season
Team_ORB_perc	Average offensive rebound percentage of the team this season
Team_DRB_perc	Average defensive rebound percentage of the team this season
Team_TRB_perc	Average total rebound percentage of the team this season
Team_AST_perc	Average assist percentage of the team this season
Team_BLK_perc	Average block percentage of the team this season
Team_ORtg	Team Offensive Rating
Team_DRtg	Team Defensive Rating
Opponent	Abbreviation of the opponent team
Opp_FG	Average number of field goals made by the opponent this season
Opp_FGA	Average number of field goals attempted by the opponent this season
Opp_FG_perc	Average percentage of field goals made by the opponent this season
Opp_2P	Average number of two-point field goals made by the opponent this season
Opp_2PA	Average number of two-point field goals attempted by the opponent this season
Opp_2P_perc	Average percentage of two-point field goals made by the opponent this season
Opp_3P	Average number of three-point field goals made by the opponent this season

Opp_3PA	Average number of three-point field goals attempted by the opponent this season
Opp_3P_perc	Average percentage of three-point field goals made by the opponent this season
Opp_FT	Average number of free throws made by the opponent this season
Opp_FTA	Average number of free throws attempted by the opponent this season
Opp_FT_perc	Average percentage of free throws made by the opponent this season
Opp_ORB	Average number of offensive rebounds by the opponent this season
Opp_DRB	Average number of defensive rebounds by the opponent this season
Opp_TRB	Average number of total rebounds by the opponent this season
Opp_AST	Average number of assists by the opponent this season
Opp_STL	Average number of steals by the opponent this season
Opp_BLK	Average number of blocks by the opponent this season
Opp_TOV	Average number of turnovers by the opponent this season
Opp_PF	Average number of personal fouls by the opponent this season
Opp_PTS	Average number of points made by the opponent this season
Opp_TS_perc	Average true shooting percentage of the opponent this season
Opp_eFG_perc	Average effective field goal percentage of the opponent this season
Opp_3Par	Average 3-point attempt rate of the opponent this season
Opp_FTr	Average free throw attempt rate of the opponent this season
Opp_AST_perc	Average assist percentage of the opponent this season
Opp_BLK_perc	Average block percentage of the opponent this season
Team_0PP	Average number of zero point plays by the team this season
Team_1PP	Average number of one point plays by the team this season
Team_2PP	Average number of two point plays by the team this season
Team_3PP	Average number of three point plays by the team this season
Team_4PP	Average number of four point plays by the team this season
Team_TP	Average number of total plays by the team this season
Team_TMP	Average number of team possessions by the team this season
Opp_0PP	Average number of zero point plays by the opponent this season
Opp_1PP	Average number of one point plays by the opponent this season
Opp_2PP	Average number of two point plays by the opponent this season
Opp_3PP	Average number of three point plays by the opponent this season
Opp_4PP	Average number of four point plays by the opponent this season
Opp_TP	Average number of total plays by the opponent this season
Opp_TMP	Average number of team possessions by the opponent this season
t!	omeadvantage] Team plays at home (TRUE) or does not play at home (FALSE)
Team_Wins	Number of games the team won this season
Team_Losses	Number of games the team lost this season
Opp_Wins	Number of games the opponent won this season
Opp_Losses	Number of games the opponent lost this season

Section IV

Miscellaneous Applications of Statistical Methods in Systems Assurance

12 New Exponential Dispersion Models for Count Data – a Review and Applications

Shaul K. Bar-Lev
Holon Institute of Technology,Holon, Israel
E-Mail: barlev@stat.haifa.ac.il. ORCID: 0000-0002-1770-7091.

Ad Ridder
Vrije Universiteit Amsterdam, Netherlands
E-Mail: ad.ridder@vu.nl. ORCID: 0000-0002-2414-0548.

12.1 INTRODUCTION

Modelling univariate count data by parametric families of counting distributions was initially based on the Poisson and the negative binomial distributions. Later on, many other count models were developed in four paths: a) Generalizing existing models (as generalized Poisson or negative binomial distributions); b) Mixing distributions (as Poisson-inverse Gaussian); c) Discretization of continuous distributions (as discrete gamma distribution); d) Adding extra probability mass on zero (as zero-inflated Poisson). All these models share one thing in common – their probability mass functions (PMFs) can be expressed explicitly in a closed form – thus making them computable. All such parametric models generate, however, a limited set of counting distributions.

In contrast, we introduce a method enabling the generation of an enormous number of new classes of counting distributions belonging to natural exponential families (NEFs) and their associated exponential dispersion models (EDMs), whose probability distributions cannot be expressed explicitly. Still, they can be calculated numerically by the availability of existing powerful computing software.

Our review study is mainly based on a fundamental paper Letac & Mora, 1990 on cubic variance functions (VFs) by Letac and Mora (1990, *The Annals of Statistics*, V.18), which presents a systematic, rigorous, and comprehensive study of natural exponential families (NEFs) on the real line and their characterization through their VFs. In particular, their Proposition 4.4 provides sufficient conditions for a VF to be associated with an NEF of counting distributions supported on the nonnegative integers. This proposition also presents a formula that allows (at least numerically) to compute the counting measure μ which generates the NEF. In our opinion, the main reason why their formula was not used is that in the eighties and nineties of the last century, there were no powerful mathematical programs that would allow the complex and cumbersome calculations of the mass of μ on $\mathbb{N}_0$.

Indeed, several papers (Awad *et al.*, 2022; Bar-Lev & Ridder, 2019, 2021, 2022, 2023; Bar-Lev *et al.*, 2023, 2024), have applied the aforementioned proposition. This paper reviews shortly these papers that present new classes of NEFs and EDMs. These classes of counting distributions have not been used before, simply because they have not been known before. Such classes possess, however,

DOI: 10.1201/9781032664330-12

some desirable properties: all are infinitely divisible, over-dispersed, zero-inflated, skewed to the right, and leptokurtic. Applications of these classes are demonstrated on model fits for count data, GLM, risk insurance aspects, and mortality projections and show superiority over well-used other probability models (such as the Poisson and negative binomial). We will mainly focus the review on the following classes of VFs of NEFs: the ABM, LM, Takacs, Abel, arcsine and large arcsine, where the last four are cubic.

The aim of this survey paper is mainly methodological. We introduce an abundant set $\mathscr{D}$ (basically an infinite set) of NEFs and EDMs that have not been known before. This set is introduced in the concluding section of this survey paper. Members of $\mathscr{D}$ are easy to construct, as indeed demonstrated there. The ABM, LM, and all four cubic functions establish only specific examples of $\mathscr{D}$. The analysis of count data sets is not at all the main core of the article. Nevertheless, we analyze a small number of them just to demonstrate the applicability of the presented models. The seven reviewed articles were written in the following chronological order. We first introduced the ABM class (Awad *et al.*, 2022) in terms of Bayesian aspects. We then introduced the ABM and LM classes via a classical frequentist approach (Bar-Lev & Ridder, 2021, 2023). The four cubic VFs were then introduced with applications to risk and some actuarial measures (Bar-Lev & Ridder, 2019, 2022; Bar-Lev *et al.*, 2023). The more recent paper is Bar-Lev *et al.*, 2024. It introduces the $\mathscr{D}$ class and its properties.

For this review, we introduce in Section 12.2 some preliminaries on NEFs, EDMs, and their associated VFs along with Proposition 4.4 in Letac & Mora, 1990. Section 12.3 introduces the above six VFs. Section 12.4 presents applications of the ABM and LM to model fit and demonstrates the applicability of the ABM class to mortality projections via the Bayesian approach. Section 12.5 presents applications of the four cubic VFs to insurance risk aspects. Some short concluding remarks are given in Section 12.6.

12.2 PRELIMINARIES ON NEFS, EDMS, VFS AND PROPOSITION 4.4 OF Letac & Mora, 1990

12.2.1 NEFS, EDMS, VF AND MEAN VALUE PARAMETERIZATION

Natural Exponential Family (NEF).

Let μ be a positive Radon measure on $\mathbb{R}$ with closed convex support C_μ. Consider the set

$$D_\mu = \left\{\theta \in \mathbb{R} : L_\mu(\theta) = \int_{\mathbb{R}} e^{\theta x}\,\mu(dx) < \infty\right\},$$

and assume that $\Theta_\mu = \operatorname{int} D_\mu$ is nonempty, where $\operatorname{int} A$ stands for the interior of a set A.

Definition 12.1. *The NEF F generated by μ is defined by the set of probability measures*

$$F = F(\mu) = \left\{P(\theta,\mu)(dx) = e^{\theta x - k_\mu(\theta)}\,\mu(dx) : \theta \in \Theta_\mu\right\},$$

where

$$k_\mu(\theta) = \log L_\mu(\theta),$$

is the cumulant transform of μ.

The measure μ is called the generating kernel, or just kernel, of the NEF. The distributions $P(\theta,\mu)$ are parameterized by the natural parameter $\theta \in \Theta_\mu \subset \mathbb{R}$. The cumulant transform is strictly convex and real analytic on Θ_μ. The derivatives $m = k'_\mu(\theta)$, and $V = k''_\mu(\theta)$ are the respective mean and variance of $P(\theta,\mu)$. As k_μ is strictly convex, the inverse mapping of $m = k'_\mu(\theta)$ is well defined, and we denote

$$\theta = \psi_\mu(m), \quad m \in M = k'(\Theta)$$
$$V(m) = k''_\mu(\psi(m)), \quad m \in M.$$

The pair (V,M), called the variance function (VF) of F, uniquely determines F within the class of NEFs (see Morris, 1982). So, (V,M) is a transform, not of a single distribution, but of the whole family. A natural question arises: which pairs (V,M) can serve as a VF of an NEF? Numerous papers in the literature deal with this question by imposing necessary and/or sufficient conditions for such a pair to be a VF of an NEF. One of the first attempts for this was introduced by Bar-Lev, 1987, who showed that for any polynomial of the form

$$V(m) = \sum_{i=0}^{n} a_i m^i, \quad \text{for } a_i \geq 0, \sum_{i=0}^{n} a_i > 0,$$

implying that $(V,\mathbb{R}^+)$ is a VF of an NEF. This structure constitutes a vast class of VFs containing all VFs discussed in this paper.

Exponential Dispersion Model (EDM).

An EDM associated with an NEF $F = F(\mu)$ is defined as follows. Firstly, the Jørgensen set is defined by (Jørgensen, 1987, 1997)

$$\Lambda = \left\{p \in \mathbb{R}^+ : L_\mu^p \text{ is a Laplace transform of some measure } \mu_p\right\}.$$

Note that Λ is nonempty as it is closed under addition due to convolution. The Jørgensen set is also called the dispersion parameter space in the terminology of EDMs and generalized linear models (GLMs). An interesting result related to Λ is (Jørgensen, 1987),

$$\Lambda = \mathbb{R}^+ \text{ iff } \mu \text{ is infinitely divisible.}$$

This would imply that the dispersion parameter space in GLM is $\mathbb{R}^+$ as in the normal case.

If $p \in \Lambda$, the NEF generated by μ_p is the set probability distributions of the form

$$F_p = F(\mu_p) = \left\{P(\theta,\mu_p)(dx) = e^{\theta x - pk_\mu(\theta)}\, \mu_p(dx) : \theta \in \Theta_{\mu_p} = \Theta_\mu\right\}.$$

In the sequel, when no confusion is caused, we shall suppress the dependence on μ of Θ_μ, k_μ, and similar functionals.

Mean Value Parameterization

Using the relations of the first and second derivatives of the cumulant $k_\mu(\theta)$ to the mean and variance of the $P(\theta,\mu)$ distributions, it is a small step to introduce a reparameterization of NEFs and EDMs, called the mean value parameterization; see for instance Bar-Lev & Kokonendji, 2017; Letac & Mora, 1990; Morris, 1982.

- *For NEFs*

  For a given VF (V,M) of an NEF F generated by a measure μ, choose any two primitives $\psi(m)$ and $\varphi(m)$ of $1/V(m)$ and $m/V(m)$, respectively, i.e.,

$$\begin{aligned} \theta = \theta(m) = \psi(m) &= \int \frac{1}{V(m)}\, dm; \\ k(\theta(m)) = \varphi(m) &= \int \frac{m}{V(m)}\, dm. \end{aligned} \tag{12.1}$$

  The mean value parameterization of a NEF F is then

$$F = \left\{P(m,\mu)(dx) = e^{x\psi(m) - \varphi(m)}\, \mu(dx) : m \in M\right\}. \tag{12.2}$$

  Note that there are infinitely many primitives $\psi m)$ and $\varphi(m)$. So, the two specific primitives should be chosen to make (12.2) probability distributions. This point will be elaborated in the sequel for the case of a counting measure μ.

- *For EDMs*
  The mean value parameterization of EDMs is defined in a manner similar to that for NEFs. Indeed, given a VF (V,M) of an NEF then the mean parameter space and variance corresponding to are
  $$M_p = pM; \quad V_p(m) = pV(m/p).$$
  The primitives of $1/V_p(m)$ and $m/V_p(m)$ are
  $$\begin{aligned}\psi_p(m) &= \int 1/V_p(m)\,dm;\\ \varphi_p(m) &= \int m/V_p(m)\,dm,\end{aligned} \tag{12.3}$$
  where
  $$\psi_p(m) = \psi(m/p); \quad \varphi_p(m) = p\varphi(m/p),$$
  and thus, the mean value parameterization has the form
  $$F_p = \left\{P(m,\mu_p)(dx) = e^{x\psi(m/p)-p\varphi(m/p)}\,\mu_p(dx) : p > 0, m \in pM\right\}.$$

12.2.2 THE COUNTING CASE AND PROPOSITION 4.4 OF Letac & Mora, 1990

A counting measure μ on $\mathbb{N}_0$ can be represented as

$$\mu = \sum_{n=0}^{\infty} \mu_n \delta_n,$$

where δ_n is the Dirac measure on $n \in \mathbb{N}_0$. The following fundamental proposition is due to Letac & Mora, 1990, which we slightly rephrased.

Proposition 12.1. *(Letac & Mora, 1990, Proposition 4.4) Let F be an NEF on $\mathbb{R}$ with VF (V,M), then*

- *F is concentrated on $\mathbb{N}_0$ with $\mu_0 > 0$ and $\mu_1 > 0$ if and only if (i) $M = (0,b)$ for some $0 < b \leq \infty$; (ii) there exists a real analytic function φ' on M such that $\varphi'(m) = m/V(m)$ on M, and such that*
  $$\lim_{m\to 0} \varphi'(m) = 1. \tag{12.4}$$
- *In this case, let φ be a primitive of φ', choose a primitive $\psi(m)$ of $1/V(m)$, satisfying*
  $$\lim_{m\to 0} me^{-\psi(m)} = 1, \tag{12.5}$$
  and define,
  $$G(m) = me^{-\psi(m)}, \quad m \in M.$$
  Then the NEF F is generated by a kernel μ whose entries are computed by
  $$\begin{cases}\mu_0 = e^{\varphi(m)}\Big|_{m=0};\\ \mu_n = \frac{1}{n!}\left[\left(\frac{d}{dm}\right)^{n-1} e^{\varphi(m)} \times \varphi'(m) \times (G(m))^n\right]\Big|_{m=0}, \quad n \geq 1.\end{cases} \tag{12.6}$$

Remark 12.1. *Condition* (12.4) *is necessary for the NEF to be concentrated on the nonnegative integers, however, it leaves infinitely many choices for the function* φ *which are all equal upto an additive constant. For our purposes, it is most convenient to impose the condition*

$$\lim_{m\to 0} \varphi(m) = 0. \tag{12.7}$$

This condition gives $\mu_0 = 1$.

Moreover, substiting conditions (12.4), (12.5), *and* (12.7) *in computing* μ_1 *according to* (12.6), *it follows that* $\mu_1 = 1$.

Accordingly, if N is a random variable taken from such an F, its probability mass function (PMF), using the mean value parameterization for NEFs, is

$$f_N(n) = \mathbb{P}(N=n) = \mu_n\, e^{\psi(m)n-\varphi(m)} \quad n = 0,1,\ldots. \tag{12.8}$$

Example 12.1. *(The Poisson NEF) We illustrate* (12.8) *with the Poisson NEF. Here, the VF is*

$$(V,M) = (m,\mathbb{R}^+),$$

so that by (12.1), $\psi(m) = \log m + a$ *and* $\varphi(m) = m + b$, *where* a *and* b *are constants that follow from conditions* (12.5) *and* (12.7) *to be* $a = b = 0$. *Substituting these in* (12.6) *and noting that for any* $m > 0$, $G(m) = 1$ *and* $\varphi'(m) = 1$, *entails that* $\mu_n = 1/n!$ *for* $n \in \mathbb{N}_0$. *Hence, we obtain the NEF of Poisson distributions generated by* μ, *and using the mean value parameterization, by*

$$f_N(n) = \mu_n\, e^{\psi(m)n-\varphi(m)} = \frac{1}{n!} e^{n\log m - m}, \quad n = 0,1,\ldots; m \in \mathbb{R}^+.$$

Note that

$$\sum_{n=0}^{\infty} f_N(n) = e^{-m} \sum_{n=0}^{\infty} \frac{\left(e^{\log m}\right)^n}{n!} = e^{-m} \sum_{n=0}^{\infty} \frac{m^n}{n!} = 1,$$

which shows that $f_N(\cdot)$ *is indeed a PMF.*

The generalization to EDMs is trivial, noting that for any value of the dispersion parameter p the corresponding NEF is given by the variance function $V_p(m)$ from which the primitives ψ_p and φ_p are determined, see (12.3), and the generating kernel $\mu_p = \{\mu_p(n), n = 0,1\ldots\}$ is computed as in (12.6) where the functions are adapted to its p-versions. In this way, the PMF of a counting variable taken from an EDM has mean value parameterization

$$f_N(n) = \mathbb{P}(N=n) = \mu_p(n)\, e^{\psi_p(m)n-\varphi_p(m)} \quad n = 0,1,\ldots. \tag{12.9}$$

Note the complexity of the μ_n's in (12.6). Indeed, no explicit expressions can be obtained for μ_n, except for the Poisson (see Example 12.1), and the negative binomial VFs. Otherwise, numerical techniques are required for its computation which we will elaborate in Section 12.3.1.

12.3 THE ABM, LM AND FOUR CUBIC VFS DEDUCED FROM PROPOSITION 12.1

Four cubic VFs that satisfy the premises of Proposition 12.1 were derived in Letac & Mora, 1990, and two, the ABM and LM were derived in Bar-Lev & Ridder, 2021, 2023. We start by presenting the four cubic VFs. These belong to the Abel, Takacs, strict arcsine, and large arcsine EDMs. We should note that the respective VFs were derived in Letac & Mora, 1990 whereas the corresponding primitives $\psi_p(m)$ and $\varphi_p(m)$ were derived in Bar-Lev & Ridder, 2021, 2023; Bar-Lev *et al.*, 2024. These are given by,

Abel EDM:

$$V_p(m) = m\left(1+\frac{m}{p}\right)^2, \quad m>0, p>0,$$

and

$$\psi_p(m) = \log\frac{m}{m+p} - \frac{m}{m+p} + \log p;$$
$$\varphi_p(m) = \frac{mp}{m+p}.$$

Tacáks EDM:

$$V_p(m) = m\left(1+\frac{m}{p}\right)\left(1+\frac{2m}{p}\right), \quad m>0, p>0,$$

and

$$\psi_p(m) = \log m(m+p) - 2\log(2m+p) + \log p;$$
$$\varphi_p(m) = -p\log\frac{m+p}{2m+p}.$$

Strict arcsine EDM:

$$V_p(m) = m\left(1+\frac{m^2}{p^2}\right), \quad m>0, p>0,$$

and

$$\psi_p(m) = \log m - \frac{1}{2}\log(m^2+p^2) = -\frac{1}{2}\log(1+(p^2/m^2));$$
$$\varphi_p(m) = p\arctan(m/p).$$

Large arcsine EDM:

$$V_p(m) = m\left(1+\frac{2m}{p}+\frac{2m^2}{p^2}\right), \quad m>0, p>0,$$

and

$$\psi_p(m) = \log mp - \frac{1}{2}\log\left(2m^2+2mp+p^2\right) - \left(\arctan\frac{2m+p}{p} - \arctan 1\right);$$
$$\varphi_p(m) = p\left(\arctan\frac{(2m+p}{p} - \arctan 1\right).$$

A few remarks should be pointed out concerning these EDM classes.

Remark 12.2.
- *The associated distributions of the four cubic VFs above appear in Letac & Mora, 1990. They were derived by using intricate applications of the Lagrange formula and not by* (12.6) *– a task that seems to be impossible.*
- *The distribution associated with the Abel EDM is commonly known as the generalized Poisson distribution (Consul, 1989; Consul & Jain, 1973).*
- *The shown Takács and large arcsine EDMs are special cases of the ones described in Letac & Mora, 1990, where the variance functions have an extra parameter* $a>0$. *We give these here with* $a=1$.
- *The Laplace transform of each of the four cubic EDMs does not have explicit form – a fact that makes them not useful practically. Fortunately, their corresponding primitives* $\psi_p(m)$ *and* $\varphi_p(m)$ *are well expressed by nice algebraic functions. This fact makes them applicable and emphasizes the importance of the mean value parameterization.*

The ABM and LM classes are given by (see Awad *et al.*, 2022; Bar-Lev & Kokonendji, 2017; Bar-Lev & Ridder, 2021, 2023):

The ABM class:

$$V_p(m) = m\Big(1+\frac{m}{p}\Big)^r,\ m>0,\ p>0,\ r=0,1,\ldots, \tag{12.10}$$

where the corresponding primitives $\psi_p(m)$ and $\varphi_p(m)$ are (excluding the $r=0$ case, and defining $\sum_{j=1}^{0}=0$),

$$\psi_p(m) = \log\frac{m}{m+p} + \sum_{j=1}^{r-1}\frac{1}{j}\Big(\frac{p^j}{(m+p)^j}-1\Big), \tag{12.11}$$

and

$$\varphi_p(m) = \begin{cases} p\log\frac{m+p}{p} & (r=1);\\ \frac{p}{r-1}\Big(1-\big(\frac{p}{m+p}\big)^{r-1}\Big) & (r\geq 2).\end{cases} \tag{12.12}$$

It is important to note that the ABM is a class of VFs or EDMs. Indeed, the first three cases, $r=0,1$, and $r=2$ correspond to the Poisson, negative binomial, and Abel EDMs. All other members are new and have not been used before for statistical aspects.

The LM class:

$$V_p(m) = \frac{m}{(1-\frac{m}{p})^r},\ 0<m<p,\ p>0,\ r=0,1,\ldots, \tag{12.13}$$

and

$$\psi_p(m) = \log\frac{m}{p} + \sum_{i=1}^{r}(-1)^i\frac{1}{i}\binom{r}{i}\Big(\frac{m}{p}\Big)^i,$$

and

$$\varphi_p(m) = \frac{p}{r+1}\Big(1-\big(1-\frac{m}{p}\big)^{r+1}\Big).$$

Note again: the LM is a class of VFs or EDMs. The only known case is $r=0$, which corresponds to the Poisson NEF.

12.3.1 COMPUTING THE ABM AND LM DISTRIBUTIONS

Here we elaborate on computing the PMF $f_N(n)=\mathbb{P}(N=n)$ of the ABM when taken from its EDM class. We assume that mean $m>0$, dispersion $p>0$ and power $r\geq 2$ is given ($r=0,1$ are the trivial cases of Poisson and negative binomial). From the mean parameterization form of $f_N(n)$ as displayed in (12.9), it suffices to compute the generating kernel $\mu_p(n), n=2,3,\ldots$ ($\mu_p(0)=\mu_p(1)=1$, according to Remark 12.1). The value $\mu_p(n)$ is computed by working out its expression (12.6) in Proposition 12.1. Written differently,

$$\mu_p(n) = \frac{1}{n!}\Big(\frac{d}{dm}\Big)^{n-1}e^{H_n(m)}\Big|_{m=0},$$

where

$$\begin{aligned} H_n(m) &= \varphi_p(m)+\log\varphi_p'(m)+n\log G_p(m)\\ &= \varphi_p(m)+\log\varphi_p'(m)-n\big(\psi_p(m)-\log(m)\big). \end{aligned}$$

Note that $H_n(0)=0$, see conditions (12.4), (12.5), and (12.7). Apply the chain rule for $\mu_p(n)$:

$$\Big(\frac{d}{dm}\Big)^{n-1}e^{H_n(m)} = \sum\frac{(n-1)!}{k_1!k_2!\cdots k_{n-1}!}\Big(\Big(\frac{d}{dx}\Big)^k e^x\Big)_{x=H_n(m)}\prod_{j=1}^{n-1}\Big(\frac{H_n^{(j)}(m)}{j!}\Big)^{k_j},$$

where the sum is over all nonnegative integer solutions of the Diophantine equation $\sum_{j=1}^{n-1} jk_j = n-1$, and where $k = \sum_{j=1}^{n-1} k_j$. This leads to (see for further details Bar-Lev & Ridder, 2021, 2023),

$$\mu_p(n) = \frac{1}{n} e^{H_n(0)} \sum \prod_{j=1}^{n-1} \frac{\left(H_n^{(j)}(0)/j!\right)^{k_j}}{k_j!}$$
$$= \frac{1}{n} e^{H_n(0)} \sum \prod_{j=1}^{n-1} \prod_{t=1}^{k_j} \frac{H_n^{(j)}(0)/j!}{t}.$$

Using the expressions of ψ_p and φ_p as presented in (12.11) and (12.12), we can write for the ABM class the H_n function in the form

$$H_n(m) = q_0 + \sum_{i=1}^{r-1} q_i(m+p)^{-i} + q_r \log(m+p),$$

for some known coefficients $q_i, i = 0, \ldots, r$. The derivatives of H_n are sums of derivatives of $(m+p)^{-i}$ and $\log(m+p)$, which are easy to work out. Then by defining

$$h_n(i,j) = \begin{cases} \frac{1}{j!}\left(\left(\frac{d}{dm}\right)^j (m+p)^{-i}\right)\Big|_{m=0} = (-1)^j \binom{j+i-1}{j} p^{-i-j} & i = 1, \ldots, r-1; \\ \frac{1}{j!}\left(\left(\frac{d}{dm}\right)^j \log(m+p)\right)\Big|_{m=0} = (-1)^{j-1} \frac{1}{j} p^{-j} & i = r, \end{cases}$$

we obtain for the kernel $\mu_p(n)$ in case of the ABM class that,

$$\mu_p(n) = \frac{1}{n} \sum \prod_{j=1}^{n-1} \prod_{t=1}^{k_j} \sum_{i=1}^{r} \frac{q_i h_n(i,j)}{t}.$$

For an implementation in C++, code is available at `https://github.com/adriaanridder/abm`.

The computation of the LM class measure goes similarly.

12.4 APPLICATIONS OF THE ABM AND LM CLASSES

In this section, we present applications of the ABM and LM to model fit for real count data (Section 12.4.1), and we demonstrate the applicability of the ABM class to mortality projections via the Bayesian approach (Section 12.4.2).

12.4.1 MODEL FIT FOR REAL COUNT DATA

Consider the data set in Table 12.1 of automobile insurance claims between 2012 and 2014 at a Turkish insurance company which has been explored in Sarul & Sahin, 2015 for fitting purposes. These data show zero-inflation (meaning a high number of zeros), and overdispersion (meaning variance larger than the mean). Namely, the fraction of zeros is 79%, and the index of dispersion is 1.26. Data of this nature are typically in areas as insurance, public health, behavioural sciences, traffic accidents, epidemiology, psychology, etc.

Fitting a theoretical statistical model to the data is a key element for decision makers in these industries; for instance, for actuaries to determine a pricing policy. Here we consider models that are based on parametric families of discrete distributions, of which the Poisson, negative binomial (NB), their zero-inflated versions, or their Hurdle versions (PH) are the traditional ones. Many alternative parametric counting distributions have been proposed recently, and analyzed for their fitting properties. To name a few with two parameters (and one with three),

- PIG: Poisson-inverse Gaussian distribution Willmot, 1987.

Table 12.1
Data set 1: Insurance claim data.

value	0	1	2	3	4	5	≥ 6
frequency	8544	1796	370	81	22	1	0

- LD: a logarithmic distribution Gomez-Deniz *et al.*, 2011.
- GDP: a geometric discrete Pareto distribution Bhati & Bakouch, 2019.
- EDL: exponentiated discrete Lindley distribution El-Morshedy *et al.*, 2020.
- PT: Poisson-Tweedie mixture distribution Kokonendji *et al.*, 2004. Note that the PT has three parameters, while the other distributions above have two parameters.

We have considered all these distributions for fitting data set 1 of Table 12.1, next to our distributions that are members of the ABM and LM EDM classes of Section 12.3 given by their variance functions (12.10) and (12.13), respectively. Parameters of all distributions are obtained by maximum likelihood estimation. Performance measures of the fitting models are AIC (Akaike's Information Criteria), χ^2 value, RMSE (Root Mean Squared Error), and KLdiv (Kullback-Leibler divergence). Their definitions are summarized below.

Firstly, let $x_0, x_1, \ldots$ be the frequencies of counting values $0, 1, \ldots$, with K being the largest of these values, and n being the total number of observations. The data implies its empirical probability mass function by

$$p_k^{(\mathrm{emp})} = \frac{x_k}{n}, \; k = 0, \ldots, K.$$

Consider a (theoretical) probability model of a discrete random variable N on $\{0, 1, \ldots\}$ with probability mass function

$$p_k^{(\mathrm{mod})} = \mathbb{P}(N = k), \; k = 0, 1, \ldots.$$

The performance of such a model fitting the data, is expressed through the following measures.

- The likelihood function:

$$L = \prod_{k=0}^{K} \left(p_k^{(\mathrm{mod})}\right)^{x_k}.$$

- Akaike Information Criterion (AIC):

$$2b - 2\log L,$$

  where b is the number of parameters.
- χ^2 value; taken into account sufficient observations in the categories, which we have executed as follows. Since our data sets show zero-inflation with a rapidly decreasing tail, we take $s + 1 \leq K$ categories, such that the categories $0, \ldots, s-1$ refer to the observed values $x_0, \ldots, x_{s-1}$, and category s groups all remaining values. Doing this as well for the fitting model, we arrive at

$$\chi^2 = \sum_{k=0}^{s-1} \frac{\left(x_k - np_k^{(\mathrm{mod})}\right)^2}{np_k^{(\mathrm{mod})}} + \frac{\left(\sum_{k=s}^{\infty} x_k - n\sum_{k=s}^{\infty} p_k^{(\mathrm{mod})}\right)^2}{n\sum_{k=s}^{\infty} p_k^{(\mathrm{mod})}}.$$

Table 12.2
Performance of models fitting data set 1 (Table 12.1).

model	AIC	χ^2	RMSE	KLdiv
NB	**14066**	0.3842	3.670	0.0002811
PH	14086	29.92	30.11	0.001194
PIG	14070	1.312	12.79	0.0004522
LD	**14066**	0.3686	3.728	0.0002812
GDP	**14066**	0.3581	3.836	0.0002812
EDL	14067	2.012	8.729	0.0002998
PT	14068	0.6778	4.873	**0.0002784**
ABM	14069	**0.1399**	**3.515**	0.0003076
LM	14070	0.3386	6.596	0.0003641

Table 12.3
Data set 2: Length of stay data.

value	0	1	2	3	4	5	6	7	≥ 8
frequency	3541	599	176	48	20	12	5	1	4

- Root mean squared error (RMSE):

$$\sqrt{\frac{1}{K+1}\sum_{k=0}^{K}\left(x_k - np_k^{\text{mod}}\right)^2}.$$

- Kullback-Leibler divergence (KL) :

$$\sum_{k=0}^{K} p_k^{(\text{emp})} \log \frac{p_k^{(\text{emp})}}{p_k^{(\text{mod})}}.$$

The performances of the various models fitting the claim data of Table 12.1 are tabulated in Table 12.2.

Notice that some traditional or modern models perform good in one or two performance criteria for this specific data set, while one of our EDM models is overall competitive (or best). This is what we see typically in all our fitting experiments. We have executed an extensive numerical study on 20+ data sets, that were considered in the literature for fitting models. Moreover, a traditional or modern model might perform well on one set and bad on another, whereas always one of our EDM models is good. For instance, consider data set 2 in Table 12.3 of the length of stay after hospital admission in the USA among the elderly population. The data consists of 4406 individuals covered by Medicare, the USA public insurance program, and were collected in 1997 Deb & Trivedi, 1997.

Again, the data of Table 12.3 show zero-inflation ($p_0^{\text{emp}} = 0.8038$), and over-dispersion ($D = 1.882$). The fit of the same distributions as before resulted in performances as shown in Table 12.4, supporting our claim.

Table 12.4
Performance of models fitting data set 2 (Table 12.3).

model	AIC	χ^2	RMSE	KLdiv
NB	6023	7.709	7.386	0.001735
PH	6122	140.9	30.31	0.01294
PIG	6020	4.683	9.994	0.001421
LD	6026	11.43	9.004	0.002147
GDP	**6019**	3.539	**5.001**	0.001273
EDL	6037	23.13	13.88	0.003299
PT	6021	3.494	5.885	0.001264
ABM	6021	3.105	5.434	**0.001235**
LM	6021	**3.086**	5.932	0.001240

An interesting question is to find underlying properties of the data making our models (ABM and LM) a suitable fitting tool. We conjecture that our models are better suited for fitting data that show, besides high zero-inflation and overdispersion, also large skewness and kurtosis. Indeed, data set 1 in Table 12.1 has frequency zeros 0.7901, dispersion 1.260, skewness 2.574, and kurtosis 7.916; data set 2 in Table 12.3 has frequency zeros 0.8037, dispersion 1.882, skewness 3.963, and kurtosis 22.80.

For further testing the conjecture, we considered simulated data sets with these properties, and we found empirical evidence for our claim. As an illustration, we report two of these simulation experiments. The first experiment considers a probability distribution that is constructed on a finite support having the required properties of zero-mass and the first four moments. The probability distribution was obtained by applying the maximum entropy method with these constraints Jaynes, 1957. In the second experiment a probability distribution is constructed under the constraints of smaller zero-mass, skewness and kurtosis. From these distributions, we generated $n = 10000$ simulated data, summarized as data sets 3 and 4 in Table 12.5. The statistics of these data sets are

- data set 3: frequency zeros 0.7933, dispersion 1.843, skewness 3.569, and kurtosis 20.40;
- data set 4: frequency zeros 0.4624, dispersion 2.009, skewness 1.702, and kurtosis 6.126.

Thus, indeed data set 3 shows high zero-inflation, large skewness and kurtosis, whereas dataset 4 does not. Next, we applied the fitting to these two data sets, resulting in Table 12.6 that summarizes the performances of the fitted models. We see a confirmation that our models are suitable for fitting data with high zero-inflation, large skewness and kurtosis (set 3), and those other fitting models give better performance when the data does not have these properties (set 4).

The statistics are:

12.4.2 MORTALITY PROJECTIONS

The ABM class was applied in Awad *et al.*, 2022 via a Bayesian approach in the Lee-Carter model for mortality projections for predicting mortality rates in the US, Ireland, and Ukraine. This application was made based on the national mortality datasets of these three countries. They showed superiority to those in use (i.e., Poisson and negative binomial).

More specifically, in their seminal paper Lee and Carter Lee & Carter, 1992, introduced a model which is one of the most well-known and widely applied models for forecasting mortality rates.

Table 12.5
Simulated data.

value	0	1	2	3	4	5	6	7	8
data set 3	7933	1420	406	132	63	23	16	4	3
date set 4	4624	2387	1390	752	412	211	103	74	47

Table 12.6
Performance of models fitting data sets 3 and 4 (Table 12.5).

	data set 3				data set 4			
model	AIC	χ^2	RMSE	KLdiv	AIC	χ^2	RMSE	KLdiv
NB	14213	9.345	16.07	0.0006630	29874	11.64	51.23	0.004598
PH	14418	261.0	77.45	0.01091	30393	968.9	221.5	0.03056
PIG	14217	6.023	19.79	0.0007982	29996	99.08	149.3	0.01065
LD	14219	15.92	20.82	0.0009119	29873	11.12	52.12	0.004557
GDP	14209	2.193	4.991	0.0004428	29876	10.58	40.52	0.004700
EDL	14238	40.31	34.22	0.001882	**29853**	**2.380**	**13.91**	0.003545
PT	**14211**	2.077	4.793	**0.0004246**	**29853**	5.065	13.93	**0.003434**
ABM	14212	**1.970**	**4.609**	0.0004455	29909	32.72	84.72	0.006262
LM	14214	2.090	8.644	0.0005517	29964	69.79	124.7	0.008983

Within this model, the time series of the log mortality rates, $\log m_{xt}$, of each age is described by an age-specific intercept α_x plus a common trend k_t for all age groups multiplied by an age-specific coefficient β_x,

$$\log m_{xt} = \alpha_x + \beta_x k_t + \varepsilon_{xt}.$$

The error term ε_{xt} is assumed to be distributed with a mean 0 and variance σ_ε^2, reflecting influences missed by the model. The age and time-specific mortality rate m_{xt} is calculated as (D_{xt}/E_{xt}), where D_{xt} denotes the number of deaths in a population at age x, $x = 1, 2, \ldots, P$, and at time $t, t = 1, 2, \ldots, T$, and E_{xt} is the exposure to the risk of death. To ensure the identifiably of model parameters, constraints are imposed such that the sum of β_x over age is 1 and the sum of k_t over time is 0. To forecast mortality rates into the future, a simple random walk with drift is proposed for k_t as

$$k_t = k_{t-1} + \theta + w_t.$$

The homoscedastic error assumption of the Lee-Carter model (LC) was criticized for its limiting impact on predictions Brouhns *et al.*, 2002; Danesi *et al.*, 2015; Idrizi, 2018. This led to the introduction of the Poisson log-bilinear LC-type model Brouhns *et al.*, 2002, which, in contrast, is intrinsically heteroscedastic, namely:

$$D_{xt} \sim \text{Poisson}(\mu_{xt});$$
$$\mu_{xt} = E_{xt} m_{xt}.$$

Here, the number of deaths is directly modelled by a Poisson distribution, whose parameter is estimated by maximum likelihood estimation. This alternative approach gained momentum and alternative discrete distributions were proposed. In particular, a binomial distribution was proposed by Wang & Pengjun, 2005, and a negative binomial distribution was suggested by Delwarde *et al.*, 2007; Renshaw & Haberman, 2008. See Azman & Pathmanathan, 2020 for further discussion of these distributions within the GLM framework.

The Lee-Carter model has been widely used for many purposes Shair *et al.*, 2018, such as forecasting mortality reduction factors, assessing the adequacy of retirement income, population projections, and the projection of mortality trends for the oldest-old (older than 80, and in some sources 85). This age group is of considerable interest to policymakers as it is destined to grow as a proportion of the entire population and can outstrip existing infrastructures' capacity Buettner, 2002. This is a fairly recent phenomenon. In Canada, for instance in Légaré *et al.*, 2015, the 21st century brought about the most significant gain in life expectancy at age 85 (7.79% for women and 9.93% for men). Clearly, policies need to be devised that can meet people's special needs in what is called the fourth age Baltes & Smith, 2003, and accurate mortality projection for this age group is a must. Therefore, Awad *et al.*, 2022, focused on the adequacy of potential underlying discrete distribution functions to produce accurate mortality projections using the LC model for this age group.

In essence, while the above cited papers relied on popular and commonly used discrete distributions, one cannot say that one particular distribution is universally superior. Indeed, it is entirely plausible to assume that there is a winning distribution for any given population or even for a specific population at a certain age range. Ideally, one should consider a rich class of family of counting distributions, much richer than the two already suggested, and use the data to pick the most suitable distribution for the population under study. This paper proposes an infinitely countable set of families of counting distributions, where the Poisson, negative binomial and Abel families of distributions are special cases. The aim in Awad *et al.*, 2022 is to study this family, incorporate it into the framework of the LC model, and use real data to seek the most suitable distribution for mortality projection. While there is little doubt that the distributions discussed above could prove adequate for specific populations or age groups, other distributions within the suggested family could have the upper hand.

As an alternative to parameter estimation via the singular value decomposition used in the classical LC model or the MLE in the cases discussed above, Awad *et al.*, 2022 adopted the Bayesian approach which offers advantages succinctly expressed in Antonio *et al.*, 2015: a. The calibration and forecast steps are combined, which leads to more consistent estimates of the period effects; b. The Bayesian approach provides a natural framework for incorporating parameter uncertainty in mortality forecasts, which is relevant-for example-in the new insurance regulatory framework of Solvency II. The Bayesian approach allows adequate handling of small populations and missing data. Like Czado *et al.*, 2005 and Pedroza, 2006, the authors of Awad *et al.*, 2022 harness the power of the Markov Chain Monte Carlo methodology to estimate the model parameters and execute mortality projection. They noted that the interest in Bayesian solutions in the context of mortality projections has recently gained momentum Ellison *et al.*, 2020; Graziani, 2020.

Suppose the number of deaths D_{xt} in a population at age x and time t is distributed as follows:

$$\begin{aligned} D_{xt} &\sim \mathrm{ABM}(p,r)(\mu_{xt}); \\ \mu_{xt} &= E_{xt}m_{xt}; \\ m_{xt} &= \exp(\alpha_x + \beta_x k_t), \end{aligned}$$

where $\mathrm{ABM}(p,r)$, $r = 1,2,\ldots$ denotes the ABM NEF with parameters r and p, and

$$\begin{aligned} \alpha &= \left(\alpha_{x_{\min}},\ldots,\alpha_{x_{\max}}\right)'; \\ \beta &= \left(\beta_{x_{\min}},\ldots,\beta_{x_{\max}}\right)'; \\ k &= \left(k_{t_{\min}},\cdots,k_{t_{\max}}\right). \end{aligned}$$

Bayesian estimation of the unknown parameters α, β, k and p are based on the joint posterior distribution function of α, β, k and p given (E_{xt}, D_{xt}), when $x = x_{\min}, x_{\min}+1, \ldots, x_{\max}$, and $t = t_{\min}, t_{\min}+1, \ldots, t_{\max}$.

The superiority of using the ABM class lies in the ability to choose a model amongst candidate models (i.e., varying r within the *ABM* (p, r) class). Since no one single model necessarily fits every population and every age group well, the ABM class could allow picking, as an example, the Poisson for members of the population aged under 50, the negative binomial for those aged 50 to 80, and another member of the class to the oldest-old. The suggested criteria for preferring one member of the class over another is the mean squared projection errors. This advantage is gained at the cost of additional complexity, which is justified given the financial benefits associated with more accurate modelling. They concluded that it is no longer appropriate to assume a single distribution for the whole process of mortality projection. Instead, for every country and every relevant range of ages, a desirable approach is to pick a member of the ABM class that provides the best mortality projection.

At this point, we should note an important fact. Since a Bayesian approach was used in Awad *et al.*, 2022, there was no need to compute the generating measure $\mu = \sum_{n=0}^{\infty} \mu_n \delta_n$ of the ABM class (as μ is canceled when computing the posterior distribution). Accordingly, any other VF satisfying the premises of Proposition 12.1 can also be used in the Lee-Carter model as an alternative to the Poisson or the ABM class.

12.5 APPLICATIONS OF CUBIC VFS TO INSURANCE RISK MEASURES

In this section, we consider the usage of the counting measure N in risk models. We restrict to distributions coming from EDMs with polynomial variance functions with degree three or less.

12.5.1 CATASTROPHIC RISK

In Bar-Lev & Ridder, 2019, 2022, we consider the problem of computing tail probabilities of the distribution of a random sum of positive random variables (called claims),

$$\ell(x) = \mathbb{P}(S_N > x) \text{ for } S_N = \sum_{k=1}^{N} X_k, \tag{12.14}$$

where $X_1, X_2, \ldots$ are random individual positive claims, and $N \in \{0, 1, \ldots\}$ is the random number of claims (with $\sum_{k=1}^{0} = 0$), independent of the random claims. We assume that the random number N has a NEF counting distribution with a cubic variance function. Concerning the claims, we assume these are IID having either lognormal, Pareto, Gamma, inverse Gaussian, or NEF-positive stable distribution. We have chosen these distributions for reasons that the latter three are light-tailed distributions following a reproducible NEF distribution with a power variance function (see Bar-Lev & Enis, 1986), while the first two are subexponential distributions, commonly used in insurance models Embrechts *et al.*, 1997; Potocký *et al.*, 2014.

Problem (12.14) is called the catastrophic risk, or the loss probability. This risk factor is used in actuarial sciences for reinsurance purposes when some level of retention from the insurer is required. Specifically, large tail probabilities are vital as they reflect the risk of large losses. Exact analytic expressions of (12.14) are generally not available, and for large losses x, the rare-event probability $\ell(x)$ is difficult to compute numerically. Though, many asymptotics of these tail probabilities have been developed in the literature, which may be used as approximations. An early work is Embrechts *et al.*, 1979 for Poisson count and subexponential claims later generalized by Cline, 1987; Mikosch & Nagaev, 2001 for other (light-tailed) counts, and by Cline, 1987; Pakes, 2007 for claims with exponentially decaying tail. More on these issues can be found in Embrechts *et al.*, 1997, Section 1.3.3. Furthermore, see Jordanova *et al.*, 2017; Potocký *et al.*, 2014 for some theoretical results and numerical experiments assessing the accuracy of these asymptotics.

Table 12.7
Statistics of insurance data.

data	zeros p_0	average m	variance V
claim frequency N	0.06349	70.60	52181.5
aggregate claim size S_N		329.2	1153532.3
individual claim size X		4.663	265.3

In our study we considered Monte Carlo simulations for computing the risk factors. Moreover, we applied importance sampling to cope with the rare events of large losses. Our computations are applied to a case study of a car insurance company for reasons that in automobile insurances most policy holders do not file claims in a given year. Thus, car insurance data will be highly zero-inflated, and typically also overdispersed. Our study considers the data of a Swedish car insurance company that is publicly available in Smyth, 2011. The main difficulty of the data set is that individual claims are not given, only the number of claims n_i, and their total claim size s_i, for $i = 1, \ldots, 630$ being categories representing the larger cities. For finding good models for the generic claim number N, and individual claim size X, we first make some assumptions.

- We assume that the n_i are IID observations of N.
- We assume a single individual claim variable X across all categories, and we assume that all claims are independent.
- Also, we assume that N and X are independent.

Then, we reason as follows, given the data (n_i, s_i). The mean m_N and variance V_N of the counting variable N are estimated by the sample average and variance of the counting data (n_i). The mean m_X of the individual claim X is estimated by $\sum_i s_i / \sum_i n_i$. The variance V_S of the aggregated loss S_N is estimated by the sample variance of the (s_i) data. Finally, to estimate the variance V_X of an individual claim, we use the well-known identity for the variance of the aggregated sum S_N:

$$\text{Var}(S_N) = (\mathbb{E}[N])(\text{Var}(X)) + (\text{Var}(N))(\mathbb{E}[X])^2.$$

These computations resulted in the data statistics summarized in Table 12.7. Clearly, the claim number N is overdispersed ($V_N/m_N = 52181.5/70.60 \gg 1$), and zero-inflated ($p_0 = 0.06349 \gg 2.189 \times 10^{-31} = e^{-m_N}$). Also the claim size X is overdispersed ($V_X/m_X = 265.3/4.663 \gg 1$).

Next, we hypothesized any combination (N, X) where N assumes one of the four claim numbers N from NEF with cubic VF (Abel, Takács, strict arcsine, and large arcsine), and where X has a lognormal, Pareto, Gamma, inverse Gaussian, or NEF-positive stable distribution. As a comparison, we also assumed Poisson and negative binomial distributions for N. Parameters of all these distributions are estimated by moment matching using the statistics of Table 12.7. The fit of a (N, X) distribution pair is assessed by a parametric bootstrap method (Efron, 1979) jointly with Monte Carlo simulations. For this, we hypothesized that the data set is just a sample set of size 630 from a theoretical distribution having CDF F with parameters estimated from the data. Hence, we cannot apply directly a goodness-of-fit test such as Kolmogorov-Smirnov or chi-square. A more appropriate procedure would be to repeatedly generate random sample sets of size 630 from the hypothesized distribution, estimate parameters, and compute a test statistic (Kolmogorov-Smirnov or chi-square statistic). Then an approximated p-value is the fraction of test statistics computed from these simulated sets that are more extreme than the value obtained from the original data. This procedure has been implemented in the following way.

- We denote by $\theta = \theta(\hat{m}_N, \hat{V}_N, \hat{m}_X, \hat{V}_X)$ the tuple of the estimated distributional parameters of the specific (N,X) combination, as a function of the sample averages and variances computed from data. This means, given a data set of aggregated claims, we compute their averages and variances and from these the mean and dispersion parameters of the specific counting distribution, and the shape and scale parameters of the specific claim distribution.
- We denote by $F_S(\cdot|\theta)$ the CDF of the aggregated sum S_N, given parameter tuple θ. There is (generally) no explicit expression of the CDF. Therefore, given parameter tuple θ, we generate n (large) values of S_N by a standard Monte Carlo simulation, and we consider the empirical distribution $F_n^{\text{emp}}(\cdot|\theta)$ of these observations as a proxy of the theoretical CDF.
- Let θ_0 be the distributional parameters obtained from the means and variances in Table 12.7, estimated from the given data set. We then compute the Kolmogorov-Smirnov test statistic T_0 between the empirical CDF of the given data set, and $F_n^{\text{emp}}(\cdot|\theta_0)$.
- Recall that the data set has length 630. Next, we generate by simulation a random sample set of size 630 from the distribution $F_S(\cdot|\theta_0)$, i.e., when the aggregated sum S_N comes from the pair (N,X) with means and variances as in Table 12.7.
- Compute means $\hat{m}_N, \hat{m}_X$, and variances $\hat{V}_N, \hat{V}_X$ of the new data set, and from these the parameters, say θ_i, for the counting and claim distributions.
- Using the θ_i parameters, we generate n values of S_N by Monte Carlo simulation, and obtain the empirical distribution $F_n^{\text{emp}}(\cdot|\theta_i)$ as proxy of the theoretical CDF $F_S(\cdot|\theta_i)$. Finally, we compute the Kolmogorov-Smirnov test statistic T_i between the empirical CDF of this new data set, and $F_n^{\text{emp}}(\cdot|\theta_i)$.
- We repeat k times the latter steps of generating a new random data set and computing Kolmogorov-Smirnov test statistic.
- The approximated p-value of the fit is $\#\{i : T_i > T_0\}/k$.

Table 12.8 presents the obtained approximated p-values of these computations in which we set $n = 10000$ for the empirical distribution $F_n^{\text{emp}}(\cdot|\theta)$, and $k = 1000$ repetitions. In case of combinations with an arcsine counting variable we set $n = 5000$ because of longer computation times. We left out the cases when the counting variable is taken from Poisson or negative binomial, since the resulting p-values were all zero in all combinations. We have run the same procedure, implementing the chi-square statistic between the empirical CDF of a random data set (of length 630), and the empirical CDF $F_n^{\text{emp}}(\cdot|\theta)$ as proxy of the theoretical CDF $F_S(\cdot|\theta)$. We used the same data sets and empirical distributions for computing both the Kolmogorov-Smirnov and the chi-square statistics. For the chi-square statistics, the data are partitioned into bins of equal size (50) that are joined when the expected values were too low (5). The results of these experiments are presented in Table 12.9.

As expected, the Poisson and negative binomial distributions for the total number N gave poor fits (zero p-values) in all combinations. On the other hand, when N has any distribution with cubic variance function, the hypothesis of a good fit can not be rejected, in all combinations (p-values ranging from 20% to more than 70% depending on the combination). To illustrate, we present in Figure 12.1 two histograms of the observed versus simulated values of the aggregated loss S_N; one of the Abel-inverse Gaussian combination, and one of the negative binomial-inverse Gaussian combination showing the bad fit of the latter. Moreover, in Figure 12.2 we picture the Q-Q plots of the observed data S_N versus simulated S_N for the combination $(N,X) \sim$ (large arcsine, NEF stable), and for the combination $(N,X) \sim$ (negative binomial, NEF stable), with the same conclusion of a bad fit of negative binomial counting variable.

12.5.2 CONDITIONAL TAIL RISKS

In Bar-Lev *et al.*, 2023 we consider two other risk factors in the collective risk model involving the total loss S_N in (12.14). Let $F_X(x), x \in \mathbb{R}$ be the cumulative distribution function of claim X, clearly

Table 12.8

Approximated p-values of the fitted models using parametric bootstrapping and Kolmogorov-Smirnov test statistic.

	Claim				
Count	Lognormal	Pareto	Gamma	Inv Gaussian	Pos Stable
Abel	0.422	0.696	0.226	0.382	0.243
Takács	0.421	0.715	0.223	0.353	0.221
Strict arcsine	0.579	0.290	0.610	0.768	0.661
Large arcsine	0.647	0.691	0.211	0.424	0.559

Table 12.9

Approximated p-values of the fitted models using parametric bootstrapping and chi-square test statistic.

	Claim				
Count	Lognormal	Pareto	Gamma	Inv Gaussian	Pos Stable
Abel	0.341	0.515	0.293	0.304	0.506
Takács	0.299	0.625	0.503	0.489	0.571
Strict arcsine	0.428	0.658	0.612	0.723	0.661
Large arcsine	0.401	0.622	0.505	0.444	0.436

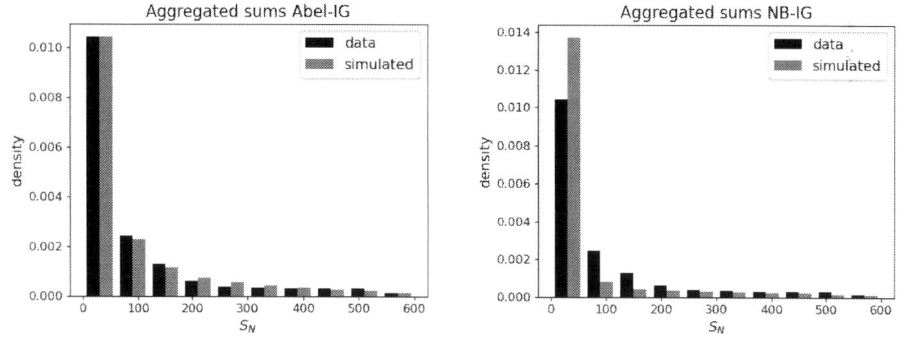

Figure 12.1 Histogram of the observed S_N data versus 5000 simulated samples, using Abel (left) and negative binomial (right) variable N.

$F_X(0) = 0$. Recall the Value-at-Risk at level $\alpha \in (0,1)$ of a claim X, defined by

$$\mathrm{VaR}_\alpha(X) = F_X^{-1}(1-\alpha) = \inf\{x : F_X(x) \geq 1-\alpha\} = \inf\{x : \mathbb{P}(X > x) \leq \alpha\}.$$

The Value-at-Risk measures the threshold level in which the probability that the claim exceeds this value is at least a given probability level $1-\alpha$. We denote is by $q = \mathrm{VaR}_\alpha(X)$.

Now we define the following two conditional tail risk measures.

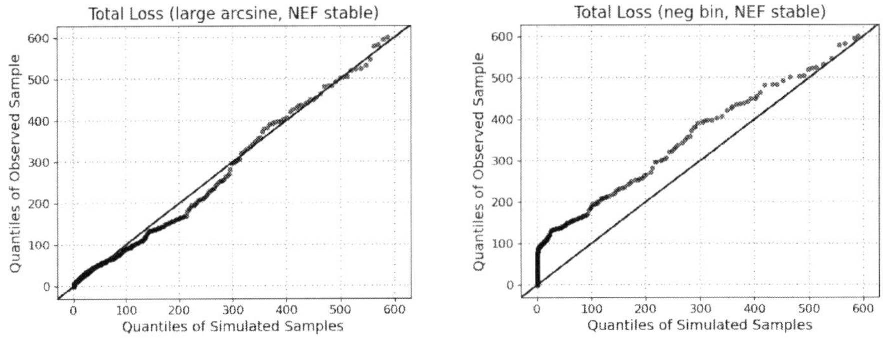

Figure 12.2 Q-Q plots of the observed S_N data versus 5000 simulated samples, using large arcsine (left) and negative binomial (right) variable N.

- $\rho_1 = \mathbb{E}\left[S_N \middle| \bigcap_{k=1}^N \{X_k > q\}\right]$.
  This measure represents the expected total loss S_N under the condition that each of the claims is greater than its VaR.
- $\rho_2 = \mathbb{E}\left[S_N \middle| \bigcup_{k=1}^N \{X_k > q\}\right]$.
  This measure represents the expected total loss S_N under the condition that at least one of the claims will face an extreme loss event.

Using ρ_1 and ρ_2, we are able to measure the risks when facing extreme loss events, i.e., events greater than a lower threshold, which we consider to be the VaR for a given percentile α. Analyzing different types of risk measures makes it possible to gain a deeper insight into the multifaceted nature of risks, and their modelling. It is important for making good decisions in various domains, such as finance, and insurance. Each risk measure offers a unique perspective on the system of dependent risks, emphasizing different aspects of extreme events and, thus, the overall loss. Analyzing different risk measures leads to a deeper understanding of the underlying data and probability distributions, facilitating improved risk modelling and prediction. We consider the conditional tail risk measures to quantify extreme loss events. We compute these risk measures for a range of combinations of distributions for the counting variable N, and the claim variable X; moreover, we consider both independent and correlated claims $X_1, X_2, \ldots$.

The combinations (N,X) for counting and variables are again

- Counting variable N has Poisson, negative binomial, Abel, Takács, strict arcsine, or large arcsine distribution. As mentioned earlier, these are all distributions (except one) coming from NEF's with polynomial VF with degree up to 3. The exception is the binomial distribution which is not relevant to risk models.
- Claim variable X has lognormal, Gamma, inverse Gaussian, Pareto distribution, or a distribution coming from the NEF generated by a positive α-stable variable. These distributions are commonly implemented in risk models. Moreover, Gamma, inverse Gaussian, and NEF positive α-stable are distributions from NEF's with power VF.

The purpose of these computations is to assess the effect on the risk measures by the various combinations. The main finding of our results is that when we fix a distribution of the claim variable, the computed values of the ρ_1 risk measure are very close for all our counting distributions, while these of the ρ_2 risk measure show much more variability. Thus, ρ_1 demonstrates robustness and insensitivity to the choice of counting distribution made by the actuary, whereas the ρ_2 risk measure exhibits sensitivity to the chosen modelling of counting distributions. This intriguing outcome raises important questions regarding the critical role played by selecting appropriate counting distributions when applying a specific risk measure.

12.6 CONCLUDING REMARKS

In this paper, we reviewed a set of NEFs and EDMs supported on the nonnegative integers. All of which are zero-inflated, overdispersed, skewed to the right, and leptokurtic. This set includes two major classes. The first includes the ABM and LM classes with VFs whose PMFs can only be numerically calculated. The second includes four NEFs having cubic VFs. We illustrated that the ABM and LM classes perform better than traditional distributional models with respect to model fit for real count data. However, more theoretical aspects and simulations need to be added to demonstrate that these perform uniformly better. However, such a demonstration goes beyond the scope of this survey paper

The second class of NEFs with cubic VFs analyzed insurance claim data and modelled the accumulated claim (or total loss) during a certain period as a random sum of positive random variables representing the individual claims. The data showed that both the random sum and the random claim size have variances as large as the cubic powers of their means. We used the NEF modelling to fit distributions with cubic VFs to the insurance data. In this way, we considered the four discrete counting variables for fitting the random sum and three positive continuous distributions for fitting the claim size, all coming from NEFs. Using these distributions we have computed several risk measures involving conditional tail expectations of the total loss. The computations are based on Monte Carlo simulations for which we have developed algorithms to sample from these NEFs with cubic VFs. In Bar-Lev & Ridder, 2019, 2022, we give an asymptotic analysis of these distributions, which results in accept-reject based sampling algorithms. Our claim size distributions are commonly known, and sampling algorithms can be found in the literature.

Finally, we mention a new line of investigations that we have reported in Bar-Lev *et al.*, 2024, in which we present an infinite set $\mathscr{D}$ of EDMs satisfying Proposition 12.1, thus supported on the nonnegative integers. The class is defined by VF of the form $V(m) = mf(m)$, where $f(m)$ is absolutely continuous and admits a power series of the form $f(m) = 1 + \sum_{n=1}^{\infty} a_n m^n$ with a radius of convergence $R > 0$ and $a_n \geq 0$ for all $n \geq 1$. This class contains all of the above VFs (i.e., the ABM, LM, and the four cubic VFs). It is abundant as any such an absolutely continuous can be used to form a VF $V(m) = mf(m)$ associated with a counting NEF or EDM. It is certainly infinite as is the set of absolutely continuous functions of the above form. Some examples are:

1. Sibuya VF of the form

$$V_a(m) = mf(m) = m\frac{1-(1-m)^a}{am}, \quad M = (0,1).$$

2. Homographic VF of the form

$$V_a(m) = m\frac{1-am}{1-m}, \quad M = (0,1).$$

3. Rational VF of the form

$$V_c(m) = \frac{m(1+2cm-m^2)}{1-m^2}, \quad M = (0,1),$$

and

4. Exponential VF of the form

$$V(m) = me^m, \quad M = (0,\infty).$$

All members of $\mathscr{D}$ are zero-inflated, overdispersed, skewed to the right, and leptokurtic. They consist mainly of probabilistic models whose probability functions cannot be explicitly expressed but can be calculated numerically – a problem now solved by the availability of existing powerful computing software. We trust that the $\mathscr{D}$ class will play a significant role in a) model fit for real count data sets, and b) describing the random error distribution in GLM with covariates (as alternatives to the Poisson and negative binomial models).

ACKNOWLEDGEMENT

We are grateful to the two reviewers for their helpful and constructive suggestions.

REFERENCES

Antonio, K., Bardoutsos, A., & Ouburg, W. (2015). Bayesian Poisson log-bilinear models for mortality projections with multiple populations. *European Actuaruarial Journal*, *5*, 245–281.

Awad, Y., Bar-Lev, S. K., & Makov, U. (2022). A new class of counting distributions embedded in the Lee-Carter model for mortality projections: A Bayesian approach. *Risks*, *10*, 111–127.

Azman, S., & Pathmanathan, D. (2020). The GLM framework of the Lee-Carter model: A multi-country study. *Journal of Applied Statistics*, *49*, 752–763.

Baltes, B. P., & Smith, J. (2003). New frontiers in the future of aging: From successful aging of the young old to the dilemmas of the fourth age. *Gerontology*, *49*, 123–135.

Bar-Lev, S. K. (1987). Discussion on paper by B. Jørgensen, "Exponential dispersion models". *Journal of the Royal Statistical Society, Series B*, *49*, 153–154.

Bar-Lev, S. K., & Enis, P. (1986). Reproducibility and natural exponential families with power variance functions. *The Annals of Statistics*, *14*, 1507–1522.

Bar-Lev, S. K., & Kokonendji, C. C. (2017). On the mean value parameterization of natural exponential families – a revisited review. *Mathematical Methods of Statistics*, *26*, 159–175.

Bar-Lev, S. K., Letac, G., & Ridder, A. (2024). A delineation of new classes of exponential dispersion models supported on the set of nonnegative integers. *Annals of the Institute of Statistical Mathematics*, *76*, 679–709.

Bar-Lev, S. K., & Ridder, A. (2019). Monte Carlo methods for insurance risk computation. *International Journal of Statistics and Probability*, *8*, 54–74.

Bar-Lev, S. K., & Ridder, A. (2021). New exponential dispersion models for count data – The ABM and LM classes. *ESAIM: Probability and Statistics*, *25*, 31–52.

Bar-Lev, S. K., & Ridder, A. (2022). The large arcsine exponential dispersion model – properties and applications to count data and insurance risk [article-number 3715]. *Mathematics*, *10*.

Bar-Lev, S. K., & Ridder, A. (2023). Exponential dispersion models for overdispersed zero-inflated count data. *Communications in Statistics - Simulation and Computation*, *52*, 3286–3304.

Bar-Lev, S. K., Ridder, A., & Shushi, T. (2023). *A simulation analysis for the role of counting distributions in estimating extreme loss events of insurance claims* [Submitted].

Bhati, D., & Bakouch, H. S. (2019). A new infinitely divisible discrete distribution with applications to count data modelling. *Communications in Statistics - Theory and Methods*, *48*, 1401–1416.

Brouhns, N., Denuit, M., & Vermunt, J. K. (2002). A Poisson log-bilinear regression approach to the construction of project life-table. *Insurance: Mathematics and Economics*, *31*, 373–393.

Buettner, T. (2002). Approaches and experiences in projecting mortality patterns for the oldest-old. *North American Actuarial Journal*, *6*, 14–29.

Cline, D. B. H. (1987). Convolutions of distributions with exponential and subexponential tails. *Journal of the Australian Mathematical Society. Series A*, *43*, 347–365.

Consul, P. C. (1989). *Generalized Poisson Distributions: Properties and Applications*. Marcel Dekker.

Consul, P. C., & Jain, G. C. (1973). A generalization of the Poisson distribution. *Technometrics*, *15*, 791–799.

Czado, C., Delwarde, A., & Denuit, M. (2005). Bayesian Poisson log-bilinear mortality projections. *Insurance: Mathematics and Economics*, *36*, 260–84.

Danesi, L. I., Haberman, S., & Millossovich, P. (2015). Forecasting mortality in subpopulations using Lee-Carter type models: A comparison. *Insurance: Mathematics and Economics*, *62*, 151–61.

Deb, P., & Trivedi, P. (1997). Demand for medical care by the elderly: A finite mixture approach. *Journal of Applied Econometrics*, *12*, 313–336.

Delwarde, A., Denuit, M., & Partrat, C. (2007). Negative binomial version of the Lee-Carter model for mortality forecasting. *Applied Stochastic Models in Business and Industry*, *23*, 385–401.

Efron, B. (1979). Bootstrap methods: Another look at the jackknife. *Annals of Statistics*, *7*, 1–26.

Ellison, J., Dodd, E., & Forster, J. J. (2020). Forecasting of cohort fertility under a hierarchical Bayesian approach. *Journal of the Royal Statistical Society, Series A (Statistics in Society)*, *183*, 829–856.

El-Morshedy, M., Eliwa, M. S., & Nagy, H. (2020). A new two-parameter exponentiated discrete Lindley distribution: Properties, estimation and applications. *Journal of Applied Statistics*, *47*, 354–375.

Embrechts, P., Goldie, C. M., & Veraverbeke, N. (1979). Subexponentiality and infinite divisibility. *Zeitschrift für Wahrscheinlichkeitstheorie und verwandthe Gebiete*, *49*, 335–347.

Embrechts, P., Klüppelberg, C., & Mikosch, T. (1997). *Modeling Extremal Events for Finance and Insurance*. Springer.

Gomez-Deniz, E., Sarabia, J. M., & Calderin-Ojeda, E. (2011). A new discrete distribution with actuarial applications. *Insurance: Mathematics and Economics*, *48*, 406–412.

Graziani, R. (2020). Stochastic population forecasting: A Bayesian approach based on evaluation by experts. *Developments in Demographic Forecasting*, *49*, 21–42.

Idrizi, O. (2018). The heteroscedasticity impact on Actuarial Science: Lee-Carter error simulation. *European Journal of Engineering and Formal Sciences*, *1*, 1–12.

Jaynes, E. T. (1957). Information theory and statistical mechanics. *The Physical Review*, *106*, 620–630.

Jordanova, P., Nefedova, Y., & Stehlík, M. (2017). Risk process approximation with mixing. *Applied Mathematical Modelling*, *41*, 285–298.

Jørgensen, B. (1987). Exponential dispersion models (with discussion). *Journal of the Royal Statistical Society, Series B*, *49*, 127–162.

Jørgensen, B. (1997). *The Theory of Exponential Dispersion Models* (*76*). Chapman; Hall.

Kokonendji, C. C., Dossou-Gbété, S., & Demétrio, C. (2004). Some discrete exponential dispersion models: Poisson-Tweedie and Hinde-Demétrio classes. *Statistics and Operations Research Transactions*, *28*, 201–214.

Lee, D. R., & Carter, L. R. (1992). Modelling and forecasting the time series of us mortality. *Journal of the American Statistical Association*, *87*, 659–671.

Légaré, J., Décarie, Y., Deslandes, K., & Carriére, Y. (2015). Les oldest-olds canadiens : Une population en pleine croissance, mal connue et à risque de manquer de services adéquats [https://ir.lib.uwo.ca/pclc/vol3/iss2/1]. *Population Change and Lifecourse Strategic Knowledge Cluster Discussion Paper Series/Un Réseau stratégique de connaissances Changements de population et parcours de vie Document de travail, 3, Article 1.*

Letac, G., & Mora, M. (1990). Natural real exponential families with cubic variance functions. *The Annals of Statistics*, *18*, 1–37.

Mikosch, T., & Nagaev, A. (2001). Rates in approximations to ruin probabilities for heavy-tailed distributions. *Extremes*, *4*, 67–78.

Morris, C. N. (1982). Natural exponential families with quadratic variance functions. *The Annals of Statistics*, *10*, 65–80.

Pakes, A. G. (2007). Convolution equivalence and infinite divisibility: Corrections and corollaries. *Journal of Applied Probability*, *44*, 407–424.

Pedroza, C. (2006). A Bayesian forecasting model: Predicting U.S. male mortality. *Biostatistics*, *7*, 530–550.

Potocký, R., Waldl, H., & Stehlík, M. (2014). On sums of claims and their applications in analysis of pension funds and insurance products. *Prague Economic Papers*, *3*, 349–370.

Renshaw, A., & Haberman, S. (2008). On simulation-based approaches to risk measurement in mortality with specific reference to Poisson Lee-Carter modelling. *Insurance: Mathematics and Economics*, *42*, 797–816.

Sarul, L., & Sahin, S. (2015). An application of claim frequency data using zero inflated and Hurdle models in general insurance. *Journal of Business, Economics and Finance*, *4*, 732–743.

Shair, N. S., Muhammad, A. S., Mohammad, J. M. S., & Zaini, M. A. A. (2018). Projected Malaysian lifetable: Evaluations of the Lee-Carter and Poisson log-lilinear models. *International Journal of Modern Trends in Social Sciences*, *1*, 60–72.

Smyth, G. K. (2011). Third party motor insurance in Sweden. Australasian Data and Story Library (OzDASL).

Wang, D., & Pengjun, L. (2005). Modelling and forecasting mortality distributions in England and Wales using the Lee-Carter model. *Journal of Applied Statistics*, *32*, 873–885.

Willmot, G. (1987). The Poisson-inverse Gaussian distribution as an alternative to the negative binomial. *Scandinavian Actuarial Journal*, *3-4*, 113–127.

13 Data Transformations and Transfer Functions: An Application to Water and Soil Pollution

Milan Stehlík
Institute of Statistics, University of Valparaíso, Valparaíso, Chile,
Facultad de Ingeniería, Universidad Andrés Bello, Valparaíso, Chile,
University of Applied Sciences Upper Austria, Campus Wels, Austria
E-Mail: mlnstehlik@gmail.com. ORCID: 0000-0003-2749-5990.

Kiran J. Bal
Independent Researcher, Vienna, Austria.

Kumud Pant
Department of Biotechnology, Graphic Era University, Dehradun, India.

Ludy Núñez Soza
Departamento de Matemática, Facultad de Ciencias, Universidad de Tarapacá, Arica, Chile.
ORCID: 0000-0003-4858-1900.

Silvia Stehlíková
Independent Researcher, Valparaíso, Chile

13.1 INTRODUCTION

In this chapter we use transformations in order to model heterogeneity and heavy-tailedness of data on rivers measured in Uttrakhand, India. By heterogeneity of data we mean parametric non-homogeneity, as developed in Stehlík *et al.*, 2022. From the perspective of water in rivers it can relate to chemical pollution and sedimentation. In section 13.2 we provide the methodology for analysis of the contamination. In section 13.3 we provide background about drinking and river water in India, together with special approach to heterogeneity Stehlík *et al.*, 2022 and extremes analysis in different regions in Uttrakhand such as Dehradun and Haridwar and its rivers (Bahukhandi *et al.*, 2017; Jalan *et al.*, 2009; Pant & Purohit, 2019; Sharma, 2010; Stehlík *et al.*, 2016a). The description and preliminary analysis with empirical densities and empirical counts and box plots of this dataset is given in Stehlík *et al.*, 2023. The dataset has vector of variables, including pH value, Biological Oxygen Demand (BOD), Chemical Oxygen Demand (COD), Conductivity, Temperature, Dissolved Oxygen (DO) and Total Dissolved Solids (TDS). First, the amount of missing data is revealed and the variable Conductivity has a particulary high amount of missing values.

DOI: 10.1201/9781032664330-13

The quality of drinking and river water is vital for a country in order to provide their inhabitants with good living standards. Poor water quality has a major negative impact on the health of humans and animals. Water, which does not fulfill the minimal standard requirements has already been linked to liver and kidney problems. Furthermore, the water quality is also crucial for the environment. Therefore, river water should be monitored and its samples analyzed regularly in order to improve or maintain the quality of water. However, problem lies in the chemical compounds of water that are not easy to asses as there are multiple outliers, missing values or distributions, which are difficult to work with. This chapter implements two new approaches. One method is to measure heterogeneity for the specific structure of water data, so that we can hypothesize the sedimentation of rivers (Stehlík *et al.*, 2022). Another approach is the t-Hill estimator and plots, which allow us to make robust analysis of heavy tails of individual measurements (Stehlík *et al.*, 2016a, 2016b).

The obtained results can be useful for decision-making on river water quality, such as identifying areas with extreme levels of contamination, assessing health risks, and designing pollution control strategies, among others. Computation is made in software R (R Core Team and others, 2015) and Jamovi (jamovi (Version 2.3) The jamovi project (2023), n.d.).

13.2 METHODOLOGY AND ON THE UNDERSTANDING WATER EXTREMES WITH CAUTION

Stehlík *et al.*, 2016a states, empirical work often starts with the analysis of the data, which is why the data quality is an important factor. However, when the data is gathered from a variety of resources, the quality of the data might suffer from imperfections. Therefore, it is vital to handle the situation in a model. The established approach is to first make a log of the data sample and then assume an contaminated model (Stehlík *et al.*, 2016a)

$$G(x) = (1-a)F(x) + aH(x) \tag{13.1}$$

with the mixing coefficient $0 < a < 1$, the predominant cumulative distribution function $F(x)$ and the contaminating cumulative distribution function $H(x)$. The mixing coefficient is usually set to 0.5 or less (Stehlík *et al.*, 2016a).

If $1 - F(x) > 1 - G(x)$ applies we call the distribution F is heavier tailed than G. This leads for example to the Exponential and Pareto tails being easily comparable. For example, in a very heavy-tailed Pareto distribution or Cauchy distribution, the expectation may not exist or it can be infinite. The arithmetic mean contains the same statistical information as the individual observation from the Cauchy distribution. A better aggregation operator for such a data is the harmonic mean. The following estimators can be utilized, assuming the data follows a Pareto distribution (Stehlík *et al.*, 2016a).

The t-estimator

$$\hat{\theta} = \frac{1}{\bar{X}_H - 1} \tag{13.2}$$

with $\bar{X}_H$ being the harmonic mean, which is only calculated for the studied upper tail (Stehlík *et al.*, 2016a).

The t-Hill estimator (Jordanova & Stehlík, 2023; Stehlík *et al.*, 2016a)

$$H^*_{k,n} = \frac{1}{\hat{\theta_{k,n}}} = \left(\frac{1}{k}\sum_{i=1}^{k}\frac{X_{n-k,n}}{X_{n-i+1,n}}\right)^{-1} - 1 \tag{13.3}$$

with $k = 1, 2, ..., n-1$ and the order statistics $X_{1,n} \leq ... \leq X_{n,n}$. The data is taken over threshold $X_{n-k,n}$

Lastly, the Hill estimator (Hill, 1975)

$$\frac{1}{\hat{\theta}} = \frac{1}{k-1}\sum_{i=1}^{k-1}(\ln X_{n-i.n} - \ln X_{n-k.n}) = \frac{1}{k-1}\sum_{i=1}^{k-1}\ln(\frac{X_{n-i.n}}{X_{n-k.n}}) \tag{13.4}$$

with $k \geq 2$.

Stehlík *et al.*, 2010 deals with the assessment of heavy tailed data and its compound sums. For this reason, the classical maximum likelihood (ML), QQ and method of moments (MM), t-score moment method and other methods are compared. Estimators based on the t-score such as t-score estimators, "generalized median", "trimmed mean" type and QQ estimators are particularly interesting as they are not dependent on the existence or finiteness of moments (Stehlík *et al.*, 2010).

Firstly, the performance of different estimators is compared with the help of the distribution of exact likelihood ratio tests of homogeneity and Pareto-tail index. This approach is used instead of the classical KS test. When analyzing sums of heavy tailed data, Stehlík *et al.*, 2010 shows that t-score estimators can outperform other estimators. Stehlík *et al.*, 2010 finds with the help of simulations based on the estimation of the Pareto tail index with known λ, that the t-score moment estimator outperforms the classical estimators QQ, MM and MLE in this case. Further, when the data shows contamination, the t-score moment estimator is better than the generalized median estimator. Nevertheless, there are some cases, where the trimmed mean estimators perform better than the t-score moment estimator. The simulations based on the estimation of the Pareto tail index with unknown λ show similar results. The t-score estimator outperforms all other estimators in question (Stehlík *et al.*, 2010). Thus we need robustness in estimation of heavy tails, since small errors in the estimation of the tail index can lead to large errors regarding the estimation of quantities based on the tail index (Stehlík *et al.*, 2010).

13.3 WATER QUALITY MAP OF UTTRAKHAND

The water sector of Uttarakhand faces many problems which increases the need of a Water Quality Map according to Sharma, 2010.

Industrialization and urbanization also play a role in poor water quality. For example industrial effluents, waste by products and solid wastes are discarded into water streams. As the waste is not treated, the surface and the ground water quality is worsened. As a consequence, the water becomes undrinkable (Sharma, 2010).

Another reason for poor quality drinking water is the chemical fertilizers used by the agricultural sector. As those chemicals are used excessively, they are absorbed in the soil and later get mixed in with ground water. Due to rain or other weathering processes, the chemicals can find their way into open water streams as well. This process contaminates available water sources (Sharma, 2010).

Missing sewage disposal facilities causes the fecal waste and domestic waste water to be disposed in open water sources. This leads to poorer water quality. Because of the lack of sanitation facilities in hilly regions and the slope factor, human and animal excreta blend in with water sources and deteriorate the water quality. Additionally, hilly regions, contain a variety of minerals. During monsoon season those minerals are mixed in with water sources, this is another reason for contamination Sharma, 2010.

Furthermore, before Sharma, 2010 the water was not analyzed in detail or regularly. There are regions, which are analyzed only sporadically. Also, the lack of any State Level Water Quality Analyzes Laboratory is responsible for the difficulty in developing a Water Quality Map. This causes the analysis of all 13 districts of Uttrakhand to be irregular (Sharma, 2010). Moreover, the effect of seasonal variation of monsoon season is also taken into account. Therefore, three samples of each site are collected during monsoon and the average value is reported.

13.3.1 STUDIES OF RIVER WATER QUALITY USING RIVER BANK FILTRATION IN UTTRAKHAND, INDIA

The widely used RBF process in Europe is known to overcome various water quality problems. RBF has established itself as an effective and well-accepted technique for the treatment of surface water. The water goes through compromised layers of sand and gravel, which contain underground water and is heavily used as a pre-treatment of drinking water. The process is explained in more detail in the following text. According to Tyagi *et al.*, 2013 River Bank Filtration makes use of overlapping physical, chemical and biological processes in order to serve as a natural filter.

The physical process consists of hydrodynamics and mechanical processes. Hydrodynamics involve

- advection
- dispersion
- transport
- and dilution

while the mechanical processes involve filtration. Natural filtration of fine sediments happens when particles are trapped in pore spaces (Tyagi *et al.*, 2013).

The physiochemical process depends on the prosperity, concentration and behaviour of metals and other inorganic compounds, the duration of water in the aquifer, temperature and pH value of water and its oxygen (Tyagi *et al.*, 2013).

This process involves

- filtration
- sorption-desorption
- solution-precipitation
- redox reaction
- complexion
- acid-based reaction
- hydrolysis
- biochemical
- microbial biodegradation reactions
- dilution

The biological process consists of

- degradation of organic matter
- mineralization of secondary substrates (Tyagi *et al.*, 2013).

The aquifier is responsible for slow rate filtration. Thus, the RBF water is of higher quality. Lastly, the water is taken out with the help of RBF wells, which are placed near the river bank (Tyagi *et al.*, 2013).

The efficiency of the RBF process is conditional on:

- surface water quality
- permeability of the riverbank
- river level
- sediment transportation variability
- residence time of the water
- temperature (Tyagi *et al.*, 2013)

Further, some downfalls of RBF are the recurring cost of running and maintenance, the limited capacity of a treatment unit and high demand of water in domestic areas. Monsoon seasons cause an even higher turbidity in water. As a consequence, lower quality water is being supplied (Tyagi *et al.*, 2013).

Tyagi *et al.*, 2013 analyzes 19 water quality characteristics of the water sampling sites at Srinagar, Augustmuni, Satpuli, and Karanprayag in Uttarakhand. Piper Diagram, Chadha's Diagram, Pearson correlation matrix and Water Quality Index (WQI) have been used to further understand the water data. In order to calculate WQI the weight arithmetic index method and the standard values of BIS are used. The overall WQI is defined as

$$WQI = \frac{\sum W_i Q_i}{\sum W_i} \tag{13.5}$$

with W_i being the unit weight for each water quality parameter. It is calculated as

$$W_i = \frac{K}{S_i} \tag{13.6}$$

K is proportionately constant and defined as

$$K = \frac{1}{\sum [\frac{1}{S_i}]} \tag{13.7}$$

with S_i being the standard permissible value of the ith parameter. The quality rating

$$Q_i = 100[(\frac{V_i - V_0}{S_i - V_0})] \tag{13.8}$$

with V_i being the estimated concentration of the ith parameter in the analyzed water and V_0 the ideal value of the parameter in pure water. All ideal values are 0, excluding pH being 0.7 (Tyagi *et al.*, 2013).

According to the analysis in Tyagi *et al.*, 2013 the data shows evidence of qualitative improvement in water quality, which is also supported by the water quality classification. The analysis shows significant improvement in drinking water quality and reduction of turbidity, bacterial contamination, iron and pH (Tyagi *et al.*, 2013). One important observation is that RBF naturally relates soil and water contaminations.

13.3.2 I-DIVERGENCE BASED STATISTICAL INFERENCE FOR HETEROSCEDASTICITY AND COMPOUNDS FOR ARSENIC CONTAMINATION IN CHILE

Because of expert expectation of strong heterogeneity and heavy tailedness in the data from Uttrakhand, we have been motivated to apply developed technologies from previous studies in Arica and Parinacota, Chile, including recent article Stehlík *et al.*, 2022. When working with real data, one can easily experience difficulties. Real data is often not normally distributed. The variables might be skewed and not independent and identically distributed. That kind of data violates the assumptions of a standard ANOVA. Stehlík *et al.*, 2022 comes up with a new approach to analyze real data. For the new measure, the correlation between two statistics radial R_N and spherical S_N is evaluated. The two statistics are a result of decomposition of the Kulback-Leibner divergence. The level of dependence of those two statistics is evaluated by several correlation measures such as Pearson, Spearman and Kendall's tau under the assumption of Gamma distribution. The proposed algorithm is discussed in Section 13.3.5.

In Stehlík *et al.*, 2022 the arsenic contamination of drinking water is analyzed. The data is provided by the Regional Ministry of Health of Arica and Parinacota. The conducted regions are located in arid climate zones resulting in limited water sources. As a consequence, the Rural Potable Water System and the Precarious System are introduced for transporting drinking water in rural areas. In this region, most of the water is collected from mountain ranges, rivers from volcanic areas and underground watersheds. Consequently, the collected water holds high levels of arsenic (Stehlík *et al.*, 2022).

Stehlík *et al.*, 2022 constructs appropriate and possible simple methods statistically detecting heterogeneity. Further, the paper also measures how often heterogeneity is detected (Stehlík *et al.*, 2022). As the majority of the results in this chapter show, there is a substantial similarity both in heavy-tailedness, also in heterogeneity.

13.3.3 RIVER DATA FROM UTTRAKHAND

The data was collected by the Uttrakhand research group from a government website with free access to data. In order to access to the data, there are multiple administrative burdens, which are time consuming. The data contains 9 different rivers with 21 riverpoints. Riverpoints are the area of a river, from which the water samples are taken.

The Bhella River has the riverpoints Kashipur and Lohia Bridge. The Dhela River contains the riverpoints at Moradabad Road Bridge and at Thakurdwara Aligwanj Road. The riverpoints for Kalyani River are at Industrial Area and Pantnagar Industrial Area. Kichha River has the riverpoint in Kashipur and Kosi River has the riverpoint at Bajpur Road Bridge. For the Nandaur River two riverpoints at Sitarganj Industrial Area exist. The Pilakhar River has its riverpoint after Bilaspur. River Suswa has its riverpoint at Mothrawala. Lastly, the River Ganga has the riverpoints at Laxmanjhula, Pashulok, Dudhiyaban, Harkipauri, Damkoti, Lalita Rao Bridge, Rishikul Bridge and Kumari Mandir.

Each sample of the river and riverpoint is taken monthly from January 2018 to March 2022. The water samples are analyzed for their pH value, Biological Oxygen Demand (BOD), Chemical Oxygen Demand (COD), Conductivity, Temperature, Dissolved Oxygen (DO) and Total Dissolved Solids (TDS).

13.3.4 DATA CLEANING PROCESS

The data was provided in multiple tables split by the river and the sample collecting location. Firstly, the data is imported in one master Table for easier usage. There has been data missing due to the river being dry or other reasons. As the reason for missing data is not relevant for this analysis, all the missing values have been set to "NA". Further, some data had the value of "<1", which is not numerical. For easier analysis, those values have been set to "0.5".

Lastly, two rivers were additionally analyzed for "Alkalinity", "Chlorides", "Calcium", "Magnesium", "Hardness" and "Total Coliform", but not for "Conductivity". As there are only two rivers with additional variables, the variables are hard to compare to the rest of the data. For this reason, those variables are not taken into account. Because there was no data collected for "Conductivity", this variable is set to "NA" for the analysis.

The dataset consists of 11 variables and 1071 observations. 49 samples could not be selected. There are multiple other missing values in the dataset. Some of them are due to the river being dry. However, not all of the missing values have an explanation. Table 13.1 shows the total of the missing values.

The description and preliminary analysis, empirical densities, box plots and the correlation of the variables of dataset is given in Stehlík *et al.*, 2023. The pH values range over 6 to 8.9, with a mean of 7.443. There are multiple outliers above and below the mean. The Biological Oxygen Demand has a minimum value of 1.5 and a maximum value of 80 with the mean being 11.58. Chemical Oxygen Demand has a minimum of 4 and a maximum of 304. In this case the mean is 48.51. Conductivity ranges from 140 to 1237, with a mean of 473. However, for this variable more than half of the data

Table 13.1
Missing Values.

pH	BOD	COD	Conductivity	Temperature	DO	TDS
75	75	75	573	157	75	103

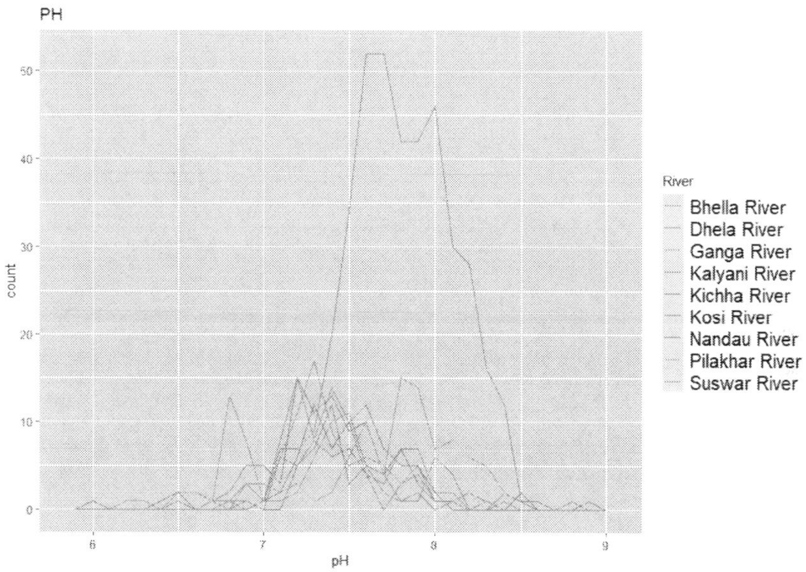

Figure 13.1 Empirical frequency plots of the individual river pH data. All empirical densities are created in Jamovi software using the implemented Kernel smoothing methodology.

is missing. For this reason the boxplot might be skewed. Temperature has no outliers and the values range from 8 to 41. The mean is 25.01. Dissolved Oxygen also has no outliers with a minimum of 0, a maximum of 9.8 and a mean of 4.443. Lastly, Total Dissolved Solids has a mean of 312, a minimum of 102 and the maximum value is 920.

Most rivers fall within the limit of IS10500:2012 of 500 mg/l. However, there are still some values above that limit present. Nevertheless, it is worth noting that, IS10500:2012 indicates the limits for drinking water and not river water.

For example, Figure 13.1 shows the densities of the rivers for the pH value. Most values of the plot start at pH value 6, peak at about 7.25 and then decline at 9. The Bhella River does have two spikes, one at about 6.75 and one at about 7.25. The Kalyani River and the Nandu River have three spikes. Kosi River has the lowest frequency of all pH values. See the heavy tailedness of all measured rivers pH values on Figure 13.2, which is also visible from kurtosis 14.8 and negative skewness −3.99, see also QQ plot on Figure 13.3.

Notice, that in the Figure 13.4 of densities of Conductivity, there is no data for Ganga River. The highest conductivity for most rivers occurs between 0 and 500. However, Kalyani River achieves its highest value slightly above 500. Dhela, Kichha and Kalyani River display conductivity, even above the value 575. For the empirical density of all measured rivers conductivity data see Figure 13.5, which is much closer to normality, see also QQ plot on Figure 13.6. The closeness to the lighter tails can be well visible also from smaller value of t-Hill estimator in the Table13.4.

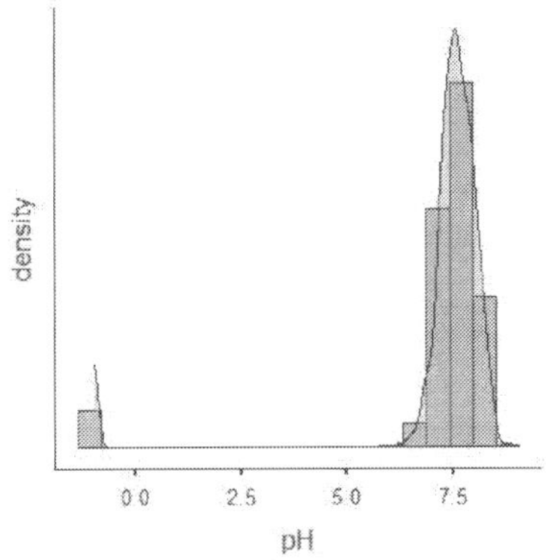

Figure 13.2 Empirical density of the all river pH data.

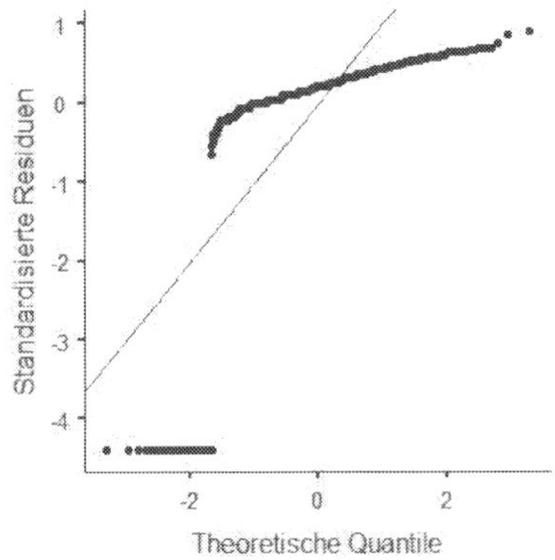

Figure 13.3 QQ plot for complete pH.

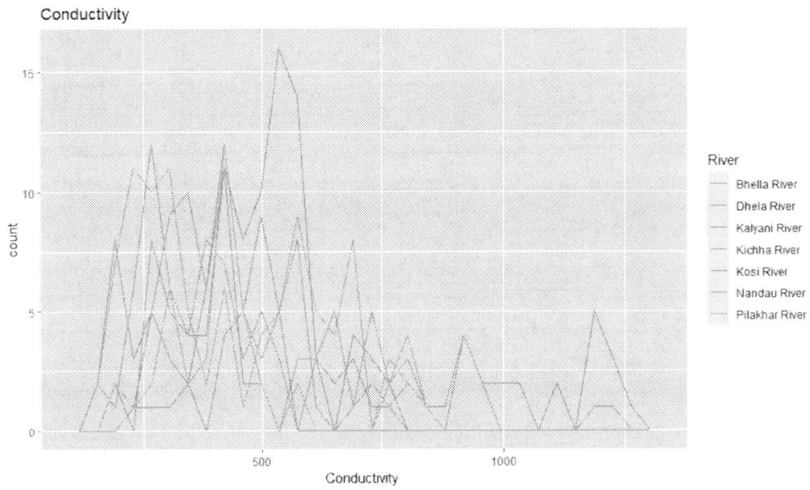

Figure 13.4 Empirical frequency plots of the individual River Conductivity.

Dissolved Oxygen and its empirical densities for individual rivers are shown in Figure 13.7. Ganga River and Suswar River peak at value 10. The remaining rivers peak below that value. Only Pilkhar River shows a little hump at value 25. The less the DO value, the cleaner the water is. Taking all river measurements, we normalize the sample, see Figure 13.8. The closeness to the lighter tails can be well visible also from smaller value of t-Hill estimator in the Table13.4 and also from QQ plot at Figure 13.9.

Figure 13.10 is dominated by the Ganga River. Much more heavy tail can be well visible also from empirical density of the all TDS data plotted on Figure 13.11 and from the value of t-Hill estimator in the Table13.4 and also from QQ plot at Figure 13.12.

Figure 13.13 illustrates the rivers and their Biological Oxygen Demand. Most spikes occur between the values 0 and 40. All rivers have the highest spikes between the values 0 and 10. From those values onwards, the lines regarding the BOD values become flat with the exception of Bhella and Suswar River. Bhella River shows a spike at BOD value 20 and 40 to 50. Afterwards, it becomes

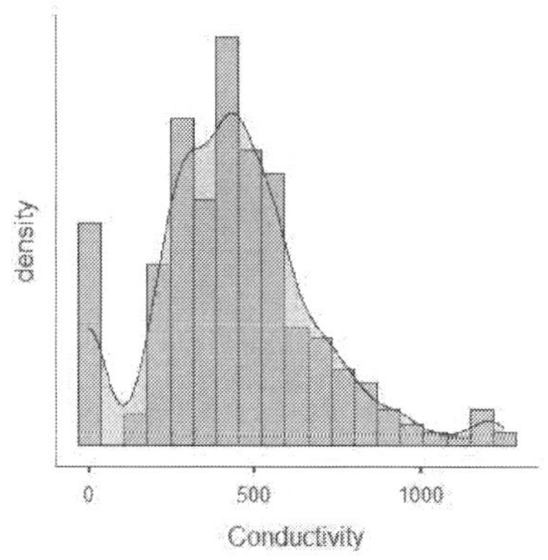

Figure 13.5 Empirical density of the all River Conductivity.

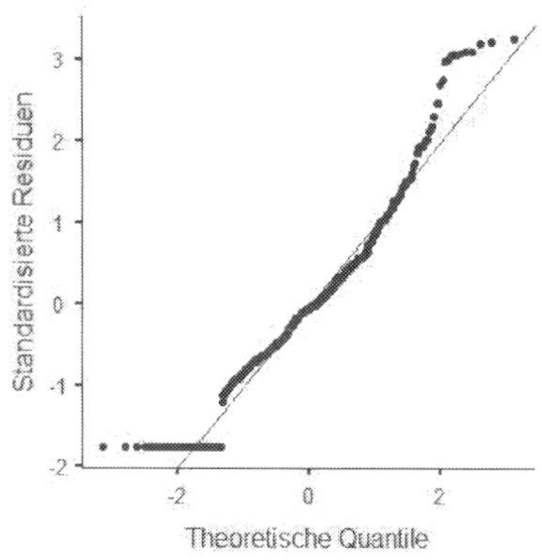

Figure 13.6 QQ plot for the all River Conductivity.

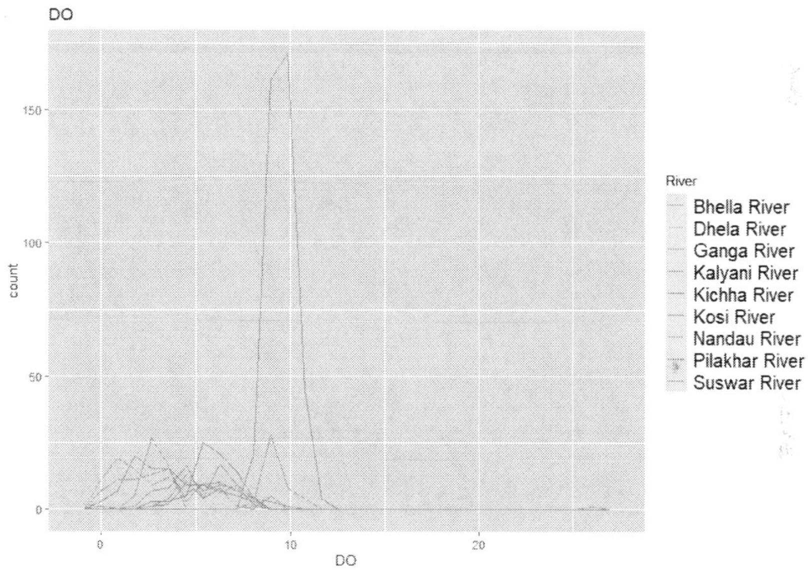

Figure 13.7 Empirical frequency plots of the individual river Dissolved Oxygen.

flat like the other rivers. Suswar River shows two spikes at about 25 and 35 before becoming flat. As we can see in the complete data empirical density (see Figure 13.14), heavy tailedness is well visible. This is justified also from the value of t-Hill estimator in the Table13.4 and also from QQ plot at Figure 13.15.

13.3.5 MEASURING OF STATISTICAL HETEROGENEITY OF PARAMETERS OF RIVER WATER

According to Stehlík *et al.*, 2022, measuring heterogeneity for real data can be difficult. Real data often do not follow a normal distribution, nor are they independent and identically distributed. Therefore, that kind of data violates the assumptions of ANOVA. Hence, the need for a novel

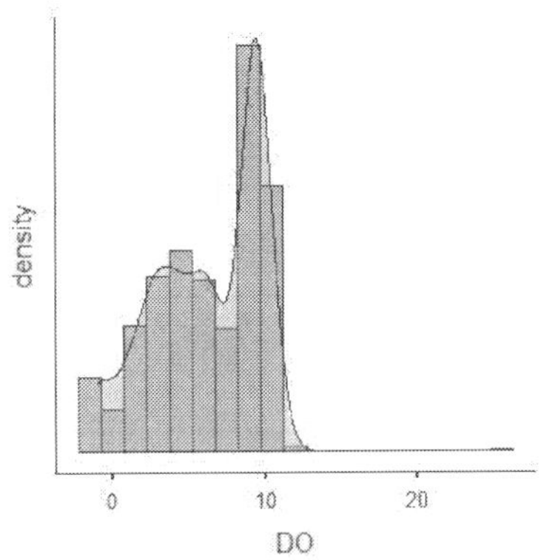

Figure 13.8 Empirical density of the all data of Dissolved Oxygen.

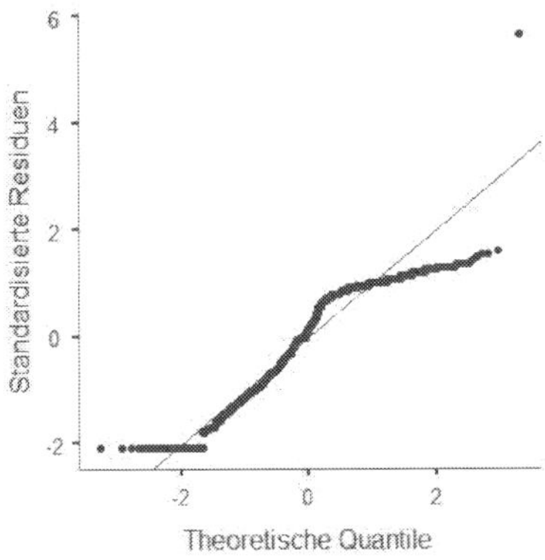

Figure 13.9 QQ plot for the all data of Dissolved Oxygen.

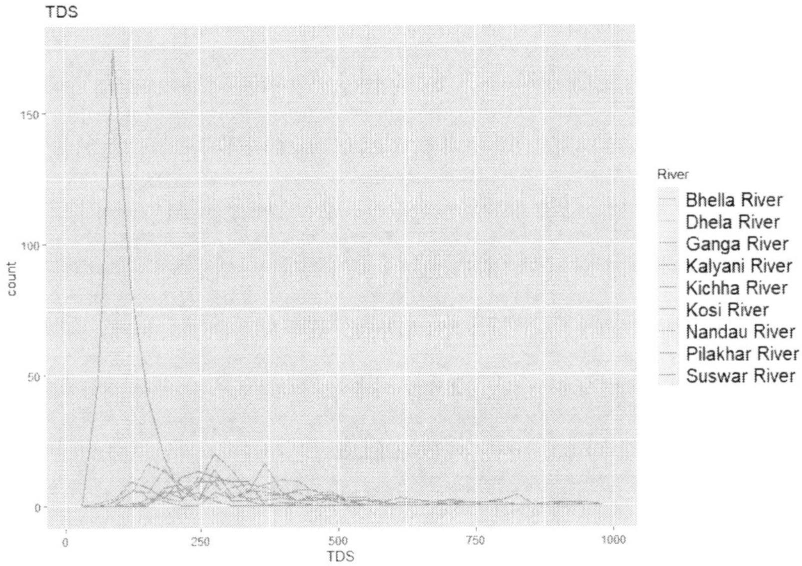

Figure 13.10 Empirical frequency plots of individual river TDS data.

approach occurs. Stehlík *et al.*, 2022 introduces an algorithm for measuring heterogeneity for data, that is not normally distributed or independent and identically distributed. The paper prioritizes finding appropriate and possibly simple methods in order to statistically detect heterogeneity in data. In order to prove the reliability of this approach, Stehlík *et al.*, 2022 also measures how often the heterogeneity is detected.

The algorithm consists of five steps. The first step separates the data into C data classes. C is dependent on the data itself and is a natural number. Then, with the help of location shift, all the data is transformed. In the second step, the shape parameter is estimated for every data class individually. Step three and four consist of computing the statistics S_{n_i} and R_{n_i}, which are derived from the Gamma distribution. Here S_{n_i} and R_{n_i} are statistics defined by the transformations developed in Stehlík, 2003, and later studied e.g. in Stehlík *et al.*, 2014. Their correlation gives levels of heterogeneity from the data; see Stehlík *et al.*, 2022.

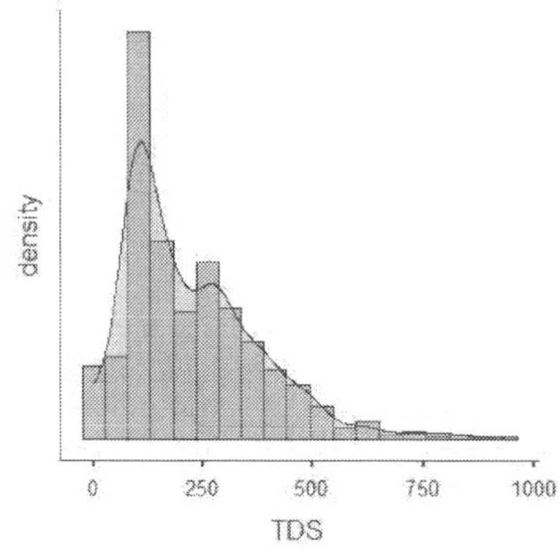

Figure 13.11 Empirical density of the all TDS data.

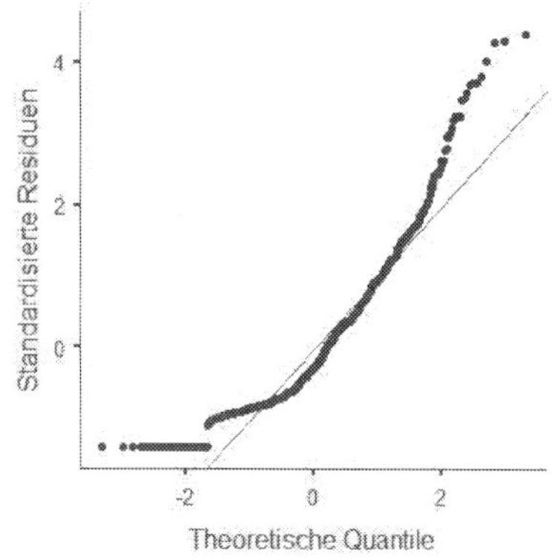

Figure 13.12 QQ plot for the all TDS data,

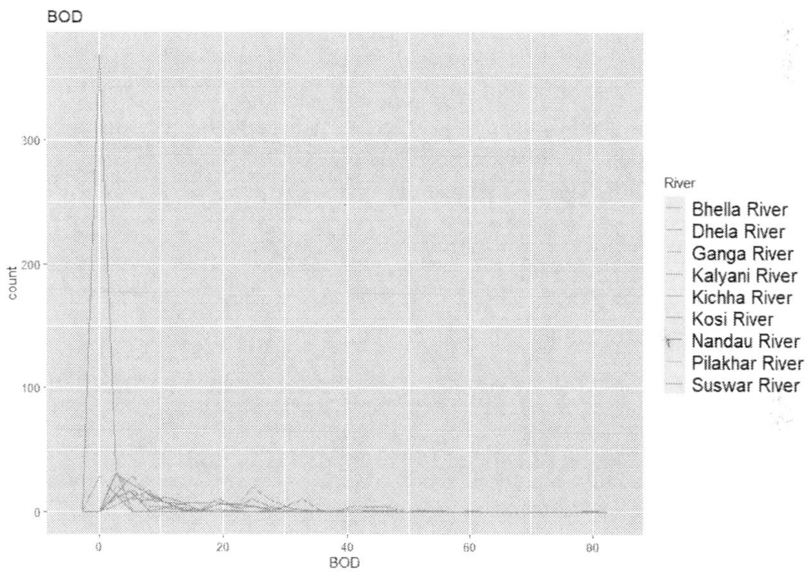

Figure 13.13 Empirical frequency plots of individual river BOD data.

σ_0 is set to 0.6. This value is set by experts with hydrological experience. In Stehlík *et al.*, 2022 the dependence measure is the correlation between R and S variables of R_n, S_n. Therefore, the correlation is also used in this chapter. However, any other dependence measure can be adapted. Lastly, the decision rule is applied to interpret the results.

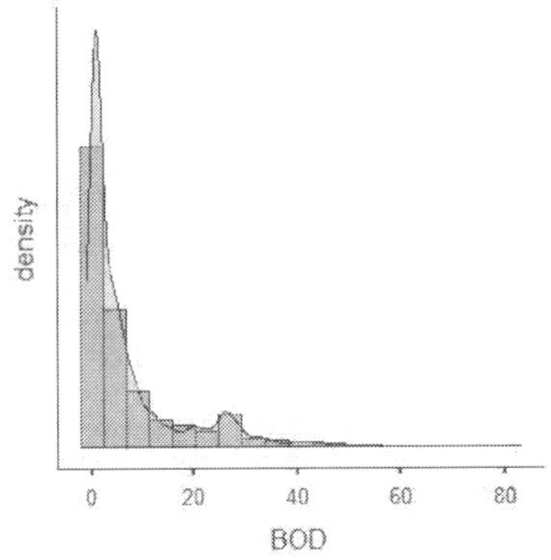

Figure 13.14 Empirical densities of individual river BOD data.

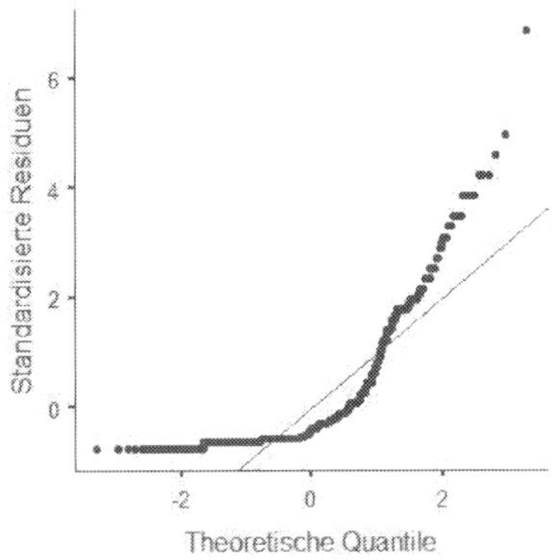

Figure 13.15 QQ plot for the all BOD data.

The algorithm for measuring heterogeneity from Stehlík *et al.*, 2022 is as follows:

Step 1: Construction of data classes

- $S(1) = \{Y > \alpha_1\}, S(2) = \{Y > \alpha_2\}, ..., S(C) = \{Y > \alpha_C\}$
- Where C is the reasonable integer lower bound constant that guarantees the minimal number of measurements in the last class; α_i are thresholds
- Transform all data in the classes by location shift
- All data y in S(i) will be transformed to $y[i] := y - i$

Step 2: Estimation of shape parameters ν in every class $S(i), i = 1, ..., C$

- $\hat{\nu} = \frac{\bar{y}^2}{s^2}$, is the moment estimator for shape parameter ν of Γ-distribution

Step 3: Compute S_{n_i} for every class $S(i)$

- $S_{ni} = n_i \hat{\nu}_i \ln(\sum_{j=1}^{n_i} y[i]_j) - n_i \hat{\nu}_i \ln(n_i) - \hat{\nu}_i \sum_{j=1}^{n_i} \ln(y[i]_j)$
- Statistics is derived from Γ-distribution
- S_n contains spherical components

Step 4: Compute R_{n_i} for every class $S(i)$

- $R_{n_i} = G_{n_i.\hat{\nu}_i}(\sum_{j=1}^{n_i} \frac{y[i]j}{\sigma_0}) - G_{n_i.\nu_i}(n_i \hat{\nu}_i)$
- In this case is $\sigma_0 = 0.6$ (expert number by hydrological experience)
- Statistics is derived from Γ-distribution
- R_n contains radial components

Step 5: Dependence measure

- Correlation of $R_n, S_n : \rho(R_n, S_n)$
- One can use other dependence measure

Decision rule:

- Low $|\rho(R_n, S_n)|$ supports homogeneity
- High $|\rho(R_n, S_n)|$ supports heterogeneity

Table 13.2
Results of the algorithm with simulated data

Simulation	2000	3000	4000	5000
$\rho(R_n, S_n)$	-0.998788	-0.998701	-0.997423	-0.994831

Table 13.3
Results of the algorithm with real data

Variable	pH	BOD	COD	Conductivity
$\rho(R_n, S_n)$	-0.905654	-0.953803	-0.88843	Incomplete Data
Variable	Temperature	DO	TDS	
$\rho(R_n, S_n)$	-0.777337	-0.995638	-0.8927117	

13.3.6 MEASURING HETEROGENEITY FOR SIMULATED DATA

Table13.2 shows the results of the previously shown algorithm for measuring heterogeneity of simulated data. The amount of simulated data varies and is shown in the respective table. For the dependence measure in step 5, the correlation of R_n and S_n is used.

The data are simulated based on the exponential distribution with rate 0.1. The same distribution is mentioned in Stehlík *et al.*, 2022. After the data is simulated, all data below the value 1 are omitted to get the logarithm of the data. If there were values below 1 present, the logarithm would be negative. Consequently, the algorithm described in Stehlík *et al.*, 2022 would not work. Building the logarithm of the data is necessary to scale the data.

As the absolute value of the correlation is relevant, the results show that all simulated data have a very high correlation. The simulations with 2000, 3000, 4000 and 5000 data all have an absolute value over 99 %. Therefore, based on Stehlík *et al.*, 2022 we can conclude that the simulated data are heterogeneous. Moreover, the simulation study shows that the results are stable as the minimal changes in the simulated data do rarely affect the results of the algorithm for measuring heterogeneity.

13.3.7 MEASURING HETEROGENEITY FOR WATER DATA

Table13.3 shows the results for measuring heterogeneity of real data. In this case, the real data is the water data with the variables pH, BOD, COD, Temperature, Conductivity, DO and TDS. As shown in Table 13.1, the variable Conductivity has 573 missing values. Due to the large amount of missing data, the construction of data classes is not possible. Therefore, there is no result for Conductivity.

The remaining variables do have enough data to perform the algorithm for heterogeneity. Likewise, to the simulated data, all values under 1 are omitted in order to build the logarithm. After the data is scaled, all five steps of the algorithm are performed. The results are illustrated in Table 13.3.

The pH value has the correlation of -0.905654, the BOD has the correlation of -0.953803, the COD value has the correlation of -0.88843 and Temperature has the correlation of -0.777337. DO and TDS also have a very high correlation of -0.995638 and -0.995638. Only the absolute values of the results are relevant. All variables show very high correlation. For this reason, all variables pH, BOD, COD, Temperature, DO and TDS are heterogeneous. The results can be associated to sediments in the mountain regions in India, Uttrakhand. Stehlík *et al.*, 2022

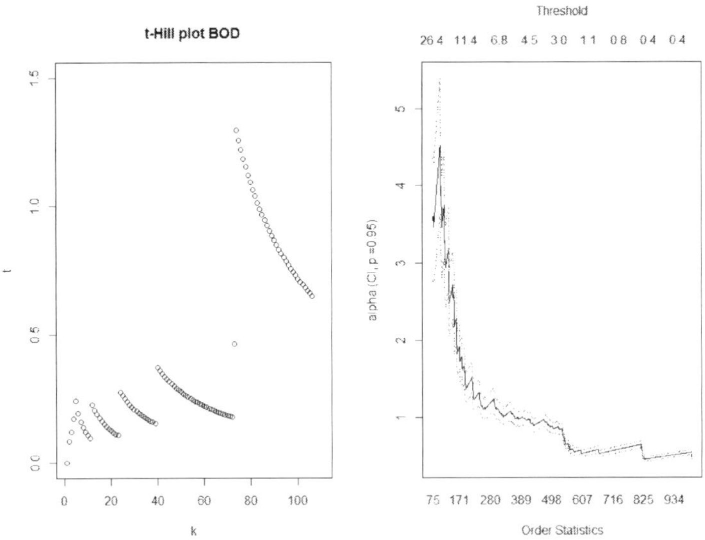

Figure 13.16 t-Hill and Hill plots for BOD.

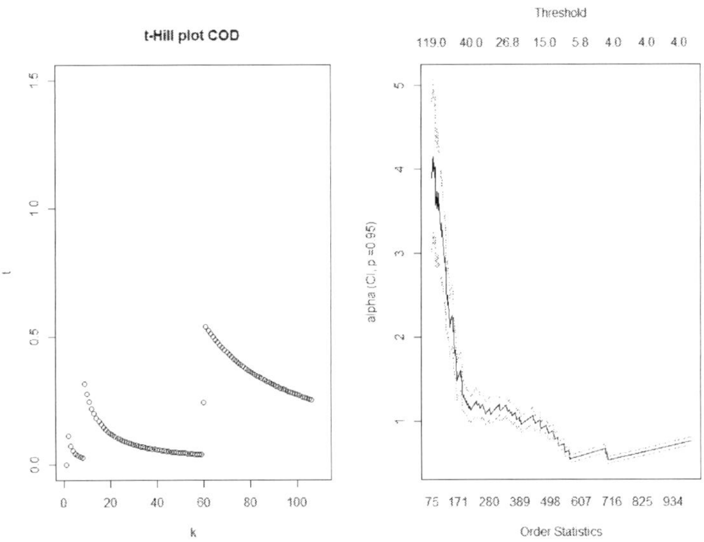

Figure 13.17 t-Hill and Hill plots for COD.

13.3.8 EXTREME VALUES

Jordanova & et al., 2016 and Stehlík *et al.*, 2016a introduce and investigate a novel method of heavy tails estimation named the t-Hill estimator. This new method relies on the method of moment for transformed scores instead of the standard maximum likelihood, like the Hill estimator. The Hill estimator as seen in Stehlík *et al.*, 2016a is based on the Pareto tail estimation by the MLE. According to Jordanova & et al., 2016, score moment estimators are robust for very heavy tailed distributions.

Jordanova & et al., 2016 shows the favourable properties of the t-Hill estimator with respect to the original Hill estimator for general settings and possibly dependent data. The infinite moving average

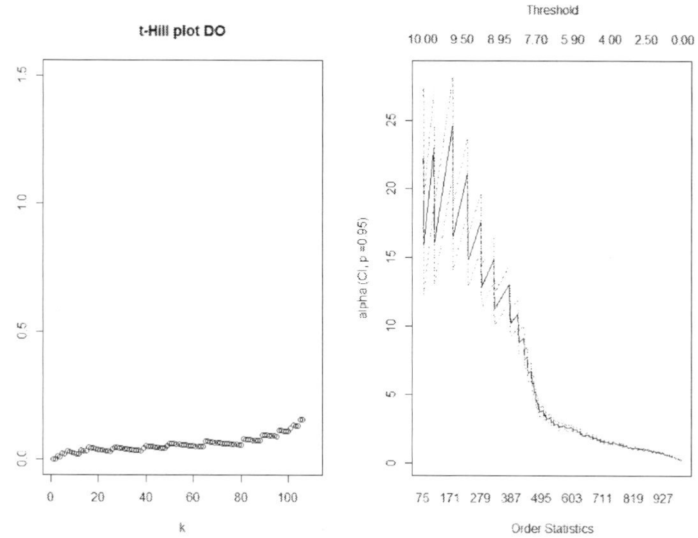

Figure 13.18 t-Hill and Hill plots for DO.

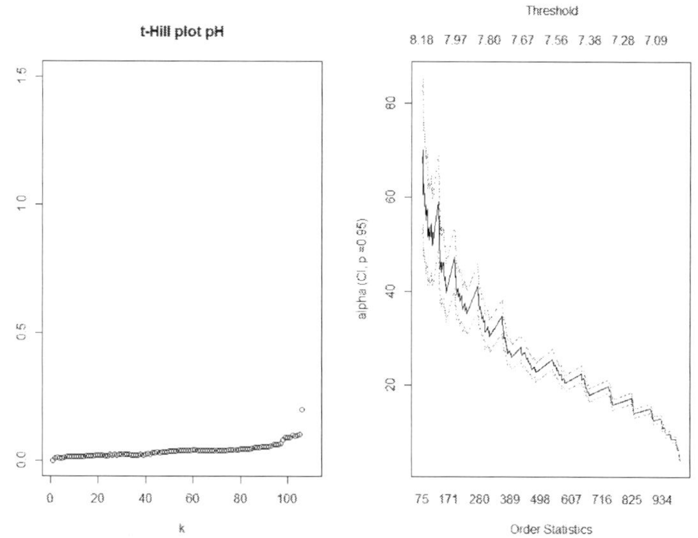

Figure 13.19 t-Hill and Hill plots for pH.

model shows weak consistency for the t-Hill estimator. Further, the special class of dependent data with large observations show that the Hill estimator is not robust.

With the help of the t-Hill estimator, extreme values are measured. Since the previous analysis indicates extreme values in the data, the t-Hill estimator is applied for further investigation. Extreme values are measured separately for all variables pH, BOD, COD, Conductivity, Temperature, DO and TDS.

We applied the t-Hill estimator to the upper ten values of the data of each variable. According to the results in Table 13.4, all estimators are below the value 0.5. The results indicate that the data of river water is very heavy tailed.

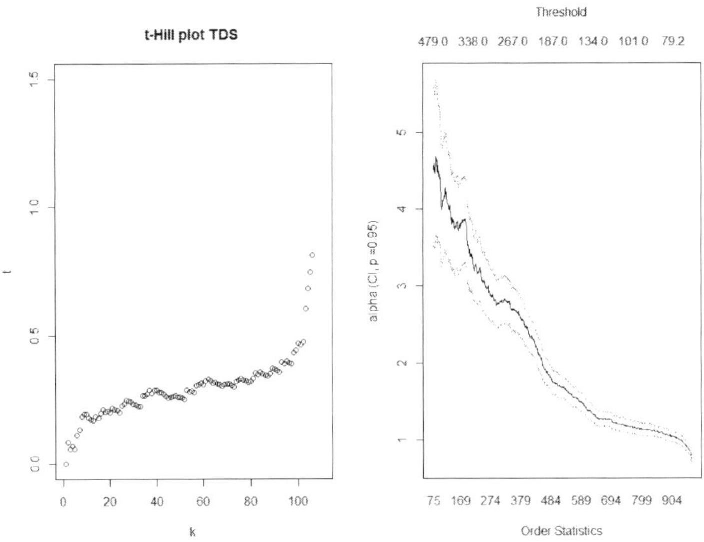

Figure 13.20 t-Hill and Hill plots for TDS.

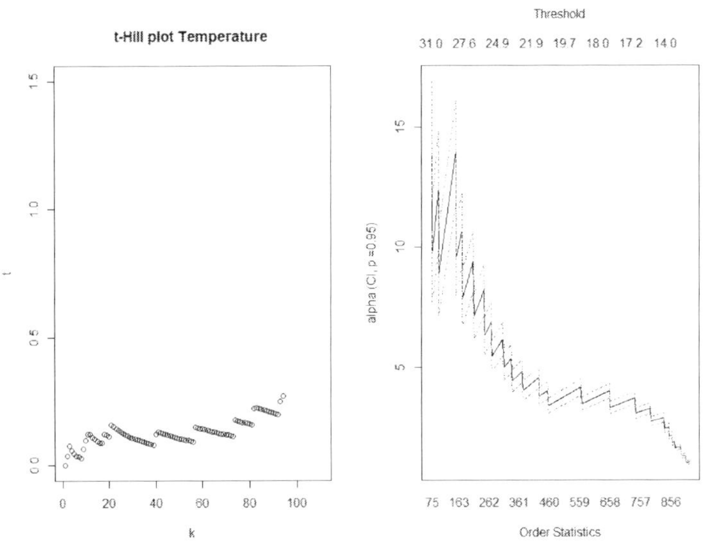

Figure 13.21 t-Hill and Hill plots for temperature.

Table 13.4
t-Hill estimator

pH	BOD	COD	Conductivity
0.02628854	0.1213034	0.08639256	0.08494529
Temperature	DO	TDS	
0.02873319	0.05233243	0.1206336	

The Pareto QQ plots inspections revealed some dissimilarities to the pure Pareto distribution. Therefore, there can be some bias in the results with respect to standard regular variation. All results of Table 13.4 show that the data is heavy tailed. Consequently, all variables are examined with the help of t-Hill and Hill plots. The t-Hill plots are created with the lower 10% of data of each variable. After omitting all "NA" values, the data is sorted. Then, the t-Hill estimator is applied to the sorted data and the results can be seen in the t-Hill plots in this section. The plots of the Hill estimator are made with the help of R software (Pfaff *et al.*, 2018).

Figure 13.19 shows the Hill and t-Hill plots for the pH values, here k is the number of order statistics in all t-Hill plots. The plot shows that the pH is stable. The little jump at the end of the points supports the fact that the data is heavy tailed.

Figure 13.16 shows more variability. As the points are clustered into seven different groups, BOD is not stable. Further, BOD is more heavier tailed than the other variables.

Similarly, COD shows a lot of variability and is not stable. Five clusters can be seen in Figure 13.17 and the variable is heavy tailed.

We omitted the t-Hill plot for Conductivity, since we have less data points because of missing values. Nevertheless, five clusters can be seen and this variable is also heavy tailed.

The Hill and t-Hill plots are seen in 13.21. Temperature is heavy tailed and stable. This is not surprising because pH and Temperature are naturally stable. Still, there are little jumps and clusters visible in the t-Hill plot.

The Hill and t-Hill plots for Dissolved Oxygen, as shown in Figure 13.18 and Total Dissolved Solids, see Figure 13.20 are stable and heavy tailed. For both variables there are jumps in plots, which divide the data into clusters.

All Hill plots first fluctuate and then flatten in the end. Moreover, all Hill plots follow a steep slope. The fluctuation depends on the visibility of clusters in the t-Hill plots. In the cases of Hill plots for BOD and COD, some fluctuation is visible at the end of the slope. Hill plots for Temperature and DO have some variation at the beginning of the plot.

It is not surprising that t-Hill and Hill plots are qualitatively different. These differences can be caused by two important issues. Firstly, naturally, real data does not follow a pure Pareto distribution. Secondly, different small sample behaviour dominated in the t-Hill and Hill plots.

13.4 CONCLUSION

The aim of the current research is to analyze the data about the quality of river water in Uttrakhand, India. The first central questions of the chapter is whether the data is heterogeneous or homogeneous. This question can answer the state of pollution in rivers based on the levels of the correlation between the transformed variables. The second main question is if and what extreme values are present in the data. With the help of the obtained results, the extend of the pollution can be illustrated.

The river data has been collected from Official Website Of UTTARAKHAND POLLUTION CONTROL BOARD Government of Uttarakhand (Official Website Of UTTARAKHAND POLLUTION CONTROL BOARD Government of Uttarakhand, n.d.). The data contains 9 rivers with 21 sample points. The samples contain results for 11 chemical components of the water. Interestingly, for the variable Conductivity, the data for two rivers are missing. As a consequence, the illustrations in the descriptive analysis are skewed. Moreover, the method for measuring heterogeneity can not be applied and the t-Hill estimator, t-Hill plot and Hill plot is analyzed with caution for this variable.

Temperature and the pH value are both normally distributed. Since this is consistent with the literature, there is no need for further investigation. The rest of the variables have a lot of values for lower concentrations of chemical compounds. The higher the concentration gets, the lower amount of values are observed. This is a good trend because high concentrations of any chemicals can be an indicator for pollution. The descriptive analysis shows that BOD and COD are highly correlated,

TDS and Conductivity show a high correlation and DO has a negative high correlation with BOD and COD. The heterogeneity for all variables except Conductivity is measured. The absolute value of the results vary from 0.77 to 0.99, indicating high heterogeneity. Hence, the water samples show a large range of values for the variables. According to the t-Hill plots, all variables are heavy tailed. There are signs of clusters in all variables except pH. Since the clusters in BOD and COD are clearly visible, more analysis is needed to examine the water better.

Summarizing, there is a need to develop an appropriate water quality methodology for the most of the measured variables. In this chapter we contributed by computing of several critical parameter estimates, by state-of-art checks, data collection, data cleaning and summarization. We also outlined importance of homogeneity checks of the underlying parameters. As also ongoing research shows that soil and water contamination are closely related, and the approach developed here present great potential for direct soil contamination studies.

13.5 ACKNOWLEDGEMENTS

Authors are grateful for Editors of this volume, Professors P. Economou, A. Rakitzis and S. Bersimis. We acknowledge project UTA MAYOR 4771-23. The team or authors are highly thankful to University of Tarapaca, University of Valparaiso and Graphic Era Deemed to be University for their support in carrying out this analysis. The team or authors are also thankful to Uttarakhand Pollution Control Board, Government of Uttarakhand, India for providing access to Water Quality data.

REFERENCES

Bahukhandi, K., Bartarya, S., & Siddiqui, N. (2017). Assessment of surface and ground water Quality of Haridwar district of Uttarakhand. *International Journal of ChemTech Research*, *31*(10), 95–118.

Hill, B. M. (1975). A Simple General Approach to Inference About the Tail of a Distribution. *The Annals of Statistics*, *3*(5), 1163–1174.

Jalan, J., Somanathan, E., & Chaudhuri, S. (2009). Awareness and the demand for environmental quality: Survey evidence on drinking water in urban india. *Environment and Development Economics*, *14*(6), 665–692.

jamovi (Version 2.3) The jamovi project (2023). (n.d.). \NoCaseChange{https://www.jamovi.org}.

Jordanova, P., Fabián, Z., Hermann, P., Střelec, L., Rivera, A., Girard, S., Torres, S., & Stehlík, M. (2016). Weak properties and robustness of t-Hill estimators. *Extremes*, (19), 591–626.

Jordanova, P., & Stehlík, M. (2023). Flexible extreme value inference. *Stochastic Analysis and Applications*, *0*(0), 1–45.

Official Website Of UTTARAKHAND POLLUTION CONTROL BOARD Government of Uttarakhand. (n.d.). \NoCaseChange{https://ueppcb.uk.gov.in/pages/display/96-water-quality-data}.

Pant, M., & Purohit, V. (2019). A Tap Water Quality Assessment of Seven Places in Dehradun. *INTERNATIONAL JOURNAL OF ENGINEERING RESEARCH and TECHNOLOGY (IJERT)*, *8*(5).

Pfaff, B., McNeil, A., & Stephenson, A. (2018). Evir: Extreme values in r. *R package version*, 1–7.

R Core Team and others. (2015). R: A language and environment for statistical computing. *Foundation for Statistical Computing, Vienna, Austria.*

Sharma, B. (2010). Development of Water Quality Map of Uttarakhand. *Report.*

Stehlík, M. (2003). Distributions of exact tests in the exponential family. *Metrika*, *57*, 145–164.

Stehlík, M., Bal, K. J., Pant, K., Stehlíková, S., & Soza, L. N. (2023). Statistical analysis for the water quality of rivers of Uttarakhand, India. *AIP Conference Proceedings*, *2953*(1), 030004.

Stehlík, M., Economou, P., Kisel'ák, J., & Richter, W.-D. (2014). Kullback–leibler life time testing. *Applied Mathematics and Computation*, *240*, 122–139.

Stehlík, M., Sabolová, R., Sečkárová, V., Núñez Soza, L., & Kisel'ák, J. (2022). I-divergence based statistical inference for heteroscedasticity and compounds of arsenic contamination in chile. *Chemometrics and Intelligent Laboratory Systems*, *226*, 104579.

Stehlík, M., Potocky, R., & Waldl, Z., Helmut amd Fabian. (2010). On the favorable estimation for fitting heavy tailed data. *Computational Statistics*, *25*, 485–503.

Stehlík, M., Stehlíková, S., & Torres, S. (2016a). Understanding water extremes with caution. *AIP Conference Proceedings*, *1738*(1), 040006.

Stehlík, M., Stehlíková, S., & Torres, S. (2016b). Understanding water extremes with caution. *AIP Conference Proceedings*, *1738*(1), 040006.

Tyagi, S., Dobhal, R., Kimothi, P. C., Adlakha, L. K., Singh, P., & Uniyal, D. P. (2013). Studies of river water quality using river bank filtration in uttarakhand, india. *Water Quality, Exposure and Health*, *5*(3), 139–148.

Index

FSC
www.fsc.org
MIX
Papier | Fördert gute Waldnutzung
FSC® C083411

Kolibri 360 GmbH
Ferdinand-Jühlke-Straße 7
99095 Erfurt, Deutschland
produktsicherheit@kolibri360.de